# Philosophy
## for
# Medical Students
## and
# Practitioners

# Philosophy
# for
# Medical Students
# and
# Practitioners

## Edward R W Makhene

Edward Makhene
2022

First Printing: 2016

This is a revised and updated version of *Philosophy for the Medical Student and Practitioner* that was first published in 2008.

ISBN 978-0-9876970-2-8

Edward R W Makhene
Mississauga, Ontario.
Canada.
edmakhene@icloud.com

Dedicated to

Uncle Willie, Dr. William. F. Nkomo, whose example inspired me to follow the humanities while I was enrolled as a student in Medicine.

The South African "black" pioneers in Medicine: Drs Bokwe, Molema, Moroka, Motebang, Sebeta, Xuma whose pictures were posted on the wall in my father's office and served as my initial inspiration, at the age of 10 years, to become a medical doctor.

My heroes, my parents J. Lebelo and N. Maleshoane Makhene who faithfully and lovingly supported and encouraged me in many ways to achieve my successive goals.

# CONTENTS

# PHILOSOPHICAL CONCEPTS IN MEDICINE

# PHILOSOPHICAL SYSTEMS IN MEDICINE

# THE TOTALITY VIEWPOINT

# REFERENCES AND INDEX

# Acknowledgments

I am grateful to my wife Eileen who has endured my immersion into libraries, bookstores, and the computer for many years, and to the scores of authors from whose works I have benefited in ideas that I have not quoted in this work.

The chemical structure of Insulin is adapted from https: www .pharmacorama.com/en/Sections/Insulin_1.php (with permission for original reproduction).

DNA structure is copied from http://ib.bioninja.com.au/standard -level/topic-2-molecular-biology/26-structure-of-dna-and-rna/dna-structure.html

Insect heart heart is copied from https://www.bioexplorer.net/do -insects-have-hearts.html/

fish heart is copied from https://www.google.com/search?q=fish +heart&client=firefox-b&source=lnms&tbm=isch&sa=X&ved=0ah UKEwja8NHfobrcAhXL6IMKHU6QASwQ_AUICigB&biw=1440&bih =594#imgdii=ZDQAGSJtuXP-MM:&imgrc=enXzbmHAmg Rz3M:

# Philosophical Signposts

# 1

# Prolegomena

## Why Philosophy in Medicine?

Philosophy provides the student of Medicine with the logic-based tools and concepts that she can use to reach beyond her thinking about medicine at the descriptive level of its subject matter to the prescriptive level. At the basic level, Medicine's task is to discover general facts and causal mechanisms of health and disease and the laws that regulate their relationship as described in the Hypothetical-Deductive Method. It inquires into the basis for the categorization of different diseases and the meaning of their cures as adapted to the cure of the afflicted patient, and it investigates the elements that comprise disease, its ontological (real entity) status, whether disease exists *per se* in nature or only as a socio-cultural construction that is relevant and relative to different communities.

At the prescriptive level where social, cultural, and holistic factors are involved, it formulates preventive and therapeutic measures for the benefit of the patient according to his personal values, and provides ethical guidance in the discussion of such problems as are presented by abortion, euthanasia, vegetative states, and others. It ventures to prescribe what should be the case in addition to what is observed to be the case, presenting a moral challenge for the doctor in relation to her patient, including study of questions relating to mind versus body, causality, free will, determinism, and all the other philosophical concepts mentioned in the first six chapters below. Reductionism has no role here.

A very important philosophical question is, should we first establish the existence of disease before we study its nature to know facts about it and how to manage it in its many manifestations, since we cannot possibly study and manage what does not exist, or should we first define it in order to be able to identify it when we come across it in people, since we cannot recognize it if we do not know any of its attributes? We don't want to pick on the wrong condition and call a disease, thus inflicting injury on persons who are treated as diseased when they are are perfectly healthy, and simultaneously waste energy and resources chasing after an *ignis fatuus* or will-o'-the-wisp.

1

# Physician privilege

Physicians occupy a unique position of authority and trust in their communities, a trust that entails the reliance of patients on them to cure their acute maladies and bodily infirmities and to support them in their chronic illnesses. This trust is based on the physician's extensive, unequalled, and intimate knowledge of the workings of the human body in health and disease and on her solemn undertaking to act as the patient's fiduciary in their relationship. However, it also renders vulnerable those who place their trust in her knowledge and authority as fiduciary, if she lacks compassion for them and treats them unprofessionally like common commodities and cash cows, or if they also cannot be trusted to provide their doctor with all the information that she needs to make a learned determination about their illnesses. So there is a spiralling (upward or downward) reciprocal relationship between doctor and patient that is largely dependent on their attitudes toward this partnership.

By singling out physicians for this honour, I am not diminishing the trust that people repose in other professionals in matters related to expertise offered by those professions: teachers, lawyers, priests, engineers, economists, and others; but when faced with life and death situations, people always turn first to the physician for help to preserve their lives, so that they can continue to enjoy the services provided by all the other professionals when they need them. Of course, the nagging questions relate to how well the modern day physician's focused training and worldly education prepare her to fulfill that role of trust which everyone else, including the other professionals, repose in her, and how well she acquits herself once she finds herself placed in that important role.

In the pages that follow, I will attempt to present philosophical perspectives that I trust will help to expand prevailing attitudes of medicine as a scientific and bio-physical pursuit of the nature and cure of disease to embrace its neglected bio-psychosocial aspects that are so crucial to the total welfare of the patient who is being cured of disease and relieved of his illness and suffering. This expansion of attitudes is one that should prepare and equip the doctor with the necessary confidence to include her patient in the planning of his own medical care and encourage his active participation in it. It should also foster in him that element of trust that has been eroded by the doctor's claim to exclusive knowledge of the facts and her oracular professional authority to which the patient is expected to bow without question. In the process, we will discuss many concepts that we will find useful in our consideration of the philosophical approach to the theory and practice of medicine.

When a patient places his trust in a doctor, he does not just rely on her to do things for him only so far as it suits her whim for her to do so; he expects her to go out of her way to ensure his welfare and preserve his life by virtue of her skills, commitment to her vocation, sincerity, and integrity. So a letdown in this case is more significant than a mere disappointment in anyone's failure to play his role in a joint task. In the latter situation, the participant may renege on his commitment for selfish reasons and prove himself to be unreliable, but in the case of a doctor, there is no ethical reason for her to neglect her duty, less still her oath of service; she has to be trustworthy every moment of the day, and if she fails to perform, it should not be for selfishness or lack of reliability but for inability to overcome overwhelming circumstances that impede her performance. Her interests should not be allowed to supersede those of the patient that she has taken into her care, otherwise she should not assume that responsibility if she is not sure that she will see it through, or else she should have a colleague help her to do so. Abandonment of her duty to her patient for any reason is unacceptable and inexcusable. At no time should any patient be placed in a position where he mistrusts his doctor and the advice that she gives him. A doctor may be unreliable for keeping to appointed times, but she should be trustworthy in her dealings with patients; they should feel that she is worth waiting for.

In another vein, there is no reason for any doctor to place herself in the disreputable position of being enslaved by political dogma and the cowardice of owning up to her culpability in bringing harm to citizens to the extent of earning the dubious distinction of being labelled Dr. Death, because he uses his skills to murder (oppressed) people or to test chemicals on them while they are immobilized outside overnight to find out if they will die from the effects of the chemicals. No doctor should merit the disgrace of being styled Dr. Beetroot for denying the people appropriate, tested treatment for their ailments (like AIDS) and substituting useless measures that will only occasion the deaths of millions of them for lack of approved and proper treatment. Also, we should not have to read news reports stating that certain doctors fail to document and report injuries of patients who come under their care for maltreatment possibly inflicted by security personnel of their country in violation of their ethical code. Doctors should not be salesmen for concerns that trade in organ donation and transplantation; they should resist the lure of money and not sacrifice their moral integrity for it. All we can tell them is that the moral standard required of doctors by virtue of their vocation, if not only by virtue of their dealings with fellow human beings, demands better than that; and as one news article stated, "the world will be a better place if doctors conduct themselves in a moral fashion". Physicians who sink to these base levels of humanity and still continue to make excuses for their immoral behaviour deserve to be doubly scorned. So also should

3

doctors who are reported to participate in despicable practices like the sterilization of women of colour against their will and the Tuskegee study that is also reported to have concealed the diagnosis of syphilis from 600 rural black men and denied them treatment while observing the course of their disease over time, with telling deleterious effects on them and their offspring. That is why some people distrust doctors.

# Meaning

Philosophy deals with the definition and analysis of concepts, their meanings, logical relations, applications, and the truth of their purported representations of reality. To understand the concept of disease, for example, is to understand the meaning-in-use of the word 'disease' and when to apply it; i.e., what conditions are necessary and sufficient for its application. Meaning is the gist of concept formation; without it the application of words like 'disease', 'illness', and 'disability' to particular existential situations cannot be clarified, even as basic science may succeed in describing fully the underlying mechanisms of these conditions. It is the function of philosophy to enquire into the meanings of these words for the persons to whom they are applied and for those who apply them; it does not contest the facts, because they can be checked empirically, but it contests the logical implications and the coherence of the multiplicity of meanings and nuances that can be applied to those facts, and how they ought to be applied within the wider context that also embraces the social and behavioural sciences. Science will explain the mechanistic aspects of disease, but it will not illuminate the humanistic aspects of what pain, suffering, and life-as-painfully-lived mean to a patient.

To understand the concept constituted by the letters d-i-s-e-a-s-e, for example, is to understand the meaning-in-use of the concept word 'disease' and when to apply it; i.e., what different conditions are necessary and sufficient for its application, since the concept can't be applied to any clinical state, nor does it isolate only one clinical state. Like all other words that have not been draped with psychological and referential import, the word 'disease' *per se* would not have suggested what we now know it to mean, and people would have been free to interpret it as they wished until it was endowed with a specific meaning after a wide variety of clinical states of the person were considered and the concept *disease* was applied to some of them. Any relativity and arbitrariness in the use of the word, which amounts to using a private language that only the user and perhaps a privileged few individuals understand, will only succeed in causing ambiguity and confusion that will endanger people's lives. So once the bounds of the word 'disease' and the concept embracing it were set, any other new conditions that were found to be consistent with the meaning of the concept as laid

4

down were termed disease states and everyone was expected to use the word 'disease' as designated and not as they saw fit as would be the case if they were using a private language. In this connection we might note that a swear word can be any combination of letters that lack meaning until it has been designated as such, and the same thing applies to other combinations of letters that are synonyms of benign words but are considered to be vulgar; e.g., 'shit', the vulgar synonym of excrement, which simply means waste matter from the body. The multiple and dissimilar conditions that constitute the family of disease states share coherence like that which exists among different kinds of games that share only a family resemblance, but are quite unlike one another, as Ludwig Wittgenstein noted in *Philosophical Investigations*.

Following the example of Charles Ogden and Armstrong Richards, we can depict the relationship between the word 'disease' and its corresponding concept and object, and hence its meaning, as follows.

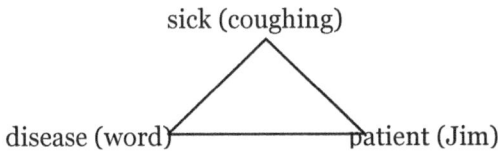

sick (coughing)

disease (word)　　　　　　　patient (Jim)

In this picture of their semiotic or semantic triangle, the arbitrary symbol d-i-s-e-a-s-e, which by itself lacks meaning, conjures up a picture of a sickness, which describes the dysfunction that a patient is experiencing within his society and the world at large. Through its linkage with its object via the associated symptom of coughing, and through many other symptoms and signs that Jim exhibits in this and different other situations, the meaningless symbolic word 'disease' thus comes to signify the disabling cough that Jim is experiencing, thereby acquiring a meaning that is now conventional and non-ambiguous, unlike what would be the case if everyone used the word to mean what he chose it to mean. (see page 26–Humpty Dumpty).

So, in the philosophical sense, the meaning-in-use of the word 'disease' will be understood to serve a designatory or referential function, but also to carry causal antecedents and consequences: pneumonia will refer to a certain constellation of symptoms and signs (physical and radiological) resulting from a contextual bacterial or viral infection from which certain consequences can be anticipated, unless therapeutic intervention alters the natural course of the disease. This category of use answers the questions: what does the concept word 'disease' (in the abstract) literally mean when the word is used in language that is not intentionally directed at an audience, e.g., disease is undesirable, and what is it a conventional sign of? Does

5

it refer to an identifiable existent or only to a concept? Furthermore, what does the doctor mean (concretely) by 'disease' when she uses that word in relation to a patient; what does she intend her patient to understand about what is going on inside of him when she uses that word; and how should the patient interpret it and react to it? e.g., John, you are afflicted with a disease called pneumonia. These statements may suggest that the doctor is talking about something that she can lay her hands on and extricate from the patient's body, but the meaning of 'disease' does not identify the ontological bearer or object of that name, because there is no such object. It is the conceptual reference of the word as it is used to designate the instantiation of a particular class of manifestations of a constellation of anatomical and functional aberrations of the living body referred to by that word; i.e., the word 'disease'$_1$ refers to a class of features that are embraced by the concept "disease", which also embraces disease$_2$ and diseases$_{3,4,5}$, etc.

Since different concepts will of necessity be generated by the same stimulus-word, e.g., disease of mind (intangible) or body (tangibles like bones, viscera, muscles, and nerves), the use of the word in the medical or lay language will determine its referential meaning in those particular domains of its application, suggesting that the word 'disease' will have as many cognitive references as it has practical applications in those and other domains. But one would expect the word 'disease' to have a universal meaning independently of the different shades imparted to it by the contexts in which it is used, in the same way as a penknife retains its character and meaning regardless of the use to which it is put, such as using it as a screwdriver, a nail cutter, or a chisel. Everyone knows what someone who talks about his penknife means, even if he does not have to refer to its many uses. Likewise, the meaning of 'disease' should not depend on its use in language. If it does, then by implication, the meaning of the word 'penknife' is incomplete without a description or demonstration of all the uses to which it can be put, including the deviant and reprehensible one of stabbing people, and that, we know, is not the case. 'A word' may not have meaning, but the word 'disease' does have meaning. So, what does the word mean and when can we apply it appropriately? Chapter 8 will deal with that question, but it is also dealt with briefly below under "definitions".

Disease as a concept entails etiology, symptomatology, and targeted treatment in a causal relationship. A causal connection explains symptoms and signs on the basis of etiology, and treatment is determined solely by etiology. The success of this conceptual approach is seen in the radicalization of the treatment of disease. "Transition from the humoral to the germ theory of disease required a major conceptual revolution, involving many kinds of conceptual change including a fundamental shift in how diseases are classified.

6

[and] . . . dramatic changes in concepts describing the infectious agents newly held to be responsible for disease."[1] As concepts of the nature of disease change, so do the modalities of its management change from biomedical to include biopsychosocial.

## Goals of Medicine

The Hastings Centre Goals of Medicine project articulated four basic goals that can be modified as follows: (1) the prevention of disease and injury and the promotion and maintenance of health and improvement of quality of life; but not in hopeless cases where more harm accrues from doggedly pursuing this effort than from ending it, and certainly not contrary to the patient's autonomous decision against heroic intervention or his request for a quick ending to his life of suffering that defies alleviation. In such cases, the empathic course would be termination of life to prevent further unnecessary injury from disease; (2) relief of pain and suffering caused by maladies that sometimes deprives the patient of good quality of life, especially in cases of incurable chronic disease, and in terminal illness, where empathetic acceleration of the patient's demise by his request can be condoned; (3) the care, in physical and mental dimensions, and the cure of those with a curable malady and the empathetic care of those who cannot be cured; (4) the avoidance of premature death and the pursuit of a peaceful death, which includes palliative care and physician-assisted death (suicide) when the patient in his particular circumstances requests it in preference to the miserable quality of life imposed on him by his suffering, realizing that death can never be held at bay for ever. "Unethical" use of modern tools of medicine to institute heroic measures in the face of a helpless situation that will only prolong a patient's suffering should not be a goal of Medicine, nor should the same measures in the case of very premature infants who will be set up for a future life of limitations from induced disabilities and wholesale dependence on others (individuals and society) for all of their daily care. This goal has regrettably fostered inflexible attitudes about preservation of life in the controversial situations of abortion and euthanasia, dependent as they are on the totality of conditions that constitute them, and about which the persons who are most vehemently opposed to them most probably know nothing, since they are not the ones enduring the burdens associated with them. Furthermore, Medicine should help with resolution of ethical dilemmas facing doctors who are required by governments to participate in executions by lethal injection and torture such as that carried out by Dr. Death. All of these goals are based on the ethics of beneficence, non-maleficence, autonomy, and justice, and on mutual respect between doctor and patient.

7

Some of the basic questions in this doctor-patient relationship will become clearer as we discuss many of the moral principle that govern the practice of clinical medicine, e.g., dealing with the patient's disease but not also with his emotions, because of the doctor's lack of empathy and her failure to communicate with him; categorizing patients for purposes of rationing health care to match available financial, technical, and manpower resources; and expending time and resources in saving the life of an extremely premature neonate in the face of the complications of cerebral hemorrhage, retrolental fibroplasia and blindness, bronchopulmonary dysplasia and chronic obstructive lung disease as a result of prolonged administration of high levels of oxygen to keep the infant alive, cerebral palsy and all the lifelong problems that will result from these complications. All of these raise the question, what has been achieved?

In addition to the above goals are the internal principles. Healing alone, which depends largely on the power of the patient's thoughts to help restore his normal functioning, is considered by some to be that sole principle. But this restriction overlooks the role of other situations that the patient might regard as basic to the furthering of his goals, such as those entailed by evolving changes in his values, which demand that he not only defer to the authority of his doctor but also insist on the accommodation of his autonomy; e.g., cosmetic surgery to improve his looks and make him socially acceptable; in vitro fertilization, contraception, sterilization, and abortion for the reproductive goals of some couples. Deferring to the authority of the doctor means allowing her to employ her own moral principles, which may be out of step with the changes that have been wrought on the patient's values by time and the march of science as it has changed and is still changing definitions of well-being and death since the advent of life prolonging ventilators and organ transplants. It also entails placing complete trust in her judgment and presuming that she will not permit the perpetuation of futile treatment on her patient or engage in any fraudulent ventures that might harm him, in contravention of her solemnly taken oath to do no harm. Too often, patients are unscrupulously exploited and exposed to unnecessary treatment by unprofessional physicians and others who have commercialized medical care for monetary gain. Reports of unnecessary or botched surgeries by qualified and unqualified surgeons, and other escapades of lay persons passing themselves off as physicians abound in the news media as proof of the greed that has overtaken the morality-based goals of medicine. But too often also doctors are placed in compromising situations by unreasonable demands from patients for unproved, unorthodox, or experimental treatments. In such cases, the patient's narrative should be patiently entertained and the risks of such treatments should be clearly pointed out. He should not be dismissed with a humiliating refusal.

In the prevailing climate of economic constraints, limits on what can be done for patients while still staying on budgetary targets have imposed new perspectives on the import of these goals to the extent that governments are abandoning patients to private health care facilities whose primary objective is profit, which turns the tables on the patients by imposing constraints on their short and long term survival from disease and disability. So it is not surprising that other people consider these procedures to be external to the goals of clinical medicine in the same light as they consider the role of doctors in administering lethal injections to condemned prisoners and limiting the access of patients to some forms of treatment under Health Maintenance Organization (HMO) rules. These external pursuits contradict the ends and internal morality of medicine as outlined in its promulgated goals set out at the beginning of this section. They argue against a universal morality of medicine, maintaining the relativistic view that morality differs with communities and the times of their existence; e.g., we live in an age where all medical resources should be rationed sparingly; hence the need for restrictive HMOs; and while homosexuality is acceptable in libertarian communities, it remains taboo in less tolerant ones even after its cerebral (amygdala) origins have been postulated. But they also allow for the fact that Medicine can sometimes be shaped by society or patients, at the risk of losing its objectives as a discipline to which patients appeal for help by virtue of its own *bona fides* about what it represents in their lives. Patients sometimes want to take over the direction in which Medicine should go to suit their whims, forgetting that although their input is important and welcomed, the doctor is still the professional, the one with the years of training behind him, and the one who is best suited to determine that direction relevant to his educated role in the health care situation. Due consideration should, however, be given to the modern Nurse Practitioner who has not yet fully replaced the physician, surgeon, obstetrician-gynaecologist, or the psychiatrist, although Advanced Paediatric Nurse Practitioners confidently profess equal expertise with paediatricians. The physician should welcome the professional input of these and other rational disciplines for the good of the patient who is the subject and object of medical care.

In the end, it seems that the morality of medicine will be decided by a dialectical progression between its internal and external components in the manner outlined in the section on Dialectic. For instance, there is no other way of resolving the conflict imposed by the principles of autonomy (relieving a patient's suffering by euthanizing him on his request) and deontology (observing the duty of preserving life at all costs), except through the dialectic of casuistry discussed on page 323, since denying him that prerogative will

9

invariably appear like sacrificing his interests and self-determination on the altar of society's interests (consequentialism), as we will see in chapter 20 under the section on euthanasia. However, where the patient's autonomy presents a threat to the welfare of those with whom he is in contact, there is no question of trying to balance it dialectically against that welfare; his autonomy will have to be overridden without question, as should happen, for instance, in the case of the HIV positive patient who demands confidentiality while taking no precautions to ensure that he does not infect others by behaving indiscreetly. The merits and demerits of each case should be considered individually within the determinations of the prevailing moral principles and the setting of realistic, value-driven goals for both doctor and patient, including avoidance of unreasonable and illegal acts. But no one should have the prerogative to endanger the lives of others for his own good; that is neither utilitarianism nor virtue, and it falls short of deontological requirements.

## Some relevant philosophical concepts

Every concept is always formulated against the background of other related concepts, and none of them exists in strict isolation in spite of all appearances to the contrary. In the search for meaning, questions will always arise about how we know that such and such is the case, given the scientific conglomerate of facts relating to the clinical conditions about which we will be making assertions. Such questions are not on the first-order level of the existence of brute facts about what we claim to know, but on the second-order level of meaning—what do we mean by saying that we know? Do we mean that we have justified, true belief? Can we be mistaken in claiming justification of our beliefs, and is it possible to know something that is not true? And what do we mean by "true"? Do we mean the invariable coherence of our experiences, the pragmatic value of their regularity for us, or the correspondence with reality of the facts derived from those experiences; and if so, then what do we mean by "reality"? Do we mean something that exists out there independently of us that we come to know by means of contact with it through our senses? If so, how do we know that our senses represent reality as it exists, undistorted, and how do we know that our interpretations of these sensory experiences are themselves not ambiguous, erratic, or vague and therefore unreliable in conveying to us correspondence between their representations and this reality? If the snare of fallibility is so ubiquitous, why should we not be skeptical about all our experiences and about reality itself? Can we ever escape from the limitations imposed on us by our sensory apparatus, or do we have extrasensory means of knowing material and mental reality?

10

In applying philosophical concepts to elucidate the meanings of medical terms and concepts, we will be looking for self-consistency in the theories that have been advanced, because those that fail to meet this basic requirement cannot be true, as we will see when we discuss the laws of logic. Coherence of several theories as a criterion of truth is problematic, because untrue theories may cohere better than true theories without thereby being true. Besides, any statement that implies, directly or indirectly, the existence and non-existence of the same condition at the same time cannot be true, since its incompatible elements make it self-contradictory—A cannot be both A and *not-A* at the same time in the same place. Furthermore we will be checking these statements and concepts for comprehensiveness, rationality, pragmatism, and simplicity in the tradition of William of Ockham's razor not to multiply entities beyond absolute necessity, and Einstein's chopper not to invent more processes than are really needed for an explanation. In the end, we will be ensuring that clinical medicine, the basic sciences on which it rests, and the social sciences with which it interacts are engaged in a non-dogmatic and honest search for truth, since their assertions affect the lives of millions of people.

Philosophy dissuades us from taking anything for granted. It fosters the habit of self-criticism as a hedge against the dogmatism and self-confidence that oftentimes takes possession of some scientists and physicians, prompting them to place their self-esteem and ambitions ahead of the quest for truth, and resulting in fabrications and falsifications of data for the sake of proving their pet hypotheses against the weight of empirical evidence to the contrary, which they misinterpret deliberately. Philosophy may not be able to expose the insincere motives of the doctor or scientist who is serving the financial interests of those pharmaceutical and other parties whose vested interests she is beholden to promote for reciprocal, financially lucrative favours, even if she claims the right to a relativistic perspective on these matters. It will, however, alert the thinking doctor to these cases of invalid reasoning from true or false premises to incongruities and false conclusions that do nothing to promote the search for truth and the best interests of her patients.

If the assumptions on which a theory is based are logically inconsistent with conclusions derivable from them, they discredit the truth of those conclusions and that theory, regardless of the validity of the arguments used in the process. Assumptions generally entail more than their immediate consequences, as demonstrated by denying remote consequences to invalidate assumptive propositions. These are the logical steps that will expose some of the warped arguments and the contrived conclusions drawn from them to justify spurious research that is often passed as authentic. In some cases assumptions have to be modified or abandoned because they cause conflict in the fit between theory and new facts as predicted The epistemic principle of closure asserts that if p is known to entail q, then q is known to be true. But q may not be true for several reasons; so one does not know q from knowing p, in spite of the entailment relation between p and q. Scientists and physicians should honestly acknowledge the shortcomings in the theories that they promulgate on the basis of these flawed relations—a spurious causal relationship between measles and autism does not justify inciting public panic.

It follows, then, that if truth is to serve our needs successfully, it should transcend relativism and be accessible to all alike, in spite of the fact that we have inherent or acquired competing individual perspectives on the world; my truth, your truth, his truth may be so tainted, prejudicial, and divergent that they fail to add up to the pure common truth that we all should be pursuing. Besides, relativism is self-refuting, because if it claims that its truth is universal, then its own claims are also relative; i.e., its truth is not universal and can be dismissed by the rest of us. Needless to say, in the process the patient suffers, because his interests are displaced, minimized, or ignored; hence the need for his doctor to be sensitive to his dependence on her respect for truthfulness, humaneness, and truth-oriented worldviews. This view is, regrettably, not the rational one taken by doctors who fabricated the irrational disease "drapetomania".

# Worldviews

Worldviews are the everyday perspectives and attitudes that we harbour, consciously or unconsciously, towards the sorts of things that we believe are in the world, how we think about them and their relationships, how they determine our goals in life under their guidance, and how we plan our whole lives in concert with them. Subtending each one's weltanschauung are the same universal principles that shore up the different values that we each expose in these worldviews. These *values and principles* influence our *attitudes* —how we think and our *perception*—how we see reality. Together they influence our *behaviour*—how we act, and determine our *performance*—what we do. If we stop to re-evaluate our worldviews, in light of our actions and their consequences, we might be persuaded to adopt different values and principles to guide our lives, especially when we realize that we are hurting more people than we are helping. Worldviews relating to the same subject vary in authenticity and plausibility with each person, but they can almost always be reconciled. An attitude to the practice of medicine as a business, rather than a vocation, will colour our perceptions of and interactions with our patients, whether we see dollars and cents in them and

12

objects to be processed as quickly as possible through our revolving door practices or persons in need of our patience (they are patients), help, compassion, and support. It is important, therefore, for medical students to start off their medical carriers with the kinds of worldviews that will position them well to assume a humane attitude toward their patients all the time. It is inexcusable for medical students to go through their training and emerge with a lack of the education that can foster the virtues of caring and compassion. Such lack is sometimes encountered in the field of medical practice, even though these virtues can be cultivated and nurtured and attitudes can be moulded to conform with the compassionate and sensitive worldviews that are expected of students in their dealings with patients at all times. Where the students are not malleable, as some will not be, or where they prefer impersonal contact, they can be urged to consider careers within the medical field or in the allied sciences that do not require compassionate contact with patients.

Philosophy also promotes and provides the means for self-assessment and criticism by, for instance, consoling the doctor in her sense of despair at not being cork sure about what she is dealing with and how to deal with it for the total welfare of her patient and his organ systems. Among other things, philosophy will teach her that nothing is absolutely certain, in spite of the staggering advances of scientific technology that can steer man into space and keep him there for months on end, and in spite of the fact that our patients expect precision in diagnosis and certainty in the ability of our therapeutic manoeuvres to restore them to good health, because we have created that impression in their minds. Certainty and the ideal are only guideposts in the doctor's ongoing courageous and hopeful striving to serve her patients, and they should be made aware of the existence of this limitation, so that they will not live with false hopes. They should know that while physicians continue to press on with efforts to understand and conquer disease and its ravages on their health, neither of them should have any illusions about the ultimate elimination of death when the best that can ever be done is only to postpone it. The philosophically minded doctor will embrace this attitude readily and not pretend to her patients that she can magically solve all their health problems with her prescription pad and pills.

At the same time that the doctor is developing a humane approach to her patients, she also needs to promote in herself a critical attitude to her construction of a pool of general knowledge that she will use in treating them and their specific clinical problems. Her search for truth about disease and illness in the patient-populated world should be devoid of dogma and irrelevancies. The examples *par excellence* of the method of critical analysis involved in this procedure are Evidence-Based Medicine, which uses generalities to solve specific patient disorders, and Precision Medicine, which fits treatment to each patient by his genetic constitution. She uses these and other principles to rationally justify the efficacy of the many diagnostic and therapeutic procedures that she will employ in her management of her patients. This mental outlook calls for a logical approach to clinical problems and case studies, and the appropriate application of available research and experimental findings from the relevant literature to the case of the particular patient that she is managing. In addition, the physician should always be ready to review her actions and improve on them in the light of current knowledge in her field of study and operation; hence the need for continuing medical education courses to enable her to critically assess and interpret any piece of old and new evidence.

Therefore the doctor's dealing with her patients should display a balance of logical and analytical approach with compassion and sensitivity. In the midst of making objective and difficult decisions guided by the evidence of mechanistic cause and effect, she should leave room for respecting the dignity of her patient as a unique, active agent with his own temperament and cultural and other values that do not conform to any of the logical formulae that she uses to solve her stock of science-based medical problems. For doing the latter she most likely does not need any input from the patient and his kin, but for dealing with the patient's stock of problems in his own context, she will certainly need the input of those very persons, and she should be accommodating and receptive to that input to avoid emotionally costly bungling on her part. Patients have defined their dignity as ability to make their own decisions, participation in the enjoyable activities of life, control of bodily functions, freedom from debilitating pain and suffering, and a degree of independence that prevents them from being a burden on their families. Those goals deserve the same kind of respect that Medicine's goals demand.

Too often doctors misdiagnose their patients because they did not pay heed to what their patients or the parents and relatives of the patients were trying in vain to communicate to them. The doctor has her routine of gathering information in the shortest possible time with standard questions that she applies to all cases, and she sees the patient's narratives of his special situations as derailments from her beaten track, besides consuming a lot of her precious time, and yet the essence of diagnosis is the patient's story of his illness and how it has affected his life. Investigative procedures are adjuncts that help to elucidate the doctor's impressions from the patient's account of what he feels and how it affects his daily functioning, coupled with her physical examination of the patient. Without the narrative to direct the focus of her examination and then her technological enquiry, she will be trying to find a needle in a haystack, guided only by speculation. Taking her cue solely from her investigations will lead her into a chase after will-o-the-wisps and chimeras that will only harm her patient in violation of the dictum: do no harm.

The acute care hospital setting is, by its legitimate nature and therefore expectedly, particularly conducive to this kind of detached attitude to patients that deprives them of their personhood in its concentration on the use of technology to make correct diagnoses and institute treatments with alacrity; but facilities for the care of chronic illnesses for which the only beneficial measure is emotional support for the patient also seem to be failing in their duty to provide solace to their residents, as if they have equally pressing concerns to attend to. This attitude is wrong and it calls for corrective action.

## Biophysical versus Biopsychosocial Medicine

By and large, the academic training of medical professionals (doctors) is strictly scientific: acquisition of facts and data in Physics, Chemistry, and Biology provide the basis for the understanding of normal form and function as represented in human Anatomy and Physiology. Pathology represents deranged form and function, which underlies all disease processes, and it helps to explain disease patterns in Medicine, Surgery, Obstetrics, and Gynaecology. Pharmacology equips the trainee with a repository of drugs to treat most pathological states. Psychiatry is left out in the cold to explain itself; it still cannot find a niche for itself in the sciences, as we will see in the discussion in chapter 5, perhaps because it is more art than science. Philosophy featured once only and for a brief period in my medical student career when I had to write a critical essay on the contrasting outlooks of orthodox and unorthodox medicine, material for which was not to be found in the medical textbooks. That was my first eye opener to the assumption of a tolerant attitude towards complementary medicine, which the medical establishment despised and to which it had not given any serious positive thought at that time. Today the practice of complementary medicine is still highly prevalent in the "old world" and is fast gaining roots in the "new world", thus attracting much attention as more and more people are becoming disenchanted with the impersonal attitudes of those who practice allopathic medicine and their failure to provide them with the answers to important questions about their total welfare that alternative medicine is trying to provide.

The model of training and practice of medicine that has prevailed for many years is the biophysical model based on the scientific basis of medicine that concentrates on explaining derangements of body function and their correction on a mechanical basis. It has maintained strict separation of mind from body while maintaining that disorders of the mind can still be attributed to molecular dysfunctions in the physicochemical constitution of the patient. According to this model, disease is explicable on the physical basis provided by the role of scientifically identifiable microbes in derailing normal physiological function, and scientific methods are all that is needed to account for and correct the adverse effects of disease. The role of psychological and social solutions has traditionally been considered to be secondary and less relevant, because they are not test tube procedures; but it is now becoming more apparent by the day that such approaches are vital to solving today's psychiatric problems that have escalated to lethal confrontations between law enforcement and psychiatric patients. The doctor's advocacy role is clearly cut out here to help save those lives.

This is the atmosphere in which doctors have been trained over the years, and are still being trained today. The approach to the sick person has always been and is still undeniably mechanistic and reductionist, finding and fixing cellular and organ dysfunctions of his mechanistic body while displacing and sometimes omitting the humane aspect that contributes equally to the recovery of the whole person from the effects of his combined mechanical and "spiritual" impairment of function. In the midst of the confusion engendered by a myriad of available and sometimes poorly understood diagnostic and therapeutic procedures, patients, no less than some doctors, rely on gleaning information about these procedures from sources that harbour vested interests. News media and Pharmaceutical companies encourage patients to request special tests and drugs ostensibly to make hitherto unknown choices available to them. Some doctors with vested interests in these enterprises even direct patients to selected diagnostic facilities for questionable procedures and motives. Health Management Organizations (HMOs) assume the prerogative of dictating to doctors what investigations and referrals they may initiate, under the direction of non-medical office staff, in the interests of saving money to provide hefty salaries and bonuses for their chief executive officers (CEOs) and boost their profits. (See Jennifer L. Lycette. AITA for Pointing Out to the Insurance Company That I'm the Expert on My Patient? - Medscape - Jul 13, 2022.)

But no one ever takes the trouble to popularize and promote concern for the patient-as-person with unfulfilled, environmentally caused physical, psychological, and emotional needs requiring attention. Thus the atmosphere of the healing art of medical care has become commercialized by the detached scientific approach, which is accessible to misinterpretation by the uninitiated in their search for quick fixes. They fall into the trap of assuming that the generalities of scientific laws and concepts can be applied unmodified to individual

16

patients who have inherently different personalities and responses, as revealed in the narratives of their illnesses in their multifarious contexts, to which some doctors do not pay any attention. A case in point is the patient (and doctor) who knows that a heart attack is characterized by crushing chest pain of left anterior coronary artery thrombosis radiating to the left arm, jaw, and neck, but does not allow for the patient's account of similarly disabling pain of right posterior coronary artery thrombosis, such as I experienced, which starts in the epigastrium and radiates to the back like the pain of indigestion, peptic ulcer, or pancreatitis, contrary to the stereotype. This is the problem of trying to fit the patient into the disease mould, when the mould should be modified to accommodate the patient's unique problem. In this age of automated EKGs and other monitoring devices in which some doctors place implicit faith, this problem of fit is a common source of confusion in diagnosis and treatment.

There is need for recognition of the patient as a complex of interaction between body, mind, and social circumstances, with each one exerting its influence on the others. The patient who lives in poverty and has difficulty making ends meet can be the subject of constant worry about the welfare of his family and thus develop peptic ulcers by well-known mechanisms, or his immune system can suffer to the extent of rendering him an easy target for invasion by microbes of all kinds, besides just being exposed by his rueful circumstances to unsanitary conditions that breed epidemics like Cholera. On the other hand the chronicity of disease can result in depression and cause the patient to neglect his social circumstances and thus initiate a vicious cycle. Hence the need for understanding the social, psychological, environmental, economic, political and other deleterious influences, in addition to the biological ones, on the genesis of the patient's clinical state to ensure a satisfactory outcome (to himself and to his physician) of his treatment, more than the mere treatment of his disease. The biopsychosocial aspect of medicine integrates concerns with both body and mind of the patient. It places his physical disease in its psychological and social contexts as they affect his perspective on it, its effects on his psyche—his hopes for functional recovery and his fears of succumbing to it, and his particular social milieu with its adequate or inadequate supportive structures geared at preventing complications from his illness. Hence the role of psychology (even placebo) and the holistic approach in the treatment of the patient's illness more than only his disease, and hence also the need for the political systems of states to pay more attention to human needs.

# Core questions.

In the preceding paragraphs terms like disease and illness, cure and heal feature in a manner suggesting that they severally or collectively constitute complementary or contrasting pairs. It is time now to pose a series of questions that every student and practitioner of medicine should be constantly asking herself in her contact with sick people. To begin with, what logical principles do we use to define disease and illness, and what tools do we have at our disposal to help us formulate the multiple concepts that we use to talk about these two conditions? We have already dealt briefly with questions relating to our knowledge of disease and illness, both by description and by acquaintance. Knowledge of these conditions by description, without first hand or personal acquaintance, is always necessarily objective, calling for understanding, sympathy, and empathy, but not of the same caliber as that originating from knowledge of them by acquaintance—having lived the experience of disease and illness and having a better appreciation of what patients are suffering. Does every doctor have to live the experience of disease before she can be more appreciative of patients' subjective feelings, or can she be trained to be more sensitive to them? Some live through one or two experiences, but many have to be trained to appreciate their patients' feelings by listening to him and by projecting them on themselves.

Next we need to know what we mean when we talk about the person who harbours disease. Is a person primarily a body with a mind or psychological experiences added to it, or is a person an irreducible complex of body and mind; and what is mind, anyway? How are body and mind related, and where do they interface? How does disease or disorder of one cause that of the other or differ from it? Can we cure the mind of disease, assuming that it does become diseased, as we can cure the body of disease? If not, why not? How do we prove the causes of mental disorder when we have no means of laying a handle on them, let alone bringing them together with their so-called effects to be able to observe changes resulting from this "spatiotemporal" approximation, as seen with bacteria and healthy living tissues where we can boldly postulate cause and effect?

If a patient has a broken leg, do we fix the leg or do we fix him and his leg? If he has alcoholic steatosis and steatohepatitis, do we reduce his disease to biochemistry and pathology, correct the disorder technologically and stop there, or do we try to help him change his habits and treat his mind while we treat his body; i.e., treat the whole person? If so, what do we need to do to treat the whole patient and his disease? If we manage to restore his form and function to normal, what we mean by normal in different situations, and how we establish it, so that we can know when the patient has recovered from his illness and is back to normal. Is it when his laboratory tests and imaging studies show conformity with established parameters of normality, or when he is comfortable doing what he used to do before, or when he is feeling well? How well?

Then there is the question of the cause of disease as it relates to the general philosophical problem of causality. We postulate that everything (except quantum phenomena) has a cause, but at no time do we ever see the cause bringing about the effect. All that we ever see or observe is one event following another, never one event causing another. Is disease caused or does it arise spontaneously? If it is caused by bacteria, viruses, fungi and other micro-organisms, have we ever seen these organisms causing disease or do we just find them in the neighbourhood of diseased tissues before, after, or at the same time as we observe disease? Can we legitimately conclude *propter hoc* (because of this) from *post hoc* (after this)? What about other factors known and unknown to us that enter into this equation of cause and effect? Anyway, what is the nature of disease (marked by an objective pathological connotation)? Is it the same "thing" as sickness (with its mutable sociological connotation) or illness (which has a subjective connotation), and how does it differ from both? Is disease a thing which people catch and which then possesses them; and does it have an ontological status, i.e., does it enjoy independent existence among other existents in the world, or is it only a concept, existing purely as a mental construct from physical and psychological phenomena? What is disability and how does it compare with disease and illness in the degree to which it impedes the normal functioning of the person? How does the biological manner in which the disease $x$ presents itself in a patient differ from how a patient as the person with disease $x$ and feelings, goals, and aspirations presents himself, and is the student or the doctor always sensitive to this phenomenological difference?

How are we able to re-assure a patient and plot a prospective course of recovery from illness for him? Is it by virtue of what we call the natural history of an untreated disease process and the interposition of our treatment to alter this natural course, or is it by using statistical data to calculate his odds, taking everything into account—his sex, age, weight, racial origin, genetic inheritance, environment etc? How much reliance can we place on these measures of prediction in the face of aberrations caused by non-linear dynamics as postulated in chaos theory? Are we never flabbergasted by frequent strange turns of events that defy our prognostications, such as the miracle babies who survive prolonged immersions or falls from prohibitive heights? Does the practice of medicine allow room for an aesthetic appreciation of persons as subjects of value apart from their sorry state as objects of ill health? If it does, that fact is not apparent from the kinds of concepts in which it dabbles, nor is it suggested by its language or by the attitudes of its practitioners. If it were, we would not hear so many complaints of the poor bedside manners of some practitioners, reflecting a lack of sensitivity to the dolorous states of some of their less socially significant patients.

Finally, there is the inevitable lapse of time towards the end of life for every person; that time when an ex-patient of mine, Mr. M, felt that the efforts of his doctor would be futile in trying to preserve his life by performing his usual rituals. Does medicine have the moral right to interfere with a patient's choice to have his life of futile suffering terminated? Does it also have the moral right to prohibit the further development of a severely disadvantaged fetus from reaching maturity? What is the role of existing ethical theories and moral considerations in helping to shape decisions related to these thorny issues? How do we see time's influence on disease, illness, and treatment? Have we emerged from voodoo and religious ascriptions of disease and illness to a better understanding of the mechanisms of disease as underlying some cases of illness, or is the pendulum swinging back in the form of more varieties of so-called unorthodox methods of diagnosis and treatment of disease and illness, because of the emotional barrenness of our biologically and physico-chemically based approach to patients' illnesses that excludes their emotional involvement? What does the future hold for orthodox Medicine?

The practice of clinical medicine as a science is rightly concerned with maintaining our elusive states of health by preventing and eliminating the many identifiable diseases that threaten our bodies; but as an art as well, it should be paying heed to the illnesses that afflict the personalities of many a sick patient whose physical anguish cannot be artificially separated from its ever present and sometimes solely overwhelming emotional component. Besides, some patients may have only mental symptoms of illness that the doctor should not neglect in her chase after the physical, because if she does not do it, the politicians will certainly not care for the mentally ill, and their health care will continue to be neglected and underfunded while they relentlessly and blindly pursue self-aggrandizing projects.

The healthy but elusive state, as the normal state that we are all chasing after, is not a fixed or steady state; it is subject to variations and fluctuations within the wide range of states that we will discuss under the heading of "normality"; nevertheless, the chase goes on unabated as increasing recognition is given to the role of evolutionary factors in the nature of disease and how it should be treated in concert with nature's own ways of dealing with it. The discussion of Darwinian Medicine in the closing chapter of this book is meant to direct attention to that discipline and its importance to clinical medicine in rounding off its quest for the whole truth. But we should never forget that "in clinical medicine [where we] encounter ill people rather than biological processes, the naturalistic view should be augmented by a phenomenological perspective. This perspective encompasses both the objective biological body and the lived body, the body as it is subjectively experienced."[2]

20

# 2

# Philosophical Tools and Concepts

## Definitions

### Role of definitions

Definitions play a pivotal role in the delineation of concepts. To be able to identify anything, we must have a working definition of what we are looking for. To go looking for a unicorn, even if it does not exist, without some description of what it will look like, if we should happen to encounter it, would be senseless. So we arbitrarily define or describe it as a horse-like animal with a lion's tail and a single horn in the middle of its forehead, and then we set out to look for it. That way we eliminate all other creatures that do not bear a resemblance to our target animal. But we cannot afford to dissipate our efforts chasing after imaginary objects like unicorns; so we need to take stock of the sense and reference of each object-specifying word that we use. We may understand the meaning of the word, but still fail to locate an object in the world to which it refers, because there is no object that corresponds to that verbal description—the concept entailed by the syntactical combination and semantic import of the words in our statement is devoid of the object referred to by the use of those words, i.e., the concept is empty.

The word 'unicorn' makes grammatical sense in the above sentence—in the context of imaginary animals—but it has no reference in real life, because unicorns do not exist. So our theoretical definition of a unicorn as a horse-like creature with a horn in the middle of its forehead will serve no practical purpose; it is only a story relative definition. If we are looking for a real live creature like a lemur, again we need a descriptive definition to help us focus on and find the real lemur, not just any small mammal. Therefore our definitions should not be the trumped up ones like that of a unicorn, but they should satisfy both necessary and sufficient conditions for the existence of what we are trying to define; for example, trauma of a degree that is more than just minimal is a necessary and sufficient condition of bone fractures, because fractures do not occur without preceding trauma as a necessary condition, and only trauma of a that degree, as a sufficient condition, will guarantee the occurrence of fractures, since not every small degree of trauma will cause bone fracture.

Many of our definitions will have ultimate reference to what is considered normal in the particular domain of their reference. The problem of normality will be discussed later, but for now we need only mention that standards will be required to establish what disease or disability is in relation to its deviation from what obtains normally. So our definitions of these conditions should be precise for correct identification, otherwise we will be talking at cross purposes all the time, and in the ensuing confusion we will place the welfare of our patients in peril by failing to identify their problems precisely. For example, if we stipulate that the feel of water at 37°C is normal, and then discover that to the hand previously dipped in water at 32°C water at 37°C feels warm, but the same water at 37°C feels cold to the same hand previously dipped in water at 41°C, we have to wonder if the water is warm or cold and if our definition of the feel of water at 37°C as normal without attached conditions still makes sense.

## Types of definitions

In dealing with disease, disability, and illness, we first need reportive definitions, also known as descriptive or lexical definitions, because they convey the commonly accepted use of words, such as the Oxford dictionary definition of disease as a serious derangement of health, or of pain as the sensation that one feels when injured. The accuracy of such definitions cannot be assured, because they derive their meanings from their everyday use, which sometimes lacks precision, and may even be as ambiguous as that disease is an affliction of the tissues that may or may not cause derangement of bodily function. Next we need stipulative, precising, contextual, or disciplinary definitions, which will set limits for new applications of commonly used words and explain sentences in which a term or word is used, such as that fever is a certain level of temperature under specified conditions, not just a feeling of warmth to the touch. These definitions force us to limit the scope of application of the definiens in a well meant effort to ensure that we are all talking about the same condition, 'A' or 'class of As', in the particular discipline of discourse by excluding the classes of Bs, Cs,....Zs from the class of As. The counter to stipulative definitions is that in laying down restrictive new rules for the use of words, they represent only the worldview(s) of the stipulator(s), which may not be widely shared; hence the unending, contentious differences of opinion that we are witnessing as to the personhood of fetuses, humans in persistent vegetative states (PVS), and anencephalics. (PVS "persons" are alive, but they lack self or environmental awareness, interaction with others, and evidence of purposeful responses to various stimuli, although their cranial and spinal nerve reflexes are preserved).

The counter to this counter is that if we are at variance about what we each mean by persons, disease, and the modes of action of drugs, for instance, we cannot concur in our prevention of disease or recognition of it when we encounter it as manifested in patients, and we cannot make appropriate plans and take appropriate concerted action to treat those who are afflicted with it. How we define conditions determines how we will react to them and how we will regard those who are afflicted with those conditions—with sympathy, empathy, culpability, or indifference. So our definitions of these conditions must be precise and have limits of applicability. Definitions that are not concise but too wide will include conditions that do not qualify, and those that are too narrow will exclude conditions that qualify as members of the class under consideration. In both cases, we will defeat our efforts at delineating our concepts precisely, and we will end up wasting time, money, and effort on aimless and misdirected quests for chimeras, besides being caught up in redundancies. Examples are given below in the definitions of seizures and fever.

Next, we need <u>exhaustive</u> definitions to include all possibilities, e.g., seizures are paroxysmal discharges of neurons resulting in altered neurological function. If they are <u>exclusive,</u> they entail alternatives, only one of which can apply or be true at any time, e.g., seizures are paroxysmal discharges of cortical neurons resulting in altered neurological function (temporal lobe seizures), or seizures are neuronal discharges arising from deep midline structures in the brain stem or hypothalamus resulting in altered neurological function (absence seizures). They may be <u>circular</u> when they include the terms that they are purporting to define, e.g., seizures are abnormal neuronal states resulting in convulsions, where convulsions=seizures; or <u>obscure</u> when they employ vague, ambiguous, or metaphorical language to explain the terms in need of clarification, e.g., seizures are spells. Our definitions should not be so loose and imprecise as to leave room for vagueness and ambiguity. People should not have to second guess our meaning, because that makes for error in judgment and treatment resulting from misunderstanding, and it can cost lives. If we define fever as a temperature of 38°C or more, we have to specify how the measurement was carried out to be able to apply it to all cases—orally, rectally, or under the armpit; and we should also specify whether the person sucked ice or drank tea before the oral measurement, or was in cardiovascular shock during the armpit measurement. Furthermore, there should be no counterexamples to the definition and its converse (a counterexample is a fact that falsifies a stated all-inclusive claim; e.g., all cases of fever are due to infection; but fever can also be caused by aspirin overdose—overdose with the very same drug that is used to lower fevers).

Finally, we have essentialist and nominalist definitions which are the converse of the definitions that we have been considering. Whereas the latter define the essential properties of subjects or objects in a manner that often demands the use of many words (sometimes approaching pleonasm) that make precise communication about diseases and their diagnostic labels extremely difficult, nominalist definitions facilitate this process by serving as pointers in the same way as the definiendum represents the definiens in as few words as possible. We use nominalist definitions as a pointed, abbreviated way of communicating about diseases like Type 1 Diabetes mellitus, instead of enumerating its essential properties of immunologically induced destruction of pancreatic β cells resulting in insufficiency of insulin secretion and excess of counter-regulatory hormones (glucagon, adrenaline), which in turn results in deranged glucose metabolism accompanied by the symptoms of polyuria, polydipsia, polyphagia, weight loss, etc. Nominalist definitions are useful labour-saving device that quickly and unambiguously pinpoint their subject.

Definitions couched in negative terms should be avoided, because they fail to refer, e.g., schizophrenia is a non-physical disease. So what is it?

## Applications of definitions

The combination of all these types of definitions helps us to delineate concepts of disease and other conditions by stating the essential or defining characteristics that any of those conditions should have to be examples of these concepts. It also helps us to identify accidental characteristics that accompany certain diseases some or most of the time. We never see the disease, but we deduce its presence from symptoms, signs, investigations, and the course of the illness, just as we never see a magnetic field but deduce its presence from its specific effects. What makes condition $x$ a disease and disqualifies condition $y$ from being a disease does not depend on the fancy or emotions of a social group but on the causal essence of disease; namely, the basic characteristics relating the etiology and symptoms of condition $x$ that also satisfy the necessary and sufficient conditions for $x$ to be a disease. For disease $x$ to occur, then, those conditions must obtain to guarantee its existence, and only those conditions can cause disease $x$; without them—micro-organisms, states of the body such as genetically lowered resistance or deficient barriers, environmental conditions—the disease cannot occur. Disease $x$ (definiendum) is defined in terms of causal conditions 1,2,3...(definiens), and the definiens (the sum of conditions 1,2,3...)

applies to every condition that can be described as the definiendum (disease $x$), and to nothing else. Conditions 1,2,3...are also the connotation or the attributes of the disease whose denotation or reference is $x$; (the connotation of $x$ is the assembly of signs and symptoms that give meaning to the concept $x$ by being both the necessary and sufficient conditions for the recognition of $x$ or ascribing symptoms 1,2,3...to disease $x$. The denotation of $x$ is the disease to which $x$ refers or the disease embraced by the concept $x$). The two categories must be freely interchangeable without altering the sense or meaning of what is being asserted by the use of either one; i.e., the concept $x$ and its reportive definition of disease $x$ or conditions 1,2,3... must be applicable to the same cases. A definition, according to this analysis, proves to be like a coin with the definiendum on one face and the definiens on the other face.

There appears to be a need for a universal definition of disease to combat the relativity and parochialism of existing societal sentiments about what constitutes disease from the perspective of each society, even though the concept of disease has paradoxically been abstracted from specific situations as they occur in their cultural contexts. But disease cannot answer to only one concept, and trying to collate different and necessarily inconsistent local concepts of disease into one universal concept will only create opportunities for equivocation on such a concept. Some people will argue that homosexuality is a disease, because it is a deviation from their normal practice. No one can argue against them if definitions of disease and normality are allowed to be parochial and subject to prejudices and ignorance; but if we have rational definitions derived from collaborative efforts of medical scientists, clinicians, and other humanists—including patients and how they view their situations—as to what might constitute disease in each one of those conditions, we can expose the fallacy in this argument and correct other illogical concepts of disease entertained by different groups of people, such as the infamous disease of drapetomania defined by Samuel Cartwright as a "disease of the mind causing [slaves] to abscond."[1] They went on to recommend that to prevent them from running away, slaves needed only to be treated like children in a manner consistent with the submissive state that they were meant to occupy for all time. This outrageous definition of that so-called disease and the cure prescribed for it by these doctors are amazing both from the point of view of science and of intellectual maturity; as amazing as Immanuel Kant's advice for flogging African servants and slaves into submission with a split bamboo cane to inflict maximum pain and suffering. Fortunately, even base disvalues and prejudices change with time and circumstances as more sensible people add their sane and rational points of view to these insane and irrational points of view.

# General Concepts

## Concept formation

Concepts are mental devices or abstractions used to delineate the scope within which the senses of words in a language pick out their referents, which may be material or syntactical objects, by broadly defining the essences that objects must possess to qualify for inclusion in a particular category, as in the concept of disease defined above (not Cartwright's misguided definition). They make sense of all our referring statements, and by their means we are able to refer to objects like tables and to recognize a table in real life when we see one of the many different tables referred to by the concept word 'table'. They are our means of reaching out to reality generally and specifically. One does not have to undertake the impossible task of describing the characteristics of every table in the world to enable people to recognize a specific table when they see it; the general concept that objects must satisfy to serve as tables will suffice for the purpose of creating the class of tables. Concepts thus enable the doctor to discriminate disease from non-disease, failing which she is unable to pick out those conditions that will most probably satisfy the necessary and sufficient criteria for inclusion under general terms like 'disease'. Her intelligent use of the word 'disease' in language constitutes the sufficient conditions for her possession of the concepts referred to by that word. Thus, knowing how to use the word 'disease' implies the ability to fit the concept of disease to her experience of the relevant conditions; it does not entail abstracting the concept from a conglomerate of these conditions, but misapplying the concept entails failure to communicate. As Lewis Carroll wrote,

> 'When I use a word,' Humpty Dumpty said, in rather a scornful tone, 'it means just what I choose it to mean — neither more nor less.'
> 'The question is,' said Alice, 'whether you *can* make words mean so many different things.'[2]

Without the use of concept words to refer to objects around us, we would have to be pointing all the time without really conveying any variety in meaning to our gesticulations beyond just drawing the attention of others to the presence of those objects—a difficult task when pointing to a variety of diseases that have no ontological status in sick persons. Concepts also help us to avoid individuating or enumerating individual diseases where use of a collective indicator will suffice. Altogether, those diseases that have common causes and clinical manifestations function are grouped into classes by their respective concepts, e.g., the concept "mucopolysaccharidoses" refers to a variety of diseases characterized by derangement in the metabolism of mucopolysaccharides as distinct from mucolipidoses in which mucolipids are at fault.

To be truly representative of our limited experience of reality, concepts must deal in the essential (constitutive and defining) properties of things thus far known to us, so that they can facilitate the categorization of information by providing stable meaning and scope to the words that describe objects, diseases, or situations for our better understanding of known and surmised aspects of the world, and for the growth of knowledge. Uninstantiated concepts that lack representative objects or specific diseases in reality are vacuous, and accidental (incidental) and fleeting properties are useless.

Our standard extrapolative method of forming concepts raises problems, which Peter Geach criticizes severely in his book, *Mental Acts*, because it consists in the retention of only similarities among objects and elimination of dissimilarities, and it fails dismally to account for the logical concepts of alternativeness and negativeness, because reality is not populated with examples of objects from which "the concept of *or* or of *not*."[3] can be formed by abstraction. In the perception of a maple leaf, for example, we are presented with particulars, e.g., colour, texture, shape, consistency, and odour, from which we form the image reflecting what we consider to be the most essential identifying characteristics of a maple leaf. We then use this image as the basis for assembling all similarly constructed and coloured objects to form the class of maple leaves. At no time are we presented initially with universal leafiness, greenness, etc., which we then combine and use to apply to particular objects, so that we can classify them as maple leafs. All our generalities, or universals in philosophical parlance, start from the recognition of particulars like the object-leaf and its qualities. They do not descend upon us like "The Forms" of Plato's world with its extra-terrestrial, unknowable "Forms" of greenness, leafiness, etc., which might not even have concrete examples in the world of particulars that we share with Aristotle, but in which all material maple leafs are presumed to participate. The formation of concepts, understood as the way we use words and could possibly use them in discourse, and "the criteria or principles by which those uses are determined"[4], includes asking the right questions to get the right answers, such as asking how many Mondays in are a week versus whether Monday is sleeping, because the concept "sleeping" cannot be applied to Monday without committing a category mistake. Mondays don't sleep.

At any time, of course, personal and domain variations in concept formation ultimately have to conform to the ideal. If the ideal concept of a rose in one world is R, individuals A, B, C, and D will each have her own concept of that rose, viz., $r_1$, $r_2$, $r_3$, and $r_4$, corresponding with her perspective, and so will the corresponding individuals in the world in which the concept of a rose is Rx. Barring any widespread delusion, we expect roses R and Rx to have similar qualitative characteristics of fragrance, colour, and beauty. We also expect that the sum of $r_1$, $r_2$, $r_3$, and $r_4$ will be close to the ideal concept R, and that the same situation will prevail in the other world with a different set of the perceivers of Rx. In both worlds we also expect differing individual concepts to still make communication and appropriate action possible within the context of their application, although a difference in the ultimate make up of R and Rx will surprise each group when they come to examine the other group's rose. One rose, R, will be substantial, and the other rose, Rx, will be ethereal, compelling each group to change the limits of the application of their concept of rose and to restrict it to only their rose in their world, because, although the language and the logic of R and Rx are the same in the two worlds, their ontology is vastly different, even contradictory. Nevertheless, communication about roses in each world will still be possible because A, B, C, and D will still be talking about the same object perceived from four different perspectives. In the case of disease, if some people's concept of it is Di, while that of other people in the same world is Dx, they will always be at odds when they think that they are talking about the same condition.

Concepts thus enable us to understand and use language semantically and coherently only within our universe of discourse in a manner that conveys commonly shared meanings to speaker and listener or to author and reader. Ludwig Wittgenstein epitomized it thus: "*The limits of my language* mean the limits of my world."[5] In stressing the fact that reality cannot defy the pervasive nature of logic by allowing the existence of logically impossible situations and objects, he added that "the limits of the world are also its limits."[6] But one would like to think that logic transcends the limits of the world to encompass the universe. An illogical universe is unimaginable, because logic does not deal in contradictions, and the concept "round square" could not possibly make logical sense anywhere in the universe. It lacks an ontological counterpart, because it consists of a physically and logically incompatible combination of predicates that cannot be simultaneously realized in the same subject or object. Our world, in spite of having round objects and square objects, does not have objects that combine these attributes by being round and square at the same time. Such a concept, with contradictory constituents, would be an absurd representation of the real world, because reality and logic as truth, do not deal in contradictions and inconsistencies. In the case of the concept like "tree", all the plants that conform to a particular form satisfy the concept, although they may be of different sizes or shapes, like firs, jacarandas, or palm trees. These wide variations do not constitute any form of contradictory application of the concept, because their essential characteristics are not logically incompatible, although their non-essential characteristics like colour, shape, and size may be so. Empirical concepts (those formed *a posteriori* or from sensory experience) formed in this case must originate from trees and what exists in the world if they are to have content and meaning, unlike rational concepts like those of round squares that originate purely in the mind (*a priori*) and can have meaning without content. Both concepts may, however, depend on the definition of their terms for their origin, but their truth is determined by causally related objects in the world that fall under them.

As we have seen, we form concepts by first abstracting similar properties from objects, and then combining them, to the exclusion of dissimilar ones. For example, from objects *xa, xb, xc* we usually form the concept "*x*" from the presence of that property in all three objects, and we disregard the dissimilar properties a, b, and c. We thus end up with a very limited and selective concept that is not representative of all aspects of the three objects, and one that cannot be reduced to its original constituents. How does one reduce *x* to *xa, xb,* and *xc* when *a, b,* and *c* are lacking in the concept "*x*"? If, on the other hand, we formulated our concept by extracting *x* (similarities) and combining dissimilarities to represent them with *Y*, then the resulting concept *xY* would adequately embrace the properties of *xa, xb,* and *xc,* and each would be derivable from *xY*, because *Y* could be reduced to *a, b,* and *c,* since it is a compound of them. In the present context, which is limited to human health and disease, the concept of arthritis, for instance, will refer to the well delineated general category of traumatic, inflammatory, infectious, or degenerative joint disorders that any medical condition, or specifically, any joint disorder will have to satisfy, if it is to be included in the scope of that concept.

## Medical concepts

A good medical example is provided by the microbiological concept of symptomless urinary tract infection (UTI) as a persistent, significant bacteriuria, especially in patients with abnormalities of the urinary tract or indwelling catheters who will require treatment even if they do not have symptoms. To be significant the bacteriuria has to be $10^5$ or more colonies in a clean catch or midstream urine specimen, and to be persistent it has to occur in at least three sequential urinary specimens. For symptomatic UTI with convincing accompanying clinical symptomatology, either a pure culture of $10^4$ bacterial colonies on a single occasion and similarly obtained, or a catheter specimen with fewer bacteria will qualify as a UTI, while as few as $10^1$ bacterial colonies in a single urine specimen obtained by suprapubic aspiration will also qualify as a UTI in the same circumstances. These considerations suggest that the limits of our concepts are largely regulated by the context of discourse and the assigned meanings of particular words and phrases employed in that context. So the limits of someone's concept of UTI will be coextensive with the limits within which she understands and uses the connotation of the diagnostic denotation UTI: clean catch midstream, catheter, or suprapubic puncture urine specimen. In the words of Wittgenstein, "the meaning of a word is its use in the language"[7], and hence the truth of a statement about UTI is a function of the use of its component terms consistent with operative rules in these contexts.

The kind of concept that we have been considering in the UTI example qualifies as a *synthetic a priori* concept of Kant, because its meaning is rendered true by its formulation from both definition and our prior experience of the world. We do not assign the diagnosis of UTI solely on the basis of what we discover the world to be like, but also on the basis of the definition that we have formulated, and to which the world must conform if it has to satisfy our definition. At the other extreme is the pure *a priori* concept, which is entirely independent of experience. For example, 2+2=4 could easily have been 2+2=5 by definition without amounting to the nonsense that our present understanding of the sum of 2 and 2 would make us believe that it is, because nothing in the external world necessitates that 2+2 should equal 4 instead of 5; only our definition does. Putting together two objects with two other objects prior to the stipulation of the definition of 2+2 as 4 a thousand times will not tell us that we have four objects as a result of this manoeuvre. Other concepts like "some", "all", and "not" are formed as the result of the use of corresponding words in language to include only a few (some), to leave out no one (all), and to be absent or lacking in specified properties (not), reflecting the nature of what exists in the world.

Clinical Medicine has adopted every possible method of constructing its concepts of disease, thereby making it possible to recognize classic, atypical, and preclinical conditions, and syndromes and their *forme frustes*. The latter are partial manifestations of a disease complex or syndrome where the less frequent clinical manifestations also replace the more typical ones in the recognition of the syndrome. These concepts of clinical medicine are not static; they evolve with the advance of knowledge in the etiology and pathogenesis of disease, shifting the emphasis from symptoms as disease to a more comprehensive picture of disease, beginning with etiological factors, through pathogenesis, to targeted treatment and prevention; e.g., Kawasaki disease started as a syndrome and has since graduated to disease status with a specific pathophysiology and mode of treatment that does not consist of treating only the syndromic manifestations but is directed at halting the progression of the pathological changes that produce those manifestations, even though its etiology is still speculative. In the case of PFAPA syndrome (Periodic Fever, Aphthous stomatitis, Pharyngitis, and Adenitis), an interconnection has yet to be made among the varied clinical manifestations for the syndrome to evolve into a disease.

The word "syndrome" will therefore define a constellation of recurrent, clinically recognizable and essential features, which consist of a pattern of non-accidentally compresent signs and symptoms marked by regular manifestations to form an equally recognizable clinical picture that suggests a particular disorder in which one or more features always entail the presence of the other accompanying features. The boundaries of the syndrome cannot be circumscribed, because they are not defined by a specific limiting etiology; hence its *forme frustes*. Although the label "syndrome" is mostly applied when the etiology and pathophysiology of the constellation are still unknown, it is also used when they are known, as in Nephrotic Syndrome (defined by proteinuria, edema, hypoalbumia, and hypercholesterolemia). This syndrome is common to many diseases with different etiologies but similar manifestations. Disorders of adrenal corticosteroid secretion are classified as Cushing syndrome, a clinical condition marked by an excess of cortisol from the adrenal gland caused by any prior condition or by extraneous cortisone, or Cushing disease only when a pituitary neoplasm is secreting excessive adrenocorticotropic hormone (ACTH) to cause the same clinical state. Such conceptualization and categorization of knowledge allows the accommodation of new information for the better understanding of the causes and natural history of disease from its pre-clinical stage to its termination in a cure or death; it also allows for the elimination of what is not disease but presents like disease; and it facilitates the logical institution of preventive measures and early intervention to terminate or alter the course of the disease to the advantage of the patient as person, which fixing only the organ, part, or system that is causing his illness will not do. Furthermore, it accommodates conditions that are non-causally related to the syndrome, e.g., atlanto-axial dislocation in Down syndrome, other syndromes that have the features mentioned above, and several other disorders that are represented in a particular syndrome.

## Specific concepts

We have determined that the basis of discourse in any branch of knowledge depends on the use of concepts. For Medicine to establish its discipline and to adopt a critical attitude toward itself, it has to employ some of these concepts that we have already discussed as, as well as some basic and pivotal philosophical concepts.

# Skepticism

Skepticism stems from the teachings of the Greek philosopher Pyrrho of Elis against the claims of Dogmatists to knowledge. He taught fallibilism: we cannot acquire knowledge of truth because of the uncertainty that clouds every object of human knowledge; therefore we should suspend judgment about knowledge and the possibility of knowing true reality and the physical laws that regulate it. His successor, Sextus Empiricus, taught an extreme form of skepticism that advocated suspension of all judgment about all knowledge. Skepticism therefore promotes a mental attitude of radical doubt about information and beliefs that are otherwise regarded as the foundations of our knowledge, because we can't claim certainty or eliminate the lurking possibility (not actual occurrence) of error in our usual perceptions, observations, inductive inferences, and memory, as discussed under Realism below. The skeptic makes this claim even as he cannot say how he knows what he is proposing with the certainty that he is demanding of us, or how he expects us to understand his demand without possessing the knowledge required for that understanding, which he is denying and which he also does not possess. If no one is justified in believing anything or knows anything, then he also is not justified in believing that he knows what he is claiming that we don't know. He claims that we cannot assume any $criterion_1$ of truth, because doing so will require a $criterion_2$ of that $criterion_1$, which will require a $criterion_3$ of $criterion_2$, and so on *ad infinitum*, and he advances a similar argument against our ability to prove virtually anything, thus pulling the rug from under his own feet by his own principles. So we can conclude against his railings that uncertainty does not nullify our ability to acquire knowledge. We cannot live in ignorance for fear of errors that we can recognize and correct by virtue of our knowledge of what is right, because even if we cannot be certain of anything, it is not the case that we cannot be certain of everything, i.e., we are not obligated to doubt everything. Saying that if we can't be certain of anything we cannot be certain of everything, i.e., we are obligated to doubt everything, does not make good sense. As long as we can adequately and rationally justify our beliefs, ensuring that we hold only correct beliefs and remembering that we don't make it the case that y is θ merely by believing it, that is enough for us to get along in the world. So we conclude against his railings that uncertainty, doubt, and error don't nullify our inclination to believe our only reliable sources of information, e.g., perception, and our ability to acquire knowledge from them.

Nevertheless, the uncertainty of the perceptual situation is such that the only knowledge that we and the skeptic can claim to possess is that arrived at by way of justified belief based on sound, irrefutable evidence, not by the inaccessible, presumed infallibility and indubitability that we would like to claim in support of judgmental certainty, because the latter is beyond our reach. Our only refuge from decimation by the skeptic is entailed in the proviso that if we can muster enough convincing justification for our beliefs about reality, and if we can demonstrate the utility of those beliefs in our survival as a species in concert with the conventions by which we regulate our lives and our relationships, then we can withstand his onslaught. For example, if we agree that a certain constellation of symptoms and signs, coupled with certain positive laboratory or other investigations, will constitute disease $x$, which we can treat successfully in specified ways, then we can ignore the doubts cast by the skeptic on the reliability of our foundational methods of gathering that information, viz., our sensory impressions, and the abiding lack of indubitable criteria to prop up knowledge acquired through them.

In our study of the basic and clinical sciences we assume that the ultimate means of accumulating the data on which they are based is reliable, in spite of the fact that we use our fallible sense organs to gain access to those data, and in spite of the fact that we cannot rely on any of our cognitive armamentaria to justify our claim to knowledge thus acquired. Besides, even if we use artificial devices to extend and sharpen the functions of our fallible senses, we still end up using the same unreliable sense organs to tell us what we see through the electron and the regular microscope, or what we hear and understand better by way of the amplification of sound or its analysis into wave forms. Barring deficiencies of form and function in our sensory apparatus, the doubt is not with the natural mechanisms of sensation, but with the ultimate veracity of the means employed to construe the thus acquired information of a reality that is independent of us and our making. If those means are not reliable, then so is the edifice of clinical medicine on which they are founded.

But such an untenable proposition cannot be entertained in the face of achievements recorded in both the art and science of medical practice. Everything happens within us, or in our brains, where all the sensory nerves converge. This situation is well represented by the analogy of the brain in the vat. This brain will never know that it is disembodied and out of touch with the real world, because the world is represented in it as authentically as it is in a normal brain. So, even if we take refuge in the corroborative evidence of other people, we are still counting on evidence from the use of their fallible collective sense organs, or from other brains in vats, which is no better than doing multiple checks ourselves and using that accumulated dubious information to justify further dubious information. In all of these manoeuvres we also have to assume that our sense organs are functioning normally; but the definition of normality requires criteria that eventually depend on the use of concepts derived from sensory impressions. We are thus trapped in the circular reasoning of forming concepts from the mental manipulation of unreliable sensory experiences of individual objects and situations, and then using these concepts to justify the normal functioning of the sense organs that produce these sensory experiences.

Who can deny the reality of hallucinations and illusions, and the deception of dreams? Every medical student has heard of the alcoholic psychotic who sees pink rats where they don't exist, the delusional patient who hears voices telling him to do strange things, and everyone knows what it is to misinterpret the sound of creaking floors in a quiet house at night as the creeping footsteps of a burglar, or the sight of a crooked stick on the grass at dusk as a snake. We are all thoroughly familiar with the apparent reality of dreams while we are having them, only to wake up to the real reality and realize that our erstwhile real experiences were phantoms. But how do we know that what we call the real reality is not also one long drawn out dream? We will never know, says the skeptic, because many, though not all, of the things and states that we thought were real turned out not to be so, and we have no way of disproving that fact.

So we should resign ourselves to paralysis of thought and action and suspend all our beliefs and judgments about the world and reality, simply because we cannot justify them conclusively—the evidence is supposed to be extremely uncertain, resting on a shaky foundation—except that they have worked successfully for us through time, even if we cannot guarantee their truth, short of claiming that truth to be self-evident. But if we do that, we pose another problem relating to the veracity or authenticity of self-evident truths beyond that of mathematical truths, such as that a part is always smaller than the whole of which it is a part. Many truths that we claim to be self-evident are not free of bias and inaccuracy, and empiricists have disqualified them from being authentic sources of truth and knowledge for that reason. Nevertheless, we have to start somewhere in our search for truth or else risk stagnation.

## Fallibilism, Foundationalism, Reliabilism

Fallibilism is the thesis in epistemology (theory of knowledge) which claims that the truth of our beliefs and judgments can never be conclusively justified, because their truth is always in doubt, thus rendering our claims to knowledge liable to error. It does not, however, subscribe to the skeptical viewpoint that all beliefs are false and nothing can be known as a result. Fallibilism recognizes that it is possible for the falsehood of our beliefs to be limited to only those beliefs derived from our sensory contact with the world, or it may

34

extend to all beliefs, including those acquired by pure reason, since error creeps in via misperception and misapprehension. This situation results from the fact that in addition to errors of expression whereby we could misrepresent what we perceive, e.g., mistaking a mirage for water, we rely on failing memory, distort the facts, or misconstrue evidence and its implications either through ignorance or through ineptitude, and we are also and worryingly subject to the possibility of perceptual errors of content about what we actually claim to know. Nevertheless, it does not nullify the consoling fact that we are able to discover and correct those errors as Foundationalists maintain. They maintain that we can rely on knowledge that is based on indubitable and infallible foundations, which are things that we know with certainty, such as our authentic internal and external sensory experiences that cannot be subjected to doubt and that therefore "guarantee the certainty of the non-foundational beliefs they support".[8] But fallibilism insists that no knowledge is perfect and no beliefs can be conclusively and rationally justified. It says all theories based on the experiences of any number of people are subject to error, because absolute certainty about empirical knowledge is as impossible as it maintains that *a priori* knowledge acquired through reason is fallible.

But at no point does Fallibilism claim that facts about the world and its content are fallible, only our meticulously accumulated beliefs of these facts may be at fault, in spite of the care and accuracy with which they have been formulated. The foundationalist goes on to say, however, that we should allow that true knowledge is possible within the empirical framework if our beliefs are found to cohere to a large extent and to yield pragmatic results, even if they are potentially fallible. Without that allowance, we expose ourselves to both mental and physical paralysis. Statements of such beliefs are justified because of their participation in a network with other statements that have been previously justified and that hang together by logical consistency. They are the only beliefs that can justify other beliefs; beliefs that do not cohere and are dispensable cannot be justified. But he seems to forget that false beliefs can also form a coherent system that does not amount to truth solely on the basis of coherence. Besides, previously justified foundational statements about beliefs cannot merit a clear pass if their own justification is still in question.

The fallibilist yields partially to this argument, granting that there can be knowledge even if it is not infallible or justified conclusively by logic, since even he, unlike the skeptic, does not claim that everything is false, only liable to falsehood. But even if most of the time we may be right in our beliefs, the lurking possibility of error still nullifies any claim we may have to absolute and infallible knowledge; we have only fallible knowledge resting on fallibly justified beliefs. Hence, the more certain we claim to be in what we perceive and know, and the wider our basic beliefs are spread, the more likely we are to be fallible; and the less certain we claim to be, and the more confined our basic beliefs, the less likely we are to be fallible. We have only fallible knowledge that is fallibly justified. We cannot divest ourselves of the allegation that the disease of today may prove to be the artifact of tomorrow as more refined methods of diagnosis become available to correct our faulty justifications and conceptions. Furthermore, the evolving state of our knowledge makes it impossible to diagnose any medical condition with certainty, always leaving open the possibility of a chaos effect whereby slight changes in original historical, clinical, and laboratory data and their interpretation can result in vastly different outcomes from those anticipated. Also, the fallibilist will argue, the line of demarcation between normal and abnormal is both indistinct and movable; depending on the total picture of the same condition presented by the same patient at different times, and by many patients with different thresholds of tolerance at the same time and in the same circumstances; and this situation also renders our judgments fallible. So how can we ever be certain about anything? Patients have been told that they have a few weeks to live, only to survive and enjoy fruitful lives for several years after the fateful predictions.

René Descartes argued for and against fallibilism in his postulation of the deceiving evil genius who implants false beliefs into his subject, making him believe that his experiences of the world are real when they are all figments of his imagination. Every one of us has woken from a dream that seemed real while it was happening but proved to be unreal now that we are awake. From this experience, one may wonder if she is not still dreaming when she thinks that she is awake and try to convince herself that if she can entertain all these doubts she must be awake. But that is cold comfort, because every experience and thought may, in fact, be part of this massive delusion of mistaking fiction for fact, thus making her very life a dream and placing certainty about anything beyond her reach and beyond rectification by appeal to the cogito (*cogito, ergo sum* = I think, hence I am) to which Descartes appealed to save himself from the ubiquitous pitfall of fallibilism. He reasoned that if he could think, then he must exist, although he discounted his body as his real self in favour of his mind that was doing the thinking, a feat that the body could not ever accomplish, even though his thinking brain was part of his body. His conclusion was that he existed as a thinking thing; therefore the skeptic and the fallibilist could not convince him that he was nothing, after all.

The reliabilist comes in at this point to assure us that we can overcome these setbacks by forming beliefs through methods that reliably result in true beliefs, i.e., ensuring reliable representation of facts and the combined sources from which we obtained the relevant information by means of sense experience or through the process of reason, memory, or testimony supported by sound evidence. But he has not provided any justification for his belief that his methods result in true beliefs; therefore his claim is without foundation, and thus not necessary or sufficient for the justification of our beliefs that the fallibilist is challenging. Hence the refuge from this accusation in evidentialist reliabilism, which can allegedly accommodate the deficiencies of evidentialism (discussed below) as well, since stringent conditions for evidentialism can be relaxed to permit justification based on memory and perception, allowing the reliabilist to utilize these relaxed conditions to justify his foundational beliefs.

## Evidentialism

Evidentialism is the view that a belief is rational or justified if and only if (iff) it is supported by relevant and sufficient evidence. When evidentialists say that a belief should or ought to be justified, they are referring to epistemic or cognitive justification as arrived at by reason. Moral suasion where the *prima facie* "ought" can be abrogated by "cannot", as in "I ought to save that child from drowning, but I cannot, because I am paralyzed", does not affect the justification process. One should undertake the pertinent process of reasoning and argumentation to weed out irrational and unfounded empirical data before concluding that the justification that one is proffering for one's belief is indefeasible. As a precondition of knowledge, evidentialism can be satisfied by high or low standards for justification of the conclusion derived from the evidence, with some of the evidence made low in status by reliable but omitted evidence that would otherwise have elevated its status.

If the standard required for justification is high, it will demand stringent evidential terms for belief of the relevant proposition and thus create an infinite regress as each successive belief demands evidence to justify it before it can be used as evidence to justify the next belief. Besides, it will exclude some useful beliefs from our repository. On the contrary, if the high standard is relaxed and a lower standard for justification is adopted, less stringent demands on one's belief about the evidence will obtain, allowing evidence such as memories and experiences (perception), to be accepted as justification for beliefs or to act as conclusive evidence in the absence of any other evidence.

The implication for medical practice is simply that the doctor should base her decisions in the care of her patient on evidence that has been gathered scientifically, as is the case with Evidence-Based Practice and Personalized / Precision Medicine; but she should not forget the phenomenological aspect of medical practice that we discussed under ethical idealism.

## Positivism

The positivist's answer to the fallibilist is that we can be certain only about the meaningfulness of statements that can be confirmed by empirical observation, and by those of mathematics and logic, all other statements being meaningless. Positivism *per se* and logical positivism into which it evolved are best applied to Medicine in the modern form of the latter as logical empiricism, which recognizes verifiable evidence obtained from observation as the only true source of knowledge, together with a limited amount of rationally acquired information, e.g., 2+2=4. On the other hand, and contrary to this contention, Bertrand Russell has described knowledge by description (seizures are paroxysmal discharges of neurons resulting in altered neurological function), and knowledge by acquaintance (John knows seizures from having them), and Gilbert Ryle has described knowing "how" (to do x) and knowing "that" (y is the case), skills that transcend the positivist criterion of the acquisition of knowledge and truth as outlined above. From the positivist circumscribed perspective, we may argue that since scientific knowledge is amenable to verification by observation, it qualifies as the only true knowledge, and since bio-physical medicine is a totally scientific discipline, its tenets should also constitute the only true knowledge; but that is not the whole story.

The aim of positivism is to discover the general laws that relate various phenomena and subject them to confirmatory or falsifying empirical tests for the purpose of giving them meaning or depriving them of meaning. Statements like ethical pronouncements, scientific theories, and scientific laws that are not subject to empirical confirmation in the way that John's seizures are, are generalizations that cannot be confirmed by any number of examples, and they are therefore meaningless by positivist criteria. Positivists claim that no one can verify the statement that all metals expand when they are heated, because it is not possible to test all existing and still-to-be metals and conclude from such comprehensive testing that all metals expand when they are heated; hence Popper's attempt to substitute for this unattainable goal by proposing the attainable criterion of falsifiability that we will encounter in chapter 13.

But the positivist doctrine itself is meaningless by implication, because it also cannot be confirmed empirically. Positivism expects the world to defy chaos effect by conforming to mechanistic predictability. In the case of disease positivists would have us say that diseases exist as entities in their own right, in determinate forms; they arise from generally known causes, follow largely predictable courses or natural histories, and therefore respond predictably and invariably to certain forms of treatment. But that is not always the case, as all doctors know, because reality is changing constantly, and what is true and confirmable now may not be so in the next instant. How often have we not heard of miracle patients whose recovery from their illnesses doctors cannot explain in mechanistic terms? Positivism fails to recognize this existential fact, thus necessitating the integration of the positivistic biomedical concept of medicine into the positivism-transcending biopsychosocial concept to give it the credibility that real life situations demand.

The basis of biomedical medicine is a reductionist, quantitative, and materialistic study and conquest of diseases that afflict people. In the tradition of Descartes' dichotomy of body and soul, it regards the body as a mechanical system of physicochemical processes whose derangements have to be repaired mechanically and instrumentally while paying less attention to the qualitative, mentalistic aspects of health, illness, and suffering as based on the sick person's role in his social, cultural, economic, and political milieu, including his mental dysfunction in that milieu. The psychiatrist will appreciate this point of view, because she deals with deviations in the mental functions of patients, and so will the drug trial scientist who witnesses the "curative" effects of placebo in some of her subjects. These are more reasons for complementing the limited biomedical considerations of disease with the wider biopsychosocial perspectives of illness for the purpose of embracing the total circumstances of the patient in the management of his ill-health. On this outlook, the patient is also encouraged to make his personal contribution to the elaboration and resolution of his problems as only he can, and is, experiencing them. Positivism regards all these vital facts about the phenomenology of illness, pain, and suffering as meaningless, because they cannot be measured, and thus lack relevance, despite their proved importance to the patient. He needs more than mere riddance if his disease.

Faced with the problem of pain in a three day old infant who cannot express his feelings when his foreskin is amputated, and lacking the means to measure the sensation caused by this disruption in the integrity of his tissues and the equanimity of his feelings, doctors erroneously conclude that this infant does not experience pain from the same kind of disruption of tissue integrity that would cause them pain. But as we will see in the section on constructivism, its comprehensive approach urges us to monitor all the parameters associated with the procedure of circumcision: heart rate, respiration, perspiration (stepped up adrenaline secretion in response to stress), muscle tension (struggle to free himself from the circumstraint), spontaneous micturition and defecation (spinal reflexes in response to pain), compound them into a comprehensive picture. Minimizing and dismissing them as of no great significance to our unfounded and blindly adhered to theories that fly in the face of contrary evidence, defeats the rational conclusion from these operations that the procedure is emotionally distressing to the little surgical patient. In effect, therefore, positivism, because of its contrary character to constructivism, isolates the event from its entire context and thus conveys a distorted view of its nature and implications. It is a form of reductionism that limits the doctor's perspective and her scope of operation by omitting the patient-person from the picture.

## Critical Realism

Critical realism is a radical advance on naïve realism, which holds that the external world as it appears is the only real world that we know and can know, independent of of our theories about noumenal and phenomenal worlds, which are postulated by Kant to represent the world as it is versus how it appears after distortion by our senses.

Our perceptual organs connect us with this never totally discoverable reality to provide the commonly shared foundations of empiricism, without which our understanding of the world would be grossly incomplete. Reliance on reason alone, without the substratum of experience, will carry us only a short distance into the world of the rational, but it will not give us knowledge of the objects and things with which we interact every minute of our lives. Accordingly, our statements about the world are rendered true or false by their representation or misrepresentation of its objective nature. But we also know that genuine sensory experiences can sometimes issue in hallucinations and illusions, which can mislead us to misrepresent the world as it is not, thus causing us to infer the existence of the independent external world, real or spurious in our judgment, from experiences within us that have no counterpart in reality. A case in point is when we infer the current existence of a star that disintegrated many light years ago from the light that it radiated then and is only now reaching us. So, we cannot always be sure that our senses represent anything when we claim to be sensing something, nor can we always trust them to represent a possibly existing reality accurately. If we do, it is because those beliefs have adequately explained many of the minute by minute collective events in our lives up to that point and also helped us to predict future events, notwithstanding known doubts about the inductive process and our habit of drawing conclusions in excess of available empirical facts.

Like realism *simpliciter*, Critical Realism affirms an objective world that exists independently of our perceptions of it, and which causes the concepts that we form to represent it. But it also acknowledges the problem of basing our knowledge on this objective relation, because knowledge derived from perception is never free from fallibility, and many of the secondary qualities that we bestow on the external world (like the green colour of grass, which the physicist says is composed of colourless atoms in motion) originate in our own thoughts under the influence of previously established beliefs about it; hence the possibility of illusions, but more seriously the possibility of formulating erroneous theories about reality. So critical realism affirms the ontology (existence and being) of the external world while it admits to doubts about its epistemological (knowable and comprehensible) foundations; but it recognizes the impossibility of perceiving reality from a detached standpoint, because our perceptions can never be free of the inherent biases that come with being centred within that reality. We are thus left with the task of reconciling two apparently incompatible perspectives, ontological and epistemological, if we intend to avoid the absurdity of multiple subjective (relativistic) views of the same reality, and this we can do only by reflecting critically on the implications of what is apparent in perception versus what is inapparent. We have here, therefore, a combination of naïve realism (which takes reality as it is perceived), critical reflection (which seeks to discover the reality behind the perceived reality), and the Kantian ontology of phenomena or appearances and noumena or hidden and perceptually inaccessible reality (which claims that such a distinction exists as fact).

In basing its claim for reality on the causal efficacy and the practical and pragmatic everyday material consequences of our actions as based on that reality, critical realism would have us realize that not all knowledge is equally fallible, even as it distinguishes between the autonomy of objectively existing reality and our perception and knowledge of it. Some knowledge has to be taken as veridical if any edifice of knowledge is to be constructed. The critical realist recognizes three levels of existence: the observable or empirical, the actual or spatiotemporal, and the real or transcendent (transfactual), which is believed to serve as the source of our empirical perceptions.

Medicine and science begin from empirical observations in search of the real nature of disease and scientific phenomena solely by way of analyzing these phenomena as they are actually presented. When a doctor observes a constellation of symptoms and physical signs that represent the presence of disease in a patient, she does not assume that eliminating these appearances will restore the patient to well-being, because she cannot achieve that without knowing the underlying factors, i.e., natural laws, that are responsible for what she observes, over and above the presence of microbes or changes in gene structure that reflect the operation of those laws. The search for those laws is the same as the search for the laws that govern every other natural process and also account for the enigmatic principle of causality that we will encounter in chapter 7.

Scientists do not have the liberty to construct their own reality, but must deal with the reality that is partly revealed to them as they strive to unravel and dissect with their research, amongst other conditions, the human genome to discover genes that cause Cystic Fibrosis, Duchene Muscular Dystrophy, Huntington's Disease, and metabolic disorders, which that research has rendered amenable to correction by gene replacement. Recombinant enzyme replacement is used in the treatment of some forms of Gaucher disease and lysosomal storage diseases. Without accepting reality as it is appears in perception but going after the unknowable, inaccessible, noumenal reality of Kant's transcendental idealism, scientists would not have anything to work with, and clinicians would not have anything to use in the application of gene therapy to produce the significant advances noted above. The use of retrovirus-transduced autologous peripherally mobilized CD34+ (progenitor) hematopoietic cells to improve T-cell function of young patients with X-linked severe combined immunodeficiency would also not have been possible. These cells correct for failure in the development, proliferation, and activation of lymphocytes due to absence or deficiency of the gamma chain protein, γc, resulting from mutations in the IL2RG gene that leaves patients with lack of T cells and natural killer cells, and non-functional B cells.

## Idealism

In the midst of prevailing doubts about objective reality, and after all these laudable achievements, the idealist undermines our shared communicative ability by assuring us that our ideas of objects and the external world are only subjective impressions, collections of sensible properties without an underlying material substance. After all, she says, when we describe oranges we describe their subjective colour, texture, weight, shape, size, taste, smell, etc., and we do not have to add "objective material thing" to complete this description and assign an external location to the orange for the sake of making it perceptually accessible to other persons. This is so, she says, in spite of the fact that these qualities exist for us only when we perceive them— their being is their perception; they are a complex of the qualities that we sense only when they are present, and qualities are known to be immaterial. John Locke claimed that qualities like the green colour of grass exist only in the mind as secondary effects, in contrast to others (primary qualities of size, shape, weight, motion, number, and solidity) that are inherent in things. But we know that both types of qualities exist as a result of the interaction between the atoms that compose matter and our perception of material things. We would not know anything about these primary qualities if we did not perceive them in the same way as we perceive secondary qualities. Besides, perception can vary them, as in the case of a large object that looks small from a distance, or a table surface that looks rectangular from one perspective and like a parallelogram from another.

The doctor's interest in metaphysical idealism as outlined above versus realism stems from the old controversy of the interaction between the immaterial mind with its psychological problems and the brain with its materialist structure, and it accords well with the doctor's acceptance of the materialism on which biophysical medicine is based, as against biopsychosocial medicine, which is clamouring for the simultaneous recognition of the physical and non-physical elements that constitute the whole patient. As we will see in our subsequent discussions of persons, there are idealist theories like the ordinary language behaviourism of Ryle, which regards as a category mistake the construal of mental terms as reflective of an internal process, because there are no occult, internal processes that precede actions in addition to externally witnessed behavioural patterns. So how do we plan a factitious disorder like Münchausen syndrome, and how else do we explain malingering if not as the deliberately induced prevailing of the mental over the physical, which the DSM-5 describes as the intentional production of false or grossly exaggerated physical or psychological problems. In contrast to Ryle, however, Geach maintains that there are *mental acts* (the title of his book). On the other hand, realist theories like the functionalism of Hilary Putnam and mixed theories like the dualism of Descartes also try to explain the nature of the person in their own way, as we will see in chapter 4. Realists acknowledge the existence of the brain as a material object, but they have a problem with doing the same for mind, and so they deny the existence of the mental world *per se,* attributing our psychological processes and mental illnesses solely to brain activity. Nevertheless, we know that some somatic disorders can be caused by mental events, such as the ulcers of worry and other physiological concomitants of emotion like those induced by fright and fear of imminent danger. So the immaterial mind does influence the material body. The perennial question is: how?

A perspective on idealism that differs from the ontological theories mentioned above relates to ethical idealism, which, in the purely philosophical sense concerns the nature of ideals and how they guide our lives into avenues that bring benefit to all people, even though by their nature they are unachievable. Ideals acknowledge the reality of what we *can* do at the same time as they challenge us to go a step further and also do what we *ought* to do. Many a time we shirk our moral obligations, because they demand consideration of others over consideration of ourselves, and we sacrifice their interests on the altar of selfishness. Ideals dare us to attempt the impossible so that when we fail to reach those heights, as fail we must, we will not fall too far down the ladder of duty and humaneness.

For the doctor, ethical idealism means having the welfare of her patients and their values as her primary professional concern. And for this purpose, her training must demand from her integrity of character that will facilitate her preparation for the cultivation of those humanistic qualities that will promote doctor-patient relationships of compassion and respect for the patient's dignity, including the avoidance of any degree of physical and psychological harm to the patient. Furthermore, it must demand courtesy, sensitivity to the patient's welfare and needs, and the appropriate professional attitude, responsibility, and behaviour that will not tolerate using the patient and his plight for any kind of personal aggrandizement, in keeping with the provisions of the Hippocratic Oath. Her response to situations that present the many dilemmas encountered in medical practice will depend on the depth of her involvement with her patient's life goals, and her appreciation of his or her autonomy versus her authority. Many a time she will have to ask herself and simultaneously answer questions like: should I assist in ending the life of this suffering patient; should I relieve this mother of this fetus for utilitarian, sound socio-economic, health, or other personal reasons, or should I dogmatically follow the deontological dictate prohibiting the ending of life regardless of the circumstances?

If she has cultivated the correct ethical ideals, she will not have any problem answering those questions and many others ethical questions related to them (see chapter 17). The practice of Medicine is a minefield of ethical dilemmas in which those who lack ethical ideals will flounder and fumble, and they will cause harm to patients while their oath of conduct admonishes them to do no harm.

## Coherentism

Coherent beliefs are those that are justified by their causal relationships and participation in a network with other beliefs that have been previously justified and are supposed to hang together by logical consistency as sufficient criteria for truth; e.g., $p$ entails $q$ iff (if and only if), given $p$, $q$ *must* be true. Coherentists claim that justification of the component propositions of a set of beliefs increases with increasing coherence of its members, thereby approaching the whole truth. (Truth is a complicated concept that I will not discuss here; I discuss it in *Perspectives on the Concept of Belief*). Necessity coherentism disqualifies from justification beliefs that do not cohere, and since justification is a sufficient condition for truth, non-cohering beliefs fail the test of truth, as per the logical entailment stated above. But these beliefs may be justified without being true, while others may be true without being justified, and all of them may cohere in their own domains without cohering with the coherent set in another domain and still be true. Besides, since complete logical coherence of any set of beliefs is not always possible to achieve, coherence where all the members of the system causally explain each other better than in any other known system provides better justification in the quest for truth.

In contrast to necessity coherentism, sufficiency coherentism only requires one to posses other compatible belief concepts to be able to hold a particular belief concept, even if none of them is coherently justified, as long as it coheres with the rest of our beliefs by means that are not irrational, like belief in witches and ghosts. The mere convergence of our independently derived beliefs by a variety of modalities and from all sources is enough to justify them, even if each one of them by itself falls short of the justification requirement. The theory admits the possibility of the existence of multiple coherent systems with which a belief is compatible, even if the systems do not bear directly on the belief under consideration, as long as they satisfy the principle of non-contradiction, e.g., asserting of $p$ that it is both $p$ and $not$-$p$, as such inconsistency amounts to nonsense.

The upshot of the foregoing discussions, which appear to be repetitive variations on one theme, is that even though we cannot guarantee infallibility in beliefs based on perception, we do not have any other means of making contact with the world to gather iron-clad evidence that we need to form infallible beliefs about the world that we know. So we have to make the best with evidence that has carried us this far. The doctor must trust the coherence that exists among her different modalities of perception: hearing (the patient's narrative), seeing (the patient's presenting signs—limp, pallor, cyanosis), touch (the cool, clammy skin of shock), taste (the Phenylketonuric patient's salty tasting sweat), smell (the smelly feet of a patient with Isovaleric Academia or Glutaric Academia Type II), even before she postulates a theory (provisional diagnosis) about her patient's illness, and clearly before she institutes a variety of investigational techniques to help her narrow down the possibilities and to arrive at a final diagnosis. Invoking the help before the initial effort never makes sense in any venture, rational people don't invite help into a vacuum. In medical practice it is a dangerous shotgun shenanigan.

## Counterexamples

Finally, we will consider briefly a useful technique for disproving any of the generalizations that we have made up to this point, and that we will make from here on. The use of the counterexample,

45

which provides even a single truthful exception to the all-embracing general statements that we make to prove that they lack universal application, is that method. The claims that As are always Bs, and As are never Bs can be undone by citing one example of an A that is not a B or a B that is an A respectively, and it is no argument to say that these exceptions prove the rule, i.e., test the rule. So when someone says abortion or euthanasia is always wrong or never right morally, one can cite cases where there is enough justification, all things considered, to overrule this bold attitude and to prove that the generalization is unjustified, as discussions in chapters 19 and 20 will show.

Counterexamples compel changes to the formulation of general statements, so that they should not claim more than they are entitled to, or else the theory supported by them becomes invalid, because it is contradicted by the truth asserted by the counter-example. As Nicholas Rescher advises in *Philosophical Standardism*, we should refrain from saying that categorically and without exception All As are, Bs but follow the safe course of saying that ceteris paribus (all things being equal), as a rule, standardly, ordinarily, as far as our experience goes and as far as statics indicate, As are Bs. And when exceptions occur, we can provide cogent explanations for them. For instance, a categorical statement that *all* cases of hematuria are caused by viral hemorrhagic cystitis can be disproved by a single case of hematuria caused by nephrolithiasis or urethritis. Counter-examples do not apply to particular statements which assert that *some* cases of hematuria are caused by hemorrhagic cystitis, because these statements allow room for other causes of hematuria, as indicated above. They also do not apply to conditional statements that *if* there there are cases of hematuria, then they are caused by viral hemorrhagic cystitis, because these hypothetical statements are clearly not categorical.

The method of counterexamples is similar to that of the falsification of general statements and hypotheses proposed by Karl Popper, whereby a scientific theory remains in force until it is falsified by predicted data that contradict and disprove it successfully, and then it is either modified or jettisoned when a more competent theory is advanced to replace it. (see Popper's method, see also Thomas Kuhn in chapter 13).

46

# 3

# Philosophical and Scientific Concepts

## Empiricism and Rationalism

Notwithstanding all the doubt about the reliability of sensory information, the empiricist still insists that the best part of our knowledge (of our internal and external worlds) derives from experience more than it does from reason. If reason plays any part in the construction of the edifice of knowledge, it is secondary to that of sensory experience without which it is useless as a foundation of knowledge. The empiricist denies the sole practical role of *a priori* knowledge in this process, because it overlooks experience by defining unassailable truths as those that necessarily transcend experience. To the empiricist, experience is, therefore, the primary and final source and criterion of knowledge, in spite of what the skeptic thinks about it. It is by sensory experience with all its ineptitudes that we make contact with the world; we learn to appreciate our environment of persons and things by establishing contact with it via our five modalities of sensation—the beautiful colours of plants in bloom, the tantalizing aroma of good food, the soothing effect of cool jazz or a piece of classical music, the soft touch of a baby's skin, etc.—relying on reason alone will not carry us far. We cannot use reason to conjure up the feeling of pains that we have not experienced, and even if we could, our descriptions would lack the limited coherence that characterizes the empirical descriptions of persons with varying levels of pain tolerance (excluding those who are pain insensitive). What reason will do for us is systematize our thinking and the presentation of our sensory experiences, so that we can dare to use them as foundations of knowledge with a high degree of probability, which is the best that we can hope for, because certainty is out of the question in this intellectual climate; but it will not replace our sensory apparatus in their work of gathering those experiences.

Within the established ethos of biomedicine, the traditional and modern roles of rationalism are to direct the doctor's concern to the single cause biomolecular mechanism of disease that proves to be consistent with generally established principles of cause and effect, and hence to its subsequent, universal management skills. Ignoring the role of empiricism in the gathering of information, rationalism says under specified conditions bacterium $x$ causes pathological tissue changes $b$, $c$, and $d$, which cause symptoms $e$ and $f$, and signs $g$ and $h$. Causally directed (rather than symptomatic) treatments $j$ and $k$ resolve the tissue changes and the symptoms and signs. Therefore all patients (bodies) who present with a similar combination of symptoms and signs can, within reasonable mechanistic limits, be rationally presumed to harbour the same bacterium and to follow the same course, and can, therefore, be treated similarly. Empiricism directs the doctor's concern to the multifactorial causation and epidemiologically observed course of a disease, and to the different evolution of symptoms and signs in one patient as compared with that in other patients with the same disease caused by the same microbe. But it also highlights the unpredictable reactions of the same person at different times and in different circumstances to the same invading microbes. The accurate compilation of data from studying these repeated varieties of the clinical manifestations of disease provides the substratum for the exercise of rationalism, which can't take place in a vacuum without being a spurious representation of reality—mentally constructed diseases that have no counterpart in real life are not diseases, and they do not warrant any attention. So another dimension is necessary for completion of this picture, connecting the presenting clinical picture with the course and outcome of the disease; and that is where empiricism fits into the total picture. A clinical medicine discipline that is based solely on rational principles is not likely to benefit patients who present with multiple manifestations of disease that require accumulated empirical experience to manage; rationalism alone is inadequate.

Empiricism thus relies more on experience and observation than on the determinations of basic science research to treat various categories of patients and to assess the impact of treatments on all aspects of their lives; but, as we will see when we discuss induction, (which entails the circularity of establishing a conclusion by using it as one of the premises of the argument; e.g., the sun has risen for the past million years, therefore it will rise tomorrow; but it may not, if a bigger celestial body should come along and change this well established pattern by obliterating it), the rationalist also depends on constructing the inductive process from empirical data to derive the principles that guide his reasoning. For example, the successful, science-based preservation of some vaccines with the addition of mercury has been questioned as a cause of autism on empirical grounds, leading to the finding of new methods of achieving the same effect without the use of the controversial preservative; and conflict has arisen in regard to the treatment of coronary thrombosis with drugs that dissolve clots successfully while they have the serious side effects of causing hemorrhages in the brain and other tissues, spurring the quest for safe compromise drugs and procedures that can produce the same beneficial effects without the attendant complications. All these empirical facts contributed to the birth of Evidence-Based Medicine (EBM) in which the aggregate of cases of individual patients with their unique clinical presentations form the basis of guidelines for managing various clinical conditions. Whether or not this is a commendable choice remains to be seen.

# Thought experiments

A thought experiment is a hypothetical reasoning or conceptual process about an imaginary scenario meant to investigate the nature of a situation for the purpose of confirming or denying a hypothesis or theory about the situation, or for delineating conditions for the proper application of a concept to that situation. A question is posed: what if? A possible scenario is constructed in answer to the question, a theory is postulated, and possible conclusions and their applications are advanced. In the process, prudent attention is given to possible irrelevancies, inconsistencies of thought, or lack of clarity that can invalidate any conclusions drawn from this conceptual exercise and its applicability to real life situations. If the result of this exercise cannot help to establish a new approach to the stated problem or is found to be too far-fetched or otherwise incoherent to challenge and refute existing theories or to confirm them, it is modified or abandoned. As well, if it fails to a) predict or project present conditions into the future, b) retrodict or speculate about a dubious distant past event and use the theory to predict a more recently past known event and establish its ultimate cause from real life situations, or c) facilitate strategy selection in the formulation of new ideas and avoidance of past errors in subsequent decision making and problem solving, it is revised until it can become applicable to real life scenarios.

An apropos example of a thought experiment is Hilary Putnam's well-known brain in a vat: an imaginary human brain placed in a vat of nutrients to keep it alive and connected to a computer by its nerve ending sites. In reverse to the famous experiments of Wilder Penfield where stimulation of appropriate sensory and motor endings in the brain resulted in subjective cognitive experiences of, and motor activities like, those found in real life situations, in this case it is the peripheral nerve endings that are stimulated to produce brain states and impressions of an outside world. For the brain in the vat, experiences of the artificially induced environment are qualitatively identical to those of real live persons; so there is no difference in the way it sees trees and people from the way that normal persons see these objects with their brains. The only difference lies in the type of causal interaction between persons and the objects of their cognition, which the brain in the vat cannot claim. But the scenario highlights the source of our doubts about the veracity of those sensory experiences that claim to represent reality to our minds, and it gives rise to skepticism about perception. Regrettably, this scenario does not allow for the fact that if we have no way of realizing that we are brains in vats, then we are not or we might be, because we can't assert a world of make-believe if we don't already know the real world via our naturally connected brains, and brains in vats don't know the real world; they have never experienced it to be able to tell the difference.

Medical science recognizes that thought experiments, like real life situations, are susceptible to false positive (type 1) errors of observing a difference where there is none and rejecting a hypothesis that should have been accepted, and false negative (type 2) errors of failing to observe a difference where there is one and failing to reject a hypothesis when its alternative is true. Statisticians call these errors rejecting the null hypothesis when it is true and not rejecting it when it is false. The null hypothesis ($H_0$) is a theory whose truth is presumed as the basis of reasoning, but which we are trying to disprove, e.g., a daily low dose of aspirin does not prevent heart attack. It asserts a lack of difference between known data and data expected from new tests. Disproving it only suggests the possible truth of its alternative hypothesis ($H_A$ or $H_1$) without confirming it. Its non-rejection does not, however, prove its truth; it only means that evidence against it and in favour of its alternative is insufficient.

In medical testing, false positive (type 1) errors falsely suggest to the doctor and the patient that an absent disease is present, resulting in unnecessary treatment, or else that a certain modality of treatment is beneficial for a patient when it is not, thus precluding the use of effective therapy and also raising false hopes of a cure. False positive tests also cause problems when persons who are not diseased are labelled as such from results of tests for rare conditions for which the tests were not designed, especially when the false positive rate of the test is high in comparison with true positives in a relatively limited population; e.g., false positive VDRL test seen in some viral infections, Systemic Lupus Erythematosus, and old age. If a test has a false *positive rate* of one in ten thousand, but only one in a million samples is a true *positive*, most of the *positives* detected by that test will be false. False negatives (type 2) errors occur in cases of patients with chest pains and negative EKGs or stress tests who subsequently suffer full blown coronary thrombosis and myocardial infarction, or in cases of negative reports of throat cultures done improperly on sufferers from group A streptococcal throat infections. They often come with costs to the patient from delayed, improper,

50

inadequate, or no treatment. Type 3 errors entail the solving of the wrong problem in answer to the wrong question. In type 4 errors the risks of illness are overestimated and the wrong treatment is given for the correct diagnosis; e.g. prescribing penicillin for a viral infection.

# Dialectic

Dialectic (from dialektikos, which means of debate/discussion) entails the use of dialogue for reasoning and the exchanging of logical arguments to resolve conflict, eliminate falsehoods, and retain only truths. In the dialectic mode of thought, contradiction is the point at which argument begins. The process has its origins in Socrates's method of question and answer to elicit knowledge of truths from persons who did not previously claim to have such innate knowledge. Aristotle regarded dialectic as inferior reasoning, based as it was, not on empirical observation, but on *a priori* knowledge. It is especially associated with Georg Hegel whose method consisted of arriving at the truth by a three-step process: stating a clearly defined concept or thesis, developing its contrary antithesis, reconciling the two, and then deriving absolute truth from the coherent synthesis of the two opposing positions. The recognized divisions of dialectic are: ontological, concerned with existing reality independently of our knowledge of it; epistemological, concerned with what we know about this reality; and relational, concerned with placing our knowledge of reality appropriately in the prevailing hierarchy of knowledge.

This concept is employed in the daily encounter between doctors and patients, where communication often goes awry, because doctor and patient are dealing with two different realities. The language of the doctor relates to her reductionist world, based on the biomedical model that she employs for thinking about her patient's illness only as measurable derangement in form and function of his organs and their related physico-biochemically determined systems, whereas the patient's world is one in which he relates his illness to his disability to function in his milieu as a result of some underlying bodily disorder. Most of the time the doctor dictates the direction and pace of the exchange and what she considers to be relevant for her pathophysiological diagnosis as her guide to treatment of the disease, thereby leaving the patient little or no scope to convey his feelings and life values in narrative that matters a lot to him. The first result of this failure in communication is the dispensing of the wrong treatment for a misunderstood illness, even though it may be the right one for the disease. The second result is the patient's non-adherence to the treatment prescribed, because it does not affect his illness, although it may transform his disease. The third result is that the disease and illness run an unrelenting course. The final result is a complete breakdown in the patient's trust of his doctor and the medical profession as a whole: they are all the same; they don't listen, because they don't care enough. We will encounter the use of this concept in chapter 16 on phenomenology.

# Logical Analysis

Logical analysis owes its prominence to Russell, Gottlob Frege, and Wittgenstein. Russell's theory of descriptions is an attempt to analyze phrases and sentences so that reference to a non-existent entity can have meaning even if it is not true of reality, e.g., the sentence "unicorns do not exist" is both true and meaningful. In the theory he states that proper names are substitutes for definite descriptions (e.g., unicorn = the horse-like creature with a single midline horn etc.), from which we can also claim that a name like "disease" refers to a non-existent entity but can be used meaningfully in a sentence as a substitute for specific descriptions of different clinical conditions suffered by patients. The use of logical analysis avoids the reification of disease without making talk about disease meaningless. In "Sense and Reference", Frege sets out to prove that proper names have both sense and reference, but some names or expressions can have sense without reference. He says, "It may perhaps be granted that every grammatically well-formed expression representing a proper name always has a sense. But this is not to say that to the sense there also corresponds a reference. . . . In grasping a sense, one is not certainly assured of a reference."[1] The sense of 'unicorn' resides in its description, but since it does not exist, it does not have a reference, it cannot be identified. On the other hand, the same sense may have different references, e.g., the expressions "the morning star" and "the evening star" both refer to the planet Venus depending on whether it is seen in the morning or in the evening. Similarly, the nonsense disease "drapetomania" (tendency of slaves to run away to freedom) has sense but no reference in reality, since it is a figment of some imagination.

This concept is discussed in chapter 12 on Logic and Scientific Method. It examines the use of language in the expression of our thoughts. By its use, obscurity and mysticism can be purged form our thinking about questions that are amenable to systematic analysis, e.g., to quote Bertrand Russell, "Suppose I say 'The golden mountain does not exist' and suppose you ask 'What is it that does not exist?' It would seem that, if I say 'It is the golden mountain', I am attributing some sort of existence to it"[2] by using it as the subject term of the subject-predicate sentence that I have uttered. In the practice of Medicine we use logical analysis to the develop and exercise skills geared to challenging and dethroning dogmatic foundations and assertions on phantom diseases by means of reasoned approaches to both scientific and clinical information on which the practice rests.

52

Wittgenstein goes even further to claim that confusion in philosophical communication is caused by imprecise use of words and language, reiterating George Berkeley's comment that "We have first raised a dust, and then complain we cannot see."[3] He uses the example of a builder asking his assistant to pass him a particular type of stone by calling "slab", "block", which the assistant understands as implying bring me a slab, or a block, in the context of the language of building. In a different context the same words would acquire and convey different meanings, depending on the language game in which they are used. From this it is possible to use words in a language game that names things that do not exist but can still be named, like unicorns, pains, diseases etc., in as much as the same word can be used in different contexts to mean different things; hence his claim that the meaning of a word is its use in language. Failure to understand how words are used in language causes miscommunication and misunderstanding, which cannot be tolerated in the practice of medicine where lives can be jeopardized by misuse of words.

All these philosophers were attempting to clear the obscurities that accompany ordinary language used to discourse about reality and setting the stage for the physician to follow suit in discoursing about disease, illness, disability, and related concepts, following Russell's contention that progress is possible only by analyzing (logically), as he did in his paper "On Denoting". The concept of logical analysis thus permeates the entire discussion in this book but its gist is discussed in chapter 13 on Logic and Scientific Method. The example par excellence of the method of critical analysis involved in the application of this concept is Evidence-Based Medicine whose basic principle is the critical investigation by physicians of the rational justification and efficacy of the many diagnostic and therapeutic procedures that they employ in their management of patients. The method calls for a logical approach to clinical problems and case studies, and for the appropriate application of available research and experimental findings from the relevant literature to the case of the particular patient that they are managing. It also entails the ability to spot fallacies in reasoning caused by use of invalid rhetorical methods that can lead to the drawing of erroneous conclusion from true premises, or from false assumptions masquerading as the truth. In addition, it calls on the physician to always be ready to review his actions and improve on them in the light of current knowledge in his field of study and operation.

## Critical Reflection

To begin with, reflection consists in rationally extracting factors that characterize both obscure and apparent similarities and dissimilarities from situations, and using this information to solve problems connected with those situations. The process results in the creation of new ideas and solutions to problems encountered in the daily conduct of life. In critical reflection the subject takes nothing for granted; even opinions from some so-called experts, which are often not free of bias. This does not mean that she should question the foundations of all our knowledge and practice where they have withstood the test of time and provided a solid base on which to base most of our disciplines—medical, astronomical, engineering, etc. But even here, we should remember that many a well-established theory has been jettisoned after decades of acceptance of its tenets and practices; e.g., flat earth, phlogiston, geocentric theories, and even Newtonian physics to which scientists are still clinging. With honesty as her sole motive, she examines all her information and opinions for authenticity, justification, and lack of biases to ensure that the conclusions that she draws from them will be reliable. She therefore opens herself to rational alternative points of view and becomes her own devil's advocate before taking a final stand on any matter to determine its public utility or why it should not become public policy, without embarking on methods that deviate grossly from accepted procedures, since such practices invariably result in needless catastrophes. It is this critical attitude that is behind all the advances that have been made in all categories of medical treatment and in all other spheres of human endeavour that we need not enumerate.

Adapted to the medical domain, this process is directed inward to the doctor's self and her actions in the context of her existential situation as the custodian of the patient's welfare who is sworn to providing him with first rate care and ensuring that no harm comes to him. It heightens awareness to the patient's needs and entails the realization that he is an autonomous agent who has personal wishes and concerns that should not be peremptorily overruled. It also demands acknowledgement of the role of other professionals who are also interestedly involved in the care of the same patients, and that their contribution to the total care of these patients surpasses any single contribution of any single physician to the exclusion or subordination of other health care professionals. It reminds us that, as the saying goes, there are many ways to skin a cat, some of which may be better than others.

The doctor must, therefore, be amenable to this process of self-criticism and evaluation that will allow her to accommodate other points of view in the management of patients on whose illnesses she does not possess the sole and final opinion. She should practice her art conventionally without being mired in procedures and attitudes that have outlived their day, and she should make use of the concepts and skills that she can gather from other disciplines like those mentioned in the rest of this book—Logic, Epistemology, Scientific Method,

Phenomenology, Darwinism, etc., and cast off the disabling burden of dogma and rigidity in her practice. She should justify her actions and the validity of the thought processes by which she concluded that they are the right ones for the patient's welfare to the exclusion of competing points of view and actions; that they qualify universally as acceptable standards of patient care. She must be honest in her critical analysis of herself. As one who engages in critical reflection, thinking critically about her thinking about her patient's problems, she will be better able to transpose her skills to the task of solving those problems correctly by systematically examining them from different perspectives and opting for the best answers.

The entire discipline of Medicine should also subject itself to similar analysis, reviewing its methods of diagnosis, investigation, and treatment of patients, revolutionizing the economics of medical care so that the welfare of the patient is not relegated to a subordinate position to other concerns like catering to corporate interests at the expense of his health. People in power have been known to resist and come out fighting against efforts to implement universal health care for the citizens in order to safeguard the economic interests of the pharmaceutical and other corporations to which they are beholden for favours. Health professionals should not be intimidated in their efforts to cater to the health needs of their patients and to shape the community and its resources to meet those needs as required.

## Critical rationalism

Critical rationalism is a philosophical method that affirms the possibility of knowledge in the face of the impossibility of claiming to ever have ultimate answers to any of the questions that we pose in our endless pursuit of truth. It acknowledges that scientific "truths" can never be certain or justified; only falsified. The process of falsification facilitates the elimination of incoherent ideas and data compounded into generalization from the inductive process, since only a single contradictory situation is enough to discredit any posited universal proposition; e.g., the existence of one white raven will nullify the universal proposition that all ravens are black. So, even if, for a while, we may believe what is false, we will be able to withdraw our belief once we find that it is unworkable, as the histories of science and medicine have proved with the abandonment of the phlogiston theory of combustion and the use of eels for bleeding sick people as a therapeutic measure.

All our scientific theories are stated in the form of universalities, but the latter are derived inductively, and therefore circularly, from an accumulation of thus far consistent, specific examples, which are compounded into (universal) laws that are then used to predict and explain future events. That all metals expand when heated is a universal law derived from observing this invariable phenomenon in many instances and then elevating the sum of these observations to the status of a law of nature. The law is then used to predict the reaction of a presumed metal, which is expected to conform to its provisions. Failure of the test results in disqualification of the presumed metal, but the law is retained. Critical rationalism says perhaps it is the law that should be modified.

Unlike scientific realists who accept the fallibility of scientific methods and the uncertainty of scientific knowledge without casting doubt on the conclusions drawn from these methods, critical rationalists espouse fallibilism and believe that there is need for more than one critical theory to explain observed phenomena. Their thinking originates from Karl Popper's substitution of a self-critical approach to the justification of a rationalism that accepted at face value purported facts garnered from experience and proved by ordinary reason. As we will see when we come to discuss Popper's method, he maintained that scientific theories are the results of conjectures that have thus far withstood falsification and refutation by the occurrence of even just one counterexample, otherwise they would have been jettisoned. They survive only because the theorist is so committed to one theory that he sees only its supporting evidence and does not want to see the evidence against it. According to Popper, if this method of formulating scientific theories is not followed, inconsistencies will become evident in the different theories that are advanced to explain the same scientific phenomenon, and the whole structure would collapse; the theories would not survive scrutiny.

The approach of scientific generalization, some people believe, overlooks the effects of chaos theory, which often scuttle our attempts to disprove any other theory, as Pierre Duhem has shown. It also overlooks the inductive but false assumption that a theory which fails a set of tests at one stage will fail those tests at another stage, regardless of the diversification of such tests meant to minimize the possibility of chance occurrences. On the other hand, successively true predictions and causal explanations of a theory re-assure us that it and its antecedents are valid, even if we can never falsify or vouch for the absolute certainty of the probability statements on which they rest, because in their claim to a degree of reliability, they necessarily do not lay claim to absolute certainty. Nevertheless, the principle of falsifiability faces the same fatal blow as that suffered by positivism, because it also cannot be falsified.

An unfortunate example of critical rationalism from the medical field is that of new drug trials. Occasionally new drugs that have not withstood or undergone independent falsification tests are licensed because the test results show only the positive effects and significant

56

p values of the drugs in study groups that are not representatively exhaustive of the population in which they will be used. The promoters of the new drugs ignore or minimize their decisive adverse effects and non-significant p values in this population for unknown reasons, with costly results on the lives of some of the patients who subsequently use those drugs—thanks to whistle blowers for exposing the fraud.

Critical rationalism does not condone this kind of procedure as dealing in truth, honesty, and the quest for untainted knowledge, because it generalizes results without making allowance for notable specific exceptions or counterexamples, which, as in this case, should falsify the generalization and disqualify the drugs for general use. The proponents of this dangerous procedure of denial produce more and more evidence in favour of their position to justify it against the dictates of reason, and in the face of these grave documented odds; but that kind of justification holds no water against existing falsifying data, according to critical rationalism. An ongoing accumulation of confirming data does not prove the case, because it relies on the principle of induction, which entails the circularity of establishing a conclusion by using it as one of the premises of the argument as in the case of anticipated sunrises. Even if we relax the stringent requirements of radical critical rationalism, because of the impracticability of attaining them, and permit these procedures on the basis of the minimum requirements of relative traditional rationalism, the claims for these drugs still fail to survive these relaxed requirements. They defy the principle that we have to work within the limits of our knowledge, acknowledging that we can never know the whole truth or be certain about the consequences of most, if not all, of our actions, and that we should never disregard counterexamples, unless we do not intend to deal in truth.

# Constructivism

Constructivism is a theory of knowledge which postulates the active participation of the subject in construing reality, usually under the influence of technology. It shuns passive engagement in reflecting on attempts to represent reality as it is traditionally believed to exist and present itself to us. For the constructivist, acquisition of new knowledge is constructed by reflection on previous learning and the subsequent integration of all new information into the already established epistemological pool, versus the passive reflection on, and imbibition of, facts that lie fallow in that pool and from which we formulate true or false representative theories of reality. In the case of infants, however, constructivism as initially applied by Jean Piaget entailed the encrypting of new information on a previously blank cognitive apparatus (mind) which John Locke called a *tabula rasa* in the empirical tradition that nothing is in the mind that was not in the senses, meaning that infants do not have prior mental data on which to construct their concepts of reality; all their data and the resulting concepts are acquired from their experience of the world, as is the case with all of us (empiricism). Therefore, for each one of us, the only reality is one that we have created with our minds from contact with objective reality. If there were no objective reality, our individual realities would be so divergent as to make communication impossible, but as things are, we communicate about the same mountains, rivers, trees, flowers, and even diseases and illnesses. So, at some point our individual realities all come together as products of one causal reality to which different interpretations are attached (see 'rose', pages 27-28).

As a learning theory, therefore, and in maintaining that the acquisition of knowledge is not a passive process, but an active method of decision making that transcends the assembly of past and present divergent bits of data to constitute new knowledge, constructivism is consistent with the dialectic synthesis. Like dialectics where interaction of thesis and antithesis produce synthesis, a new and advanced concept that did not exist before, it knits concordant and discordant facts representing relativistic points of view into a complex from which there emerges a crystallization of meaning with equal applicability to all current and future contexts embroiled in the struggle for priority of recognition—this because we have no way of proving that our individual and multiple sensory representations of an objective and unified reality in whose making we had no part are the most authentic. It also attests to the importance of accumulated, processed experience in the active construction of an edifice of knowledge.

In their denial of the objectivity of knowledge, constructivists also deny the objectivity of disease-causing processes resulting from specific bodily malfunction in favour of subjective feelings of disapproval of situations that run counter to our cultural concepts of what we consider to be the person's natural state; i.e., disease is a subjective man-made concept that reflects on our attitude to the presumed harm caused by normal biological processes with whose results we are unhappy; it does not correspond with any objective malfunctioning of the body. That is why we associate pain with disease as one of its causes, because we do not like pain and when we feel pain we look for its source and label that part of the objective-body as diseased and therefore dysfunctional. In denying the objectivity of disease, the constructivist further claims that we thereby avoid passing purely moral judgments on disagreeable conditions by attributing them to pathological sources. As Dominic Murphy observes, "constructivism . . . often looks like a thesis about how inquiry is carried on: first we identify a condition we disvalue, then we look for a biological process that causes it and say that,

58

whatever it is, it is abnormal."[4] Mere sadness has also been "pathologized" to a disease styled depression, a popular diagnosis for which millions of different kinds of antidepressants are prescribed, as is the case with Attention Deficit Hyperactivity Disorder (ADHD), a label pinned on adventurous children by teachers who want every child in their class to behave like a little angel and not deviate from the expected norm. Some people attribute the increase in the labelling to increased awareness of the disorder by both parents and physicians, but my experience also tells me otherwise. At least one child and adolescent psychiatrist believes that the prevalent use of these drugs suggests the pinning of diagnostic labels without the benefit of comprehensive assessment, and the overvaluing of prescribed drugs versus effective psychotherapeutic intervention.

As an aside, in my practice I discouraged parents from labeling their children or allowing teachers to label them as "ADHD kids", because this shotgun diagnosis stays with them throughout their school years; they are known by it in every new class that they enter, and they are given second class attention because of it. The attitude is, he is an ADHD kid, so don't spend too much time trying to bring him up to the level of the rest of the students in the class who fall well within the two standard deviations of the bell curve. He does things his way at his own pace, and we are just happy to hear the bell ring for the end of the period so that he can move on to the next teacher and be her problem, or to see the end of the academic year when he will move on to the next class with his peers. If these bright souls could only focus on the child's difficulties and forget the labels, they would expend a little more effort in helping this child to overcome or adapt adequately to his problem and prepare him well for the bigger problems that he will meet in life after school; but they don't think that far, because they have not paid much heed to the meaning of the word "education". Some of them are just doing a job; others are getting the child ready to pass his examinations (book learning); a few others, however, are to be commended for remembering the etymology of the word 'education' from the Latin 'duco', which means 'I lead'. So, let doctors lead the way if teacher's can't. The constructivists in society need to revise this prevailing labeling of children with these problems so that more heed can be paid to helping them with their specific difficulties. Sticking a general label on them and dumping them in a common pit where they have little chance of escape doesn't help them.

Constructivists reconstruct the concept of disease in tune with features that constitute a patient's disease that may be found in other diseases. Constructivism allows that when a patient presents with a hodgepodge of symptoms, some of which appear to be inconsistent with delineated diagnostic paradigms, we are sufficiently concerned to knit them together against the background of the milieu in which the patient functions into real life entities that merit our attention, instead of dismissing them as mere fabrications or attention-seeking ploys, thereby driving the victim of this unfortunate assemblage of symptoms into depression. Previously unrecognized clinical entities like Fibromyalgia, Illness Anxiety Disorder, and Chronic Fatigue Syndrome are no longer passed off as relativistic manifestations of neurosis that are out of sync with universally recognized categories of existing bio-medical diseases. This attitude of concern is consistent with acknowledgement of and respect for the diversity of human nature and personality and the variegated physiological responses of the body to the constant incursions of environmental factors into its homeostasis, as well as our attempts to accommodate many types of clinical presentations within the diagnostic and therapeutic umbrella of clinical medicine. It recognizes the contribution of the subject to the purely objective influences that determine the state of his body and its functioning, and it bestows authenticity on these objective phenomena through his subjective synthesis and expression of them. He is the one who knows how it feels for his body and mind to be thus afflicted, and the physician has no right to dismiss his feelings as unfounded or fictional, simply because such diseases have not yet been described and classified, unless she can prove that the patient is malingering or hysterical. Constructivism makes this possible even after positivism has allowed isolated consideration and predictive utilization of the facts to the exclusion of the person about whom they are facts. Hence the importance of paying heed to the patients' narrative of their illnesses as they are living it from day to day, taking note of even trivial events mentioned in passing, because all of them can lead us to the discovery of crucial cues that our instruments cannot always reveal. The patients' survival could easily hinge on such attention to the details of their narrative.

A case in point is that of one youth in my practice who suffered chronic pain in her arm after a bout of Immune Thrombocytopenic Purpura that persisted after standard treatment. Neither I nor the tertiary institution assumed that her variety of Complex Regional Pain Syndrome merited dismissal as fabrication. She received all the necessary investigation and attention to relieve her of her obscure disease and make her life comfortable, which should be the aim of all our encounters with patients, even those who do not fit the mould carved out by biomedical medicine.

# Problem solving

Scientists often reach provisional agreement solutions to problems central to their disciplines, whereas philosophers do not. Although philosophy has been practiced by outstanding intellects for over two thousand years, philosophers have not reached agreement, provisional or otherwise, on the solution or dissolution of any central philosophical problem by philosophical methods.[5]

Problem solving entails critical analysis of the form and content of the problem and situations that spawned it, and thinking clearly and rationally about what to believe about them and what course to follow in trying to resolve them. To crystallize the process of problem solving, the mnemonic "IDEAL" has been devised to represent the following steps: Identify the problem, its presuppositions, and the logical connections among its related concepts; Define and represent the problem in explicit terms, ensuring that there are no logical inconsistencies and fallacies; Explore possible solution strategies and test them on actual and ideal situations; Act systematically on the strategies that you have synthesized as per the dialectic method; Look back and evaluate your conclusions as per the hypothetical-deductive method. As we will see when we discuss logic and scientific method, the solution of problems entails generating correct hypotheses and using the theories thus derived to explain existing problems and to forecast future problems and possible solutions, or else abandoning the hypothesis because it fails to explain a succession of facts. At no point, however, can an explanatory hypothesis predict with certainty the outcomes from specific events, whether they are physicochemical events or clinical marvels expected of physicians by patients, because the numbers of variables affecting the entire course of any such events are legion, exposing the dogmatic physician to limitless possibilities of error—the concept of fallibility. We will discuss the effects of variables on the original position in the Duhem-Quine thesis and in non-linear dynamics in chapter 13. Nevertheless, if these and other limitations are acknowledged, generalizations and predictions can be made reliably from the consistent track records of certain events on a statistical basis and without the unattainable possession of all the relevant facts.

Problem solving, like logical analysis, will feature in later chapters; but, as noted above, it entails identifying and defining the nature and ramifications of a problem and its elements precisely, as empirical (factual), conceptual (related solely to meaning and logic), and normative (value driven), seeking its potential causes, outlining alternative ways of resolving it satisfactorily, and then selecting the methods that are most likely to produce both the desired and the best results while monitoring the process all the time. The success of problem solving in medicine, based on the application of the best evidence available, will depend on how well we can distinguish relevancies from irrelevancies in addressing the needs of patients; how much our approach to their problems is eventually based more on rational, than solely on, empirical considerations; how well we can

61

restore them to normal overall functioning, based on their subjective feelings of wellness; and how well we can keep them from falling victim to as many diseases and illnesses as possible. In problem solving, one has to understand the limits imposed by the scope of applicability of one's clinical data and ensure that all tentative conclusions should undergo tests of validity, applicability, and appropriateness, always with the welfare of the patient as the primary concern, and with the necessary provision for alternate courses of action in cases of failure to achieve the intended goal. There is no room for guesswork or shotgun management here.

The doctor should be aware that problem solution is notorious for generating additional problems, especially since many problems are ill-defined by virtue of the obfuscating assumptions that we attach to them, thus clouding the goals that should serve as beacons in the process of solving them. Besides, our own ingenuity creates new complicating problems from solutions of old ones, prompting the need for engaging in lateral thinking with the usual vertical thinking to facilitate the recasting and clarification of obscure problems. We can also use algorithms to break problems down into simpler constituents that facilitate further stepwise resolution. Additionally, problem clarification and solving can employ analogy as in the famous tumour problem: how does one apply high dose irradiation to a cancerous tumour without injuring surrounding healthy tissue, given the fact that a fortress surrounded by a moat can be assailed simultaneously with success by a few soldiers using the many bridges that surround it? The solution is to bombard the tumour with low intensity radiation from different directions in the same way as the soldiers use different bridges to avoid crowding into the main entrance where they can become easy cannon fodder.

# Homology/Analogy

Homology (from *homologia* or agreement) is the sharing of similar forms—genetic, biochemical, and anatomical—whose causal basis is a common ancestry and hence shared inheritance of relevant information that is necessary for the mechanisms that produce that form for performance of functions. The functions may be dissimilar, but they are characteristic for the species in question. An example is the form of the fore-limbs of vertebrates, which share a common musculoskeletal pattern that has been adapted to the needs of each species in its struggle to survive in its particular milieu. The wings of bats and birds, and the fins of whales all fit into this category, as do our middle ear ossicles and parts of fish jaws and gill arches. Our vestigial supernumerary nipples are among those homologues that have now lost their original function. Homology includes similarity of structure in the right and left limbs of the same person (or other organism), and the serial structural similarity of a person's vertebrae —an example of the philosophical evolution of the definition.

The genetic theory of homology encounters a stumbling block, however, with the observation that only a 1% dissimilarity in human and chimpanzee deoxyribonucleic acid (DNA) is known to result in the vastly different phenotypes that constitute these species. Biologists call the genes responsible for this difference regulatory or homeotic genes; but they also note that homologous structures can be produced by different genes and different developmental pathways, in as much as non-homologous structures can be produced by similar genes and developmental pathways. So, whatever the mechanism for homology is, it is not solely one of genetic programming and similarity of developmental pathways.

Similarity in function, but not in structure, between the organs or parts of different species, adapted through convergent evolution to serve the same function, constitutes analogy of similarity versus that of diversity (from *analogia* or proportion); e.g., all types of wings in all the different species that fly—birds, bats, insects—and the fins of fish and whales, which are not homologous, but enable them to navigate their way in their water habitat. (Bird and bat wings are analogous as adaptions for flight and homologous as forelimbs). Analogy may also refer to correspondence in structure between a series of parts (such as vertebrae) in the same individual. Philosophically, analogy is used in arguments to contend that a preponderance of relevantly similar kinds of features over dissimilar ones in two things means that what is true of one is also true of the other; but such truth is always tenuous because of the lack of 100% correspondence between the features of one and the other.

In addition to the foregoing two evolutionary developments, metaphysical approaches to evolutionary change are divided into (a) taxic homology, a holistic approach related to grouping of organisms by shared characteristics to establish a common origin or convergence, (b) transformational homology, relating to the sequence of events in the evolution of a specific lineage. Further philosophical evolution of the two concepts has spawned the concept of homoplasy

> to designate shared similarities that were not traceable to the most recent common ancestor of two species. Homoplasy is thus broader than analogy, because the similarity can come from any source other than recent common descent (including both functional adaptation and mere accident, selection or genetic drift) . . . [as a hedge against] identifying features as adaptive.[6]

In genetics, homology is demonstrated by comparing protein or DNA sequences. If two homologous genes share a high sequence identity or similarity, they share a common ancestor and sometimes a common function; e.g., within vertebrates, the similarity of insulins is very close. Bovine insulin differs from human insulin in three amino-acid residues, and pig insulin differs in only one, explaining why pig insulin was preferred for human use before the advent of genetically engineered insulin. The chemical structure of Insulin is shown below.

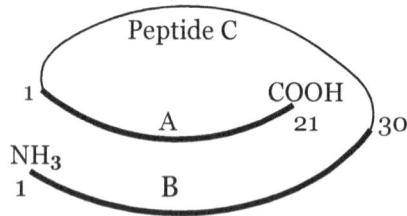

The 21 amino acids A chain and the 30 amino acids B chain are linked together by disulphide bonds at positions 1 and 30 after cleavage of the intervening C-peptide from the inactive 74-amino-acid prohormone molecule, proinsulin, to form the active product, insulin.

The concept of homology has thus proved to be useful in the treatment of diabetes with insulin, and also in bridging the gap from dissection of the frog, dogfish, and rabbit in Biology to dissection of the human body, as we learnt in our first week in the Anatomy dissection hall when we dissected the fetal pig as a prelude to dissection of the human body. Lately the skin of the pig has also been used for human skin grafting on the basis of homology between the two. Furthermore, conductors of drug trials, researchers into the genesis and treatment of cancer, and researchers on other projects also employ this same principle in experimenting on rats and other animals before they subject humans to clinical trials, and before they speculate and postulate hypotheses on the origins, behaviour, and treatment of cancers in them. In the case of the experimental use of mice, the concept has been stretched a fair bit as experience has shown that experimental information is not readily transposable from one species to the other; e.g., mice possess genes that enable them to detoxify some chemical poisons to which human beings will succumb. But their experimental use is still useful as a framework for reasoning about the human situation, in the same way as canaries are used to test for carbon monoxide in coal mine shafts to save human beings from potential lethal danger.

# Philosophy of Medical Concepts

# 4

# Personhood

## Definition of person—difficulty

The question of what constitutes a person is central to many of the concepts and decisions that have to be made in Medicine, simply because clinical medicine is concerned solely with persons as bodies and minds; but it has yet to be answered with concordance among those philosophers who have attempted to define personhood. For the doctor, the lines have been drawn by the Cartesian dichotomy of persons into accessible bodies and inaccessible minds, setting the stage for bio-mechanical medical practice, and relegating the person as conceived by Peter Strawson to a position on the back burner from where the phenomenologist is attempting to rescue it, as we will see in the pages that follow.

The multiplicity of theories of personhood, which must also elucidate the nature of mind and consciousness, only serves to compound the problems posed by the application of the concept of person to fetuses, severely mentally disabled human beings, and those in persistent vegetative states (PVS). (I will adapt the genetic USLegal biological classification of human being as "a hominid of the species Homo sapiens, a primate species of mammal with a highly developed brain" to include moral humans, except that some hominids satisfy all the other criteria for inclusion in this class but do not have the brain or the function that goes with it; e.g., anencephalics). Consciousness is also posited as the minimum necessary condition of personhood, and autonomy as its necessary and sufficient condition, but consciousness and autonomy are denied to fetuses, and autonomy is denied to the mentally disabled and to those in PVS by most "normal" people, thus depriving them of personhood and thereby creating controversy.

My concept of a person is that of a living, active, consciousness-endowed, human body, which consciously or unconsciously, voluntarily or involuntarily, displays all its integrated functions to varying degrees at different times and is capable of evaluating its actions against the backdrop of their effect on itself, others like or unlike itself, and on the milieu in which all of them exist. I discuss the concept of "self" at length in *Perspectives on the Concept of Belief*, and after a lengthy discussion of the nature persons as determined or free agents in *The Human Agent*, I conclude that ultimately, the person is what every one of us sees when he or she looks squarely in a mirror and has the guts to pass honest judgment, positively or negatively, on the one whose image appears in it. Behind that image, there is the real person who makes our world what it is for himself and for others.

# Body-mind problem

## Traditional theories

The body-mind problem arises from the assumption that a person is the result of the union of material and mental substances, the former easily accessible, and the latter wholly inaccessible. The difficulty posed by this inaccessibility of the mental aspect of persons thus creates the following puzzle:

(1) The human body is a material thing.
(2) The human mind is a spiritual thing.
(3) Mind and body interact.
(4) Spirit and matter do no interact.[1]

The human body is a physical thing that possesses mass, volume, and spatiotemporal location, whereas the human mind is not physical; it is insubstantial and not spatiotemporally locatable. Therefore the human mind and body cannot interact, because substantial entities cannot interact with insubstantial entities. We deduce the presence of mind in other persons analogously from their first person actions and pronouncements, which are third person considerations to us when compared with our first person thoughts, feelings, and actions, which are third person considerations to them. But the analogy does not even begin to get off the ground, because it is like comparing apples with oranges, and the conclusion is absurd, because we know that mind and body interact all the time, although we don't know how and where they interact, because one has location and the other does not; e.g., I can will myself to raise my arm and actually raise it without applying external force to it. Did my mind raise it, since it was not physically assisted, and it did not go up randomly and spontaneously but in response to my will or my intention? How and where did my will (mind) interact with my arm (body) to produce this result?

We owe the above impasse to Descartes who divided the human person into the sole but disparate categories of body *(res extensa)*, which is material and governed by the laws of physics, and mind *(res cogitans)*, which is immaterial intellect. He declared the mind to be

68

in control of this complex and to have the ability to exist apart from the body. But the quest for the site of interaction of these aspects of the person eluded him as much as it has eluded his successors. In the wake of the problems generated by Descartes' body-mind dichotomy, other philosophers proposed their own definitions of a person in an effort to jettison the metaphysical concept of body and mind as distinct, non-interacting ontological existents, and to look for a concept that will accord well with our experience of the freedom with which we move between the physical and mental aspects of our being. Their theories have included attempts to prove the following:

(1) a reciprocal, causal relation between body and mind, based on the being of persons as organisms endowed with sets of mental qualities (neutral monism, dualistic interactionism, or double aspect);

(2) a predetermined, parallel response of both body and mind to the same stimulus (psychophysical parallelism and original synchronization, or pre-established harmony);

(3) a basic, material bodily constitution that is more basic and ontologically prior to mind and from whose level of sophistication novel, irreducible mental and other non-physical higher order properties have emerged (emergentism);

(4) a body with mental manifestations as the necessary and sufficient accompaniments of its responses to stimuli. These theories of epiphenomenalism maintain that the brain is the cause of all our mental processes and physical actions, and our minds never exert any influence on our brains and our physical processes, presumably because mind is a product of brain activity. Psychotherapy provides a counterexample by modifying behaviour without acting physically on the brain as psychoactive drugs and electroconvulsive therapy do. These theories have all met with variable reception and rejection.

## Later theories

Later variations on them have been proposed, such as the

(1) new epiphenomenalism, which states that some bodily (brain) states are also mental states, and causal mental properties are physicochemical properties of these bodily states; so the body is not a purely material thing. This version of epiphenomenalism styled supervenience claims that every mental property (psychological characteristic) has a physical base on which it supervenes, or with which it shares a common cause and to which it is not reducible; hence every pain (mental property) is underwritten by the concomitant firing of $A$ fibres: the nerve fibres that carry sharp, localized pain sensations from skin and muscle, or $C$ fibres: those that carry poorly localized, dull pain from viscera, peritoneum, and muscle. The justification for this claim is that "the mental is 'realized'

by the physical"[2] and no mental property can have non-physical realization. The counter-example to this claim is that the same nerve fibres can fire without causing pain in an anesthetized patient. This theory is really a form of emergentism, because it bases mental states (pains) on underlying physical states (tissue damage issuing in pain behaviour consistent with the organism's neural organization) with or without causal interaction between them. It also has elements of the identity theory of mind, which claims that mental states are brain states, and therefore the mind is the brain, and all psychological processes and psychiatric conditions are physicochemical brain states; hence the utility of mood altering drugs via their chemical effects on brain chemistry

(2) identity theory, which reduces all mental states to physical states, not emergent states referred to above

(3) eliminativist theory, which rules out mental states in toto in favour of physical states.

These earlier and later theories hope to eliminate the superior ontological status accorded to the mind by Descartes, but they fail to account for the essential difference in meaning between terms referring to mental and physical processes, and they still do not explain how the neurophysiological phenomenon of the firing of $A$ and $C$ nerve fibres, which is a physical event, translates into the experience or mental event of pain. Such objective physiological (third person) descriptions omit a big chunk of the phenomenology of the experience of pain and other sensations and their accompanying emotions, which are subjective (first person) and not amenable to objective measurement or reducible to third person experiences, and which necessarily vary from person to person. We cannot possibly read a person's feelings and other experiences from analyzing the state of his neurological system, and according to Thomas Nagel, we can also not separate the functional features of pain from its phenomenological and physiological features, which entail it and each other without being entailed by it. Therefore any state that is functionally equivalent to pain would not be pain if it differed from pain phenomenologically and physically, because the mental state related to it would be different from that of pain as we know it. We can't read the feeling of pain from brain states.

In his observations concerning our conclusion that there is no necessary connection between consciousness and the physical world, because our concepts fail to reveal any such connection, Nagel further states that talk about minds, mental events, or mental processes is not talk about physical objects, events, or processes, but he allows for the reverse process whereby mental events can be traced back to and explained in terms of physical events in the brain. He goes on to assert that "we have good grounds for believing that the mental

70

supervenes on the physical—i.e., that there is no mental difference without a physical difference."3 Nevertheless, these grounds do not help to explain the mechanism whereby the supervenience occurs, as he says in another context, "mind is a biological phenomenon. So long as the mental is irreducible to the physical, the appearance of conscious physical organisms is left unexplained by a naturalistic account of the familiar type."4

(4) Logical behaviourism, which claims that mind is the totality of the actual or dispositional behavioural states of persons, without defining persons as such. It maintains that statements about mental activity can be expressed in terms of actual or possible behaviour. The immediate questions are how other people can ever get a handle on private, dispositional behaviour and whether absence of behaviour should imply absence of mind to the observer. Besides, there can be pain behaviour without pain, as in pretending, or pain without pain behaviour or any disposition to behave as if in pain, as in a state of paralysis. So behaviour and feeling are logically independent; i.e., there is no relation of entailment between them.

(5) Functionalism, which claims that mental states are functional states resulting from certain causes (sensations) and manifesting during the production of certain results (behaviour). Therefore mind is the non-physical effect of perceptual input and the simultaneous cause of behavioural output, both of which are physical. Some have compared mind with the function of a computer as the hardware (brain) produces results (actions) from the programs that are loaded into it via the software (sensory input). Hence consciousness of pain or being in pain is the result of particular sensory inputs resulting in the output of pain behaviour, which is different from saying that behaviour is the equivalent of mind, as the behaviourists maintain. So mind is not identical with behaviour, but with what produces that behaviour from raw sensation. Again, the person whose mind and behaviour are in question is taken for granted; but functionalism explains how his (mental) illness can be approached from both physical and mental perspectives in keeping with the observation that psycho-social events often have long lasting effects on brain function, e.g., cases of reports of childhood trauma causing hyper-reactivity of the hypothalamic-pituitary-adrenal axis in later life.

Stephen Priest supports the identity theory of mind, stressing that any talk about mind is talk about the capacity to think, since for something to be a mind, it must satisfy the logically necessary condition of being capable of thinking, which also happens to be the sufficient condition for being a mind. His conclusion is, therefore, that "*the mind is the brain*."5 Hence, to say that someone has a mind is the same as saying that she has the capacity to think, from which it also follows that she is in command of mental skills which she can manifest in the performance of appropriate behaviour and the display of dispositions to that kind of behaviour in the relevant circumstances, as we have already observed. The equation of mind with brain is not a necessary and sufficient relationship, but only contingent, since it might easily not have been the case. This theory does not, however, account for the fact that damage to the brain necessarily entails annulment of the capacity to think, because the mind is where the brain is. The natural and sole mental activity of the brain is thinking, just as the sole function of eyes is seeing with the brain, and that of ears is hearing with the brain. Hence, there is never a question about the point of interaction between the brain and thinking (mind), the eyes as the media through which the brain sees, or the ears as the media through which the brain hears. All these organs just do the jobs for which they are meant in the service of the person, and the improper category mistake question of whether a person is physical or mental or both does not arise on this theory.

As we will see in the next chapter, the argument is made there that related mental events reflect simultaneous brain events, which occur in response to how the person experiences the world. So the brain and its function as manifested in the person's behaviour are the rock bed of Psychiatric enquiry, since even the most recent methods of imaging will not tell us what thoughts anyone is entertaining .

# Personhood

John Locke defined his necessary and sufficient conditions for personhood as those of "a thinking, intelligent Being, that has reason and reflection, and can consider it self as it self, the same thinking thing in different times and places".[6] According to this definition, anything that does not persist through time and place and can also think and reflect by virtue of its possession of intelligence and reason, which dwell only in the mind, cannot be a person. Therefore the body alone is not a person, because it lacks those characteristics, but the mind alone can qualify as a person. Jason Stanley argues that the above interpretation of Locke's position is erroneous, because it attributes to him the implication that persons can only be "*direct* bearers of mental properties, [but] not direct bearers of their physical properties"[7], which in turn implies that they derive their physical properties from some mysterious relation to the body that has those properties. But the thrust of Locke's personhood still remains that of randomly embodied consciousnesses with psychological threads connecting its bearer's past and present experiences with her future aspirations and expectations, because only her mental properties are necessary for her particular personhood, since they make her who she is. Her body and its physical properties are contingent, and might as

72

well have been different without affecting who she is as a person—a position that is not entirely tenable, as Strawson argued.

Naturally, this definition raises questions about the self-awareness and personhood of fetuses, anencephalics, and patients in persistent vegetative states, and whether any lapse of consciousness constitutes temporary cessation of previously existing personhood and the emergence of a new personhood afterwards, such as in and after a concussion or amnesia. David Hume would have no part of this nebulous entity styled the self as reflective of personhood. He pointed out that a thorough search of oneself will not reveal an entity or self that lurks behind one's totality of experiences to define the owner of those experiences. But he did not answer the question: who is searching herself, in the same way that Locke did not define the being that he described as having those properties that he assigned to it. John Searle answered this question in his way by postulating his concept of a person as "a purely formal notion", which he then defines as "an entity" endowed with an array of mental properties: "consciousness, perception, rationality, the capacity to engage in action, and the capacity to organize perceptions and reasons, so as to perform voluntary actions."[8] It is possible that this complex of notion and entity is the same as the biological person who is constituted by the human organism with its characteristic karyotype and its potential for consciousness, the same body that is now defined as the functional being who cannot be fully described without including his consciousness as maintained by Strawson.

In his effort to discredit Descartes' metaphysical dualism of the person, Strawson has followed Benedict Spinoza in re-affirming the double aspect theory, which states that the mental and physical are two fundamental aspects of the single underlying reality of personhood, which is neither mental nor physical and cannot be reduced to either one of these categories, since it is more basic than they are. Hence, the concept of a person is not constituted by the union of two disparate, prior aspects, mental and physical, but is a primitive concept that derives from the inherent double aspect nature of personhood itself. In Strawson's words, we ascribe to ourselves physical characteristics that we also attribute to material objects, but we also attribute to ourselves qualities that we would not ascribe to material objects, such as conscious states. So we need to explain how our "states of consciousness . . . are ascribed *to the very same thing* to which these physical characteristics, this physical situation, is ascribed."[9] The answer can only be that as persons, we are primordial complexes of mind and matter that cannot be teased apart and recombined to make a person. So body and mind are property derivatives of personhood.

In Strawson's ontology, the person as basic particular predates the combination of mental and physical substances in a particular material body, which Descartes started with and then reduced it to separate, distinct, and non-interacting mental and physical components with opposite properties. A person can thus be described as weighing 40Kg, but also simultaneously as being ill (from observing his behaviour), without posing a problem about how to interpret illness in a lump of matter that cannot be described as having feelings or the capacity to display feelings of any sort, as the biological concept of personhood demands of us. Strawson's concept of a person makes possible the ascription of psychological predicates to this special lump of matter without having to commit the category mistake of assigning these properties to the biological human body that might not even warrant that ascription, as in the case of fetuses and anencephalics. But in so doing it assumes previous knowledge of who are persons and who are not, so that these predicates should be ascribable to them. It is, however, still a dualism, not of substance like Descartes', but of properties, because it recognizes two kinds of properties in the world: psychological and material predicates. It also restricts the individual roles of brain substance and of memory as being sufficient for personal identity through time, in contrast to Locke who proposed only memory as sufficient for that purpose.

Alasdaire MacIntyre's constructivist theory identifies a person with his particular life narrative, a narrative that must of necessity differ in details, values, and goals from that of any other person, and that includes physical and mental descriptions of his sequential life experiences—ideally stretching from birth to the point before death. This is the narrative that the patient often unsuccessfully attempts to convey to the doctor. It doesn't discriminate between body and mind as to questions of priority, unlike the Cartesian dichotomy expressed in his famous dictum: "I think, therefore I am"[10] (cogito, ergo sum), which places mind well above body as essentially a thinking thing with the ability to exist without the body. It presents the patient as one whole subject with physical and mental burdens, beyond the complex of physical and biochemical characteristics that present him to, and represent him in, the external world where he is the object of regard (or gaze, in the terminology of Jean-Paul Sartre) by others.

Steven Edwards[11] similarly delineates his essential aspects of personal existence that will identify the personalities of persons as (1) existence of persons in space and time, relating to the subjective and objective experiences of all the places they have been and the times involved; (2) their embodiment or their submersion in and dependence on their intact or defective bodies as representing their personal and subjective abilities for taking on the world; (3) self-conception of the kind of persons they are aspiring to be, depending on the values that they cherish in life; and (4) the pursuit of goals or self-projects that will raise their self-conception. Personhood is taken for granted here, but it remains central to Edwards' thesis, because changes in the mental competency, goals, and values of a patient may render his advance directives dubitable as applying to the same person by whom and for whom they were originally framed. Edwards is one step ahead of us in our effort to define the person by presuming that there already exists a "then-self" who framed the original directives and a "now-self" who is issuing new directives. The two selves may be related by bodily but not by mental continuity of the whole-self. Although personhood is taken for granted, the person thus described merits our respect if we intend to promote his autonomy and dignity in his total care.

Michael Tye epitomizes his concept of persons thus: "the subjects of appropriately complex psychological bundles. . . . brains in so far as those brains are in the appropriate physical states"[12] with beliefs and memories that can be connected over time. Without brains there cannot be any of this continuity, as demonstrated in states of split brains and in multiple personality disorders in which there is one brain (one body) but many persons with many mental histories, or one material substance and many spiritual substances. His theory accords well with the ego theory of personhood, which maintains that a person is a continuing spiritual substance whose essence is the consciousness of causally related experiences, thoughts, and mental states.

As we have already learnt, Nagel believes that the brain is a necessary and sufficient condition for personhood. Without a brain there cannot be a person, although there can be a human being, as in the case of an anencephalic infant. The intact brain is the material seat of all the identifying mental processes of a person, so that drastic changes in its structure amount to changes in personhood while minor functional disturbances, as in brief losses of consciousness, will not affect the identity of the person who maintains the same brain structure throughout the disturbances. Half a brain, however, results in change of personhood, as Warren Bourgeois also maintains. He ties up all these perspectives on personhood in his constitutive view of biological, psychological, cultural, and moral factors as being together primarily constitutive of a person, and he goes on to say that persons "have an indefinability and . . . an indivisibility and irreducibility as well. To be in the class of persons, one must be self-creating. . . . Chimpanzees may have 98% of our chromosomes, . . . but unless they are self-creating, they are not people."[13] He leaves the concept open to continual fulfillment as persons create themselves with their actions and choices, much as Sartre stated: "man first of all exists, encounters himself, surges up in the world—and defines himself afterwards. . . . He will not be anything until later, and then he will be what he makes of himself."[14] In this approach, Bourgeois also agrees with David Wiggins who believes that a person is an animal (substantive entity or organism) with qualities that are insufficient to qualify him completely, because he is always adding more qualities to his description. With all his thoughts, emotions, and actions, he thinks of himself "as having a past accessible in experience-memory and a future accessible in intention"[15]. So, to make him a person, the human being needs supervenient psychological components—which are described as additive without being self-sustaining—among which are included propositional attitudes and intentionality. We encountered supervenience above in the discussion of pain fibres.

## Corporeal foundations of personhood

The logical outcome of this exercise is that even though static bodily continuity is insufficient for personal identity, consciousness ensures its enduring continuity by virtue of the integrity of the neurophysiological structures on which it is founded. Hence, the amnesiac remains a person with full responsibility for his previous actions, even if he might claim to be the same body or organism, but not the same person as he was before his loss of memory, as John Locke's theory of personal identity will allow. His ability to also form conscious reciprocal relationships, which are not sufficient but are minimally necessary for personhood, differentiates him from the patient in a persistent vegetative state who has ceased to be a person by virtue of the loss of these abilities and the first person (subjective) perspective on the world that accompanies them, even as he continues to be the object from which those perspectives previously arose.

All told, it seems that the locatable, living human body, by virtue of its spatiotemporal continuity, is the assembly point of all the characteristics that matter in the delineation of personhood, even if the person himself is regarded by some philosophers as a series of momentary stages of the same organism as it progresses through space and time. Bodily identity is, therefore, the necessary condition of personal identity, because the inaccessible mental criteria of other persons cannot fulfill that criterion. All persons are known primarily by their bodily presence; their other characteristics come second. That is also the case in legal circles where bodily presence is extended to finger prints, DNA, and soon to iris patterns, and in medical circles where the same criteria are utilized for pinpointing personal identity.

The whole person, however, is a complex of mental and bodily characteristics that are both necessary and sufficient conditions for personal identity and that justify Jim's conclusion that other persons have the same mental life as he has, or other minds exist in the world. Outside of this framework, personhood can be known only by direct intuition; the experiences of other persons and their personhood are assumed by analogy from the coincidence of their behaviour with his when they are placed in similar circumstances; they cannot be known first-hand. It is quite evident that zygotes and fetuses do not qualify as persons under these criteria and under ontological distinctions that are drawn between persons and non-persons on the basis of self-consciousness or knowledge of oneself, especially of one's mental or internal states in a continually changing external and internal environment. One must be able to recognize and adapt to these different changes within the range of one's cognitive operation and occupational pursuits, and in keeping with one's level of intellectual development as determined by age and experience. The effect of this categorization spills over into the question of the moral status and worth of the lives of fetuses and those human beings who are excluded from personhood as a result of the severe cognitive (and physical) shortcomings that have contingently come to characterize them as the kinds of individuals they are, and that have become the ontological means of identifying them as mentally or physically disabled persons, because they lack these abilities.

Personhood cannot, however, be understood by dissecting human bodies and studying their constituents, but only by analyzing the actions of human agents who are steeped in a social milieu where each agent displays unique interaction with other agents consistent with his temperament, experiences, and personal history as Edwards and MacIntyre have proposed. Included in these encounters will be successes, failures, anticipations, disappointments, abilities, and disabilities that assume control of his life as the agent battles illnesses and adversarial relations with other agents, including environments that can be hostile or conducive to the unfolding of his personhood. Some persons will face the adversities of life courageously, because they have a history of fortitude and high self-regard that has also won them the support of others around them, unlike others who despair easily on their lives and personhoods, regarding their bodies as impediments to a life that would otherwise have been good. The latter miss out on the role of their bodies in their personhoods, because the body is only the vessel that introduces the person to and sustains him in the world, as we have noted. What matters is what that body is harbouring in the personality that inhabits it. People who lose their personhood to their bodies in life's battles also lose those battles even before they take them on, because they surrender to the infirmities of the body and allow matter to triumph over mind; they make no room for their individual consciousnesses to assert their personhoods over their failing bodies. Hence the belief that the positive attitude of mind toward illness goes a long way to helping the patient overcome his burden, while capitulation and depression may accelerate his demise.

# Consciousness

## Nature of consciousness

We will briefly discuss consciousness, since it appears to be the central concept of all the discourse about personhood. The central problem in the nature of consciousness is how several properties and perspectives (visual, auditory, tactile etc.) of a single object, and the experiences of different objects and situations can all come together to form a synchronic unity (an encompassing experience or a unified phenomenology). Many parts of an object under study and many sensory experiences of it unite to present as a single phenomenon with its phenomenology that differs from the phenomenologies of its parts, which are rarely, if ever, experienced separately. Similar considerations apply to one scene composed of many objects, or one experience consisting of many sensations in temporal succession, whose elements are never appreciated individually but always together as one whole. According to Tye, failure of these magnificent mechanisms of vertical and horizontal unity explain the breakdown in unity that invariably results in split personalities. A parallel situation is seen in cases of division of the corpus callosum for treatment of epilepsy where the challenge is to decide who the products of this procedure are in terms of the independent streams of consciousness that now reside in the two half brains that were previously synchronized as one brain with the left half excelling in language functions and the right half in spatial reasoning. Which one can claim to be the continuation of the person (consciousness) who was terminated to produce them, since each one of them now functions independently of the other?

In answer to these difficult questions, John Searle sets out to demystify consciousness for the sake of finding out the causes of consciousness and what it causes. He begins by defining it as "a natural, biological phenomenon"[16] that he likens to our daily biological processes of both a physical and a psychological or mental sort. In a nutshell, "those states of sentience and awareness that typically begin when we awake from a dreamless sleep and continue until we go to sleep again, or fall into a coma or die or otherwise become 'unconscious'."[17] The single quotation marks are necessary because the definiendum (consciousness) appears in the definiens (unconscious), constituting a circular definition—one cannot fully comprehend what lack of consciousness is until one understands what consciousness is. Lack is absence, and absence is non-specific; it does not zero in on anything, and so it does not convey any descriptive or identifying information. Therefore knowing that consciousness is not unconsciousness does not do much to further elucidate the definition of consciousness. Consciousness has also been defined as the inner, first person, qualitative, subjective state of everything that we feel, do, and suffer as a result of having a brain whose emergent property at higher macrolevels results from activity taking place at lower micro-processes. Hence it is that in a comatose state we lose function as a result of loss of consciousness

Nonetheless, in the tradition of those philosophers who also want to dispel the Cartesian body-mind dualism and its mystery of how body and mind interact, Searle goes on to state that consciousness is caused by "lower level neuronal processes in the brain; [hence it is] a feature of [the] brain [rather than] a separate entity from [the] brain".[18] In other words, it is an emergent property and a feature of the physical world of the same kind as that claimed for mind by the espousers of the theory of the emergence of mind from brain—pain is not only the firings of $A$ and $C$ nerve fibres, but also the conscious manifestation of that neurological process. The only problem in both cases is to advance beyond the stage of recognizing the correlation between brain and mind on one hand and brain and consciousness on the other hand to explaining how neurobiological processes in the brain are sufficient to cause each one of them, i.e., mind and consciousness. Postulating Cartesian mental and material aspects to the person will not explain this causal relationship. Unlike Priest's comprehensive theory, it only reduces consciousness and mental processes to the level of primary processes occurring in the brain, instead of second level effects of those brain processes in relation to the particular experiences of a subject.

We are thus forced to call consciousness neuronal behaviour when we should be saying that it is explained by neuronal behaviour without being reducible to neuronal behaviour. We are also forced to limit human behaviour to deterministic mechanisms; and yet we know that our behaviour and actions depend on the rational processing of neuronal input from our environment and our ability to freely make responsive decisions as active agents, if we are not acting under duress. This ability accounts for the free will that is claimed to remain when I subtract the fact that my leg goes up (after a tap on my patellar tendon) from the fact that I simply raise my leg (I raise my leg - my leg goes up = free will). Free will is in turn directed towards my goal by consciousness, which originates from activity of neurons in my brain, resulting in my ability to act otherwise than as I am actually doing, i.e., not raise my leg, and to assume responsibility for the choice and execution of my actions. So the question is still how this mass of matter in my head accomplishes all these feats.

Searle maintains that without being experienced by someone, consciousness does not exist. It "only exists when it is experienced as such."[19] Its unified structure makes it integral to its experience; it is the experience. As well, each experience has its own qualitative feel and mood, which is susceptible to pharmacological influence and which constitutes the basis for the use of mood altering drugs like Prozac (before questions were raised about its effectiveness *vis-a-vis* placebo). A key characteristic of consciousness is self-awareness, a person's sense of who he is against the background of everything else that exists in the world. This feature is lacking in people who are brain-dead or in persistent vegetative states, zygotes, and fetuses. In addition, all states of consciousness are defined by intentionality or directedness at something other than themselves. Something may be real, or imaginary like a fairy, but the thought of it is a target of the mind, and its representation in the mind alone makes it an intentional object. Examples of intentional states: Jim's thought that he could have pneumonia, his belief that he might have it, his fear that he has it, and his hope that he does not have it. All these states have the same content or object: "pneumonia", but different attitudes or psychological modes: "thought", "belief", "fear", and "hope", with different fits to the factual world of whether or not he does have pneumonia. His attitudes do not determine or influence the facts; they only influence his reactions to those facts. If the world is as he thinks, believes, or fears, then there exists a mind to world direction of fit that validates his mental content, as per Searle; i.e., conditions in the world answer to his mental states. On the other hand, if what he desires about the world is not possible, then there is a lack of world to mind direction of fit, because what he desires or believes does not obtain in the world, so his state of mind lacks a satisfying counterpart in the world; but the intentionality of his consciousness still holds, although its purported truth is void.

## Consciousness, body, person

Now, since consciousness is neither spatiotemporally locatable nor identifiable in the body that harbours it except as an occult property of that particular body, the human body once more becomes the individuating element in personhood. If that were not the case, consciousnesses would be floating around, ready to attach themselves to any material body that could accommodate them; and that would be a recipe for chaos and confusion. Doctors would have the impossible task of trying to keep track of their patients' pains and other feelings as they floated around and attached themselves randomly to any material body. They would not know which patient to treat, because they could not identify the owner of these flitting feelings with any consistency. Besides, personal identity cannot be defined by consciousness of itself, because to be able to attribute consciousness to the same person one first has to know that the

80

person concerned is its original possessor—the body that harboured it in the first place.

If we maintain that consciousness and its content are the defining characteristics of personhood, then we have to account for lapses in consciousness and memory, which some people suffer periodically and for prolonged periods, if not for the rest of their lives, and without which they cannot claim to be persons; and we have to account for different levels of consciousness as displayed by human beings who enjoy higher functional levels than animals. Hence we have to attribute to persons consciousness of their consciousness, which animals apparently do not possess. We also have to concede that after a brain transplant with Mike's brain, Jim may still have the same body by which he is externally recognizable as Jim, but he is not the same person, because he has a different brain, which carries conscious memories that differ from those of his old brain, and which causes him to display new patterns of behaviour that are unlike those to which his friends are accustomed. He may behave like Mike, but his friends will still call him Jim gone strange. On the other hand, Mike's friends will also call him Mike gone strange, because he looks like Mike, although he does not behave like him by virtue of harbouring Jim's brain and consciousness. So it seems that these considerations place the body at the centre of personhood, selfhood, and identity—objectively to onlookers, but not subjectively to the person concerned whose life is steered by his mind.

If we further maintain that personhood is constituted by the unity of all the states of the organism via consciousness, whether as awareness of self and environment or of what it is like to have certain phenomenal experiences, then any breakdown in the phenomenal unity of these multimodally experienced elements of the unity of consciousness will result in a breakdown in personality, resulting in mental disorder. Abnormal phenomenally conscious states cause more than the normal reactions that accompany these states in daily life. They result in distortions of perceptual experiences in the form of hallucinations and delusions, bodily sensations in the form of aches and pains without neurophysiological basis, felt reactions in the form of phobias, and felt moods in the form of anxiety and bipolar disorders. Additionally, if consciousness (or is it self-consciousness?) is the minimum necessary condition for personhood, and fetuses, anencephalics, and persons in persistent vegetative states are denied this capacity to any degree, then these categories of organisms are not persons by definition, although they might be persons by other criteria—which ones? Since such criteria are not forthcoming the best that we can do for these categories of beings is to endow them with the title of human being, as per the definition quoted on page 67 As we will see in the discussion of abortion, it is not easy to decide at what point in the subsequent development of a zygote a person comes into being, although the weight of argument tends to favour the presence of consciousness as the distinguishing characteristic without pinpointing its time of emergence in the life history of the person. On the popular view of consciousness, it means that the fetus as person is aware of and experiences itself as a subject of consciousness; that it is aware and values its awareness of its being or existence as a person— a proposition that sounds rather far fetched.

The bodily changes referred to also present their own kind of problem in the conceptualization of personal identity. Jim's body is not strictly identical with itself at time $t_1$ and time $t_2$, because in the interim many changes in its composition have taken place: cells have been lost and replaced by new cells, biochemical changes have occurred, and the Jim of age 3 months is not qualitatively identical with the Jim of age 3 years, 30 years, or 90 years, but he still claims to be numerically the same person, even if he has undergone multiple anatomical, physiological and biochemical changes such as plastic and transplant surgeries, renal dialysis, and regular insulin supplements for his failing pancreatic $\beta$ cells. Even with this progressive change of bodily characteristics over time, we also still have to say that he is the same person by virtue of retaining those properties that are essential to his existence as the same person through time, as opposed to those that are merely incidental, because he is consistently and sequentially sufficiently similar to himself.

As for the self that we have to use as the standard, it will probably be the spatiotemporally continuous body through which we made our first acquaintance with him, in spite of his qualitative changes over the years and our interrupted observation of him. If we kept him under continuous observation, we could confidently say that he is the same person that we first saw as a newborn baby, even if his behaviour over the years had changed so drastically, for whatever reason, that we felt compelled to say colloquially that he is not the same person. If we could not thus observe him, we might find it difficult to claim that the continuity of his memories, experiences, and his other interconnected psychological characteristics are sufficient to constitute his personal identity through time, especially if they were marked by major gaps; and this does not take into account the fact that Jim does not belong to his memories and experiences. These characteristics belong to him, even though we have not yet defined him; they are instantiated in the same body.

It seems that neither the body nor consciousness (the mind) is individually sufficient for personhood, although both are necessary conditions. Some philosophers, like Hume, whose denial of a self that owns all the personal properties with which we are acquainted, have tried to resolve this impasse by suggesting that the person is a mere

82

bundle of properties instantiated at a certain location, since extensive self-searching has revealed only sensations but no person who is the cementing substance of those sensations. Others have stated that the "I" of personhood is only a linguistic devise used to identify the source of origin of an utterance, and that it does not refer to the "ghost in the machine" [20] described by Ryle in his reference to the mind that was allegedly housed in the body of the Cartesian person, nor does it refer to the body itself. We do not say that our bodies are tired, but we say that we are tired. The statement does not, however, imply that the persons inside us are tired, but it may imply that our bodies are tired. When someone says that she is thinking, we understand her to mean that thinking is occurring where she is, perhaps in her brain. Does that make the brain the centre of personhood? Some other philosophers think so.

## Clinical perspective on personhood

The person concept is central to the kind of treatment that a patient will receive, whether his illness as he lives it or his disease as it "inhabits" him will receive treatment. In some situations, however, if treatment is limited to disease, then some patients will not receive treatment for their diseases, because the true natures of their diseases may not be fully known at the time, and no modality of treatment can be undertaken for something unknown; so they will continue to suffer from their untreated diseases and unrelieved illnesses. But if their treatment is directed at their persons, their suffering will be relieved while their diseases are being unraveled for specific treatment. On the other hand, if bodies are not persons, then patients, as persons, will be left behind in the effort to rid their bodies of disease.

The patient with a mental disorder presents a special problem, because his primary derangement is not always physical with a mental overlay, but it mostly starts as a primary mental problem with physical manifestations, thus emphasizing the point that persons, not bodies and minds, suffer from illness and disease. His problem also suggests that the person with mental illness who lacks objective signs and displays only persistently unorthodox behaviour such as paranoia is not afflicted with disease as an ontological entity that can be nailed down. We can see, hear, describe, and sometimes feel the effects of his physical behaviour; but how do we describe his mental state precisely without merely speculating about its nature on which we have no handle, and how do we assign personhood to him when he is found to lack the allegedly sufficient condition of autonomy, even though he has its reputed necessary conditions of body and consciousness? These issues lead some people to postulate that there are no mental diseases *per se*, in the realist sense, but only delineable diseases of the brain with mental symptoms, or else that mental illness is a mythical symptom of moral problems in living. For them, true illness is caused by disease of a body organ or tissue.

Personhood is central to questions of life and death, because considerations relating to the beginning and end of life and their social implications hinge on the concept of a person, as we will see again when we discuss abortion and euthanasia. Some of the supporting arguments are aimed specifically at the role of the brain as the source of consciousness and personhood. We are reminded that persons are essentially constituted by biological organisms lacking in self-consciousness without being identical with them, in the same way that statues are constituted of lumps of matter without being identical with them. One cannot come to be (have properties) without the existence of the other, since it derives its properties from its intimate coexistence with this other. The person and the organism are not two separate entities in the same place at the same time, but they share a joint state of being (existence) interdependently. Therefore the patient who is in a persistent vegetative state is not the equivalent of an organism that has survived the person's death—the person being the organism with a whole brain and self-consciousness derived from the same biological organism; he is also not brain-dead, but has permanently lost awareness of his environment and himself.

So what is his status? Joseph Fins succinctly answers this question by delineating and distinguishing between the two states of prolonged impairment of consciousness that can cause confusion, viz., diminution and abolition of consciousness as manifested in

(1) the minimally conscious state where patients are able to follow simple commands and engage in purposeful behaviour, and can still recover full consciousness if the process is not prolonged. In this state they "demonstrate unequivocal, but fluctuating, evidence of awareness of self and environment."[21] At the level where they are closer to recovering consciousness, they are capable of limited articulation of words or phrases, and they can also communicate through gestures and also show other evidence of intelligent attentive and intentional mental activity.

(2) The vegetative state, typically described as "a paradoxical state of 'wakeful unresponsiveness' in which the eyes are open but there is no awareness of self or environment."[22]
A vegetative state that lasts more than one month after brain injury is persistent, and one that is clinically irreversible is permanent, as seen commonly in states of hypoxic-ischemic encephalopathy (after 3-6 months) and traumatic brain injuries (after 12 months).

# 5

# Psychology and Psychiatry

## Psychology

Psychology studies the human mind via behaviour, assuming that inaccessible mental states of people are causally related to their behaviour. We assume and conclude from their actions that other persons have minds and thoughts that direct their physical activities, because they behave in ways similar to how we behave in the same kinds of circumstances, such as when we are in pain, although this is a spurious analogy, because we are not comparing two "knowns" but one "known" and one "unknown". Nevertheless, this perspective is in direct conflict with epiphenomenalism, an alternative to Cartesian dualism, which we now know favours a one way street of influence from body to mind and never from mind to body. But in the pain syndrome we have tissue disruption causing pain, which in turn causes certain motoric behaviour patterns, proving the existence of a two way street. Here we have psychological behaviourism, which tries to explain pain behaviour as response to the external stimulus of tissue disruption, matched against logical behaviourism, which starts at the mental level, inducing a disposition in the person to behave in certain ways. Nevertheless, to replace the phenomenological state of the experience of pain with dispositional behavioural states is still to misinterpret the nature of pain as an existential experience with personal implications for the subject of that experience, since dispositions exist only as unfelt possibilities that might not even be realized, while pain is real, present, and unpleasant.

In this situation, then, observable and dispositional behaviours only serve as aliases to explain mental processes without singling them out as the causes of the behaviour, because dispositional behaviour means only that if a particular condition obtains, (arsenic is ingested) certain behaviour will follow (the person who ingests it will display a certain array of symptoms), provided that an infinity of other conditions tending to the same end prevail. If even small, unfavourable initial conditions creep into the equation, the end result may prove to be vastly different from what was predicted, as chaos theory maintains, which means that the relations among the components of the system are non-linear. Another way of stating the

85

case is to say arsenic is poisonous means that if arsenic is ingested in the context of all the relevant circumstances, it will poison whoever ingests it; but as long as it is not ingested, it remains only potentially and not actually harmful.

So in behavioural science the mind remains dispositional, never carrying out any actual functions, but always remaining poised to do so only after the fulfillment of an endless concatenation of conditions, which is never. Behaviourism as a modality of diagnosis and as the only modality on which psychiatry has to depend also misses out on the narrative aspects of diagnosis, because the patient's demented state renders communication with him impractical. There is no direct access to a disordered mind as there is to the orderly mind of a diseased body; so psychiatry loses out on both the bio-mechanical and phenomenological advantages of general medicine, thus making its task an extremely difficult one. Outward behaviour may be causally related to our inner mental states, but it and dispositions to certain kinds of behaviour do not adequately replace those mental states. The inaccessibility of the mind raises more questions about the nature and nosology of mental disorders, because both depend largely on the patient's description of his feelings about the meanings of the psychic events and states that he is experiencing, if he has the capability to do so, and also on our error-prone conclusions about those feelings from his behaviour as observed. Laboratory tests have no role in directing our impressions and speculations.

In all forms of human thought, related mental events reflect simultaneous brain events, which occur in response to how the person experiences the world; and at any time they will be either a true or a false representation of what the world is like at that time. Their meaning will not be found in the nerve impulses that are going on in the brain at the time, because mind cannot be reduced to matter and its behaviour, but it will be manifested by the behaviour of the person in the context of his environment and his reaction to those events in it that impinge on him. The depressed person may indeed suffer from low serotonin output, but his life circumstances may also have a causal effect on that physiological dysfunction; so his problem will not be corrected by pharmacological means alone, as the psycho-therapist will testify. We have already discussed how mental events are related to the brain events that underwrite them, so that there can never be a mental event without a concomitant brain event, because producing mental events is the exclusive function of the brain, and at no time can the two be separated in the same person who is a complex of his $M$ (material) and $P$ (psychic) predicates. Nevertheless, psychology and its sister, psychiatry, rely mainly on characterizing mental illness by its symptomatology and less so by its underlying causal pathophysiology as well. But again, we do not expect all illnesses to conform to the same physical pattern; especially manifestly mental illnesses that are concerned with a non-physical "entity".

Equating the person with mind and consciousness means that one with a sick mind has lost his personhood, even if his body is in perfect condition; but one with a grossly grotesque or non-human body and sound mind is still a person. If the former regains full use of his erstwhile deranged faculties, he becomes the person that he was before he lost them. By this criterion, personhood is indeed a very strange and unenviable state that comes into and goes out of being with changing circumstances, leaving open the logical possibility that the world can one day be completely devoid of persons, if they should all loose their faculties simultaneously for some unknown reason, but be fully repopulated by the same persons in the wink of an eye when they regain their full faculties. Furthermore, mental disorder cannot be a disease in the same sense as a clinical state caused by invasion of body tissues by micro-organisms with consequent derangement in physiological function, because the mind is not the kind of physical existent that can be inhabited by micro-organisms. Mental disease is disease only in the sense that the term "disease" is used as a label for dis-ease affecting the whole person in the Strawsonian sense of affecting his P predicates which are an inseparable part of the primary complex that is his person. The determinants of mental disease may have a physical basis, but their output is psychological, as the functional theory of mind postulates. As qualitative deviations from accepted behavioural norms within societies, they are more difficult to define than physical determinants, which are amenable to universal quantification. Besides, the moral status of those afflicted with psychiatric illness is often degraded by the rest who cannot understand their behaviour, and who claim to be rational and capable of exercising freewill, in contrast with them.

# Mental disorder

If the counterargument to this line of reasoning is that mental disease is a deviation from and derangement in rational mental function, then the standard of rationality has to be established and justified. So far, it appears that the standard will be all of those who consider themselves sane by their own standards, thus begging the question of how the standard can be established using a population that harbours its own undiagnosed mentally deviant elements. Besides, the discipline of medicine cannot claim the same naturalistic basis for psychiatry that it does for the rest of clinical medicine, because psychiatry is basically a discipline of values, which defy test tube measurements. Perhaps, then, mental disease is a psychological and social disorder caused by failure of persons to adapt to the pressures of modern living and the sometimes insatiable demands made on those who lack the skills or the material means to respond adequately to those pressures as Thomas Szasz maintains. But Szasz goes so far as to call mental illness a stigmatizing label that is not indicative of mental disease, because the mind is not an organ of the body whose deranged structure can be investigated empirically, and so it cannot be subject to disease like the body (logical positivism).

Szasz's point of view is not, however, without counterexamples; e.g., there are some diseases, like migraine, that are thus labelled even though they cannot be investigated empirically, and there are structural deviations that do not cause disease, like Riedel's lobe of the liver. Besides, brain disease does cause behavioural dysfunction, which cannot be labelled as a mere social problem, thereby validating the concept of mental disorder as a genuine disease. As Jerome Wakefield has stated, mental disorder is "a harmful dysfunction, where 'harmful' is a value term, referring to conditions judged negative by sociocultural standards, and 'dysfunction' is a scientific factual term, referring to failure of biologically designed functioning"[1] in the sense expounded by Christopher Boorse (see page 136), and subject to the same criticisms.

Against the contention of some psychiatrists who claim the existence of significant correlations between some brain structure, genetic and immunological disorders and related schizophrenic symptoms, Szasz maintains that Schizophrenia is not a disease, because it does not manifest the statistically neuropathological or neurochemical abnormalities seen in organic brain syndromes. For Schizophrenia to qualify as a disease, it should manifest characteristics that conform to the definition of disease as objectively and universally identifiable bodily abnormality of form and function, and not be limited by the moral persuasions of those making the diagnosis, sometimes maliciously to confine those who are a nuisance to political authority figures in prisons called hospitals or in labor camps that go by the same name. As he says, "medical diseases are discovered and then given a name, whereas mental diseases are invented and then given a name, such as attention deficit disorder."[2] From this perspective follows his contention that "Psychiatric coercion [as purported treatment of medicalized conflict] is medicalized terrorism"[3], in conflict with the medical dictum: *primum non nocere* (first do no harm). So people who claim to be treating mental diseases are not really treating disease, they are treating "fake diseases"; i.e., pursuing the unethical practice of inflicting harm on their patients.

Life-style stresses are known to cause as much misery as infectious and other diseases, and in some cases they predispose to disease, e.g., hypertension and heart attacks, while physical disease can also cause mental symptoms like depression and nervous breakdowns, which adversely affect the person's life-style. Mental misery thus manifests as somatization, viz., emotionally precipitated physical symptoms, which lack underlying physical causes, and which erroneously prompt the misplaced application of physical and drug therapy. On the other hand, mental disorder could also be the result of natural chemical imbalances that underlie the genesis of certain mental illnesses characterized by symptoms similar to those of mental illness caused by overdoses of psychoactive drugs; hence the utility of drugs with effects opposite to those known to produce states resembling psychiatric illnesses, e.g., mania-like symptoms of excitement resulting from an overdose of Imipramine. So their antidotes should prove useful in the treatment of the clinical conditions associated with these imbalances, even if they are not caused by them. All these aspects of mental disorder can be best apprehended by a person-based clinical approach to the patient's problems. Without engaging the patient as person, the doctor can never appreciate his subjective mental states and she cannot diagnose and treat him appropriately. Her customary compartmentalization, bisection, and management of his personal problems as somatic and mental derangements should be replaced by a phenomenological approach that embraces comprehensiveness and empathy.

From these considerations we can pose a series of questions that are begging for answers. With regard to schizophrenic states, who the real person is: person $a$, $b$, $c$, or $d$ who constitutes the patient at various times in his life history, and each one of whom has his own different history from those of the others. In paranoia, whose voices does the patient hear; those of other persons, or the devil, or a God? Should those who hear the devil's voice be penalized while those who hear only God's voice are protected, in spite of the fact that people have committed heinous crimes in response to their claims to be hearing and obeying God's voice? Finally, during states of depersonalization or transformation of personality, who takes the blame for crimes and misdeeds committed by an identifiable body causally connected with those crimes and misdeeds, while the mind is dissociated from that body? Can the mind alone be punished for the misdeeds of the person, because it is the person or the motivating aspect of the person, while the body is exonerated, because it is not the person in question but the executor of the will of the mind; and how do we separate the two so that we do not mete out punishment to the undeserving partner of this Strawsonian unity?

A closer look at psychiatric diagnoses reveals that they are derived from first person accounts of internal feelings and impressions and third person observations, not from a combination of these with objective physical findings—a situation that rings reminiscent of the age-old philosophical problems of other minds and the Cartesian body-mind dichotomy, and the dilemma of treating matter (brain) to affect mind and change behaviour, since feelings and behaviour *per se* are inaccessible to any form of physical targeting with drugs and electroconvulsive therapy but can be reached via psychotherapy or alterations in the functions of the brain. Hence, the entire discipline of Psychiatry is a negation of Descartes' dual substances of mind and matter, leaving mind and brain as parallel manifestation of the same substance, with brain function expressing itself in the mind, and mind expressing itself in bodily movements of speech and other behaviour patterns, normal or abnormal by our standards that can sometimes mete out the same treatment to persons who share symptoms that have different etiologies. On this analysis, we can understand how the external world (socio-economic, political, cultural, and physical) exerts an influence on the mind and ultimately on the body to display behaviour patterns from which we can deduce normal or deranged mental functioning, in contradiction of the one way epiphenomenalist philosophical theory of mind that we referred to on page 69. Ultimately, psychiatric symptoms originate in mental manifestations of brain dysfunction, contrary to Szasz's contention, and the entire field of Clinical Psychiatry should be interpreted from all pluralistic points of view as mandated by the concept of systemism—see chapter 6.

## Existential Psychiatry

Psychiatry assumes a combined phenomenological and existential perspective when it studies the phenomena and conditions of mental life and its aberrations in the spatiotemporal settings of persons with mental problems. Existential psychiatry can be shown to rest on phenomenological tenets of what the experience is like for a particular person, relying as it does on the patient's present account of his contact with the world as he is continually engaged in balancing the conflict between his basic biological demands and his prevailing social situation. In this effort, he is choosing and making his world primarily to suit him and secondarily to earn him a place in the wider community; he is not merely adjusting and succumbing to the reality of his time and place. If he cannot succeed in this tug of war, he reacts by choosing the easier and more pragmatic alternative of assuming an identity that has been imposed on him by those upon whom he depends for survival, temporarily resolving his conflict.

But when the patient cannot handle these existential crises that occur in his life, such as failure to resolve the overwhelming conflict between succumbing to an identity imposed by circumstances and assertion of the self that he experiences himself to be, he often loses mental stability and undergoes distortion of personality, preferring to live in his own world. He would rather not continue to try to share the real world with other beings, even though he misrepresents it in his thoughts and actions in preferring to live it his way. Therefore to understand him and be able to help him, we have to know how he sees and experiences his world, which is vastly different from how every one else experiences and sees it. Like many "normal" persons, if he feels overwhelmed by circumstances that he can't comprehend or which he is powerless to affect, his life loses meaning and he sinks into isolation where normal being-in-the-world-with-others is now alien to him, because he cannot confidently assert his identity within that overpowering milieu or even exercise transcending care and concern for others. He feels more himself in his own circumscribed world in which he can retain his identity unchallenged and find meaning in his life through self trust in his diminished personhood. This attitude is not altogether alien to most persons who look at their own strange reactions to what goes on in the modern world as the only rational way to respond to the insanity that is overtaking it, because lack of good choices often forces us to choose from the only two or more evils that exist in the "total world 'design' into which [Clemens Benda believes we are] 'thrown'"4 by a deranged world with which we are inevitably and inextricably bound up.

In the midst of all these gyrations, however, the patient remains at all times a part of the communal world, which is also a part of him; therefore he is what his world is. In the same way as this world has a past and future, he is not only subject to past influences in his life, but also and more so to future ones. Even if what matters most is the here and now, the pull of his future and its moulding influence on his present actions and experiences will determine how he constructs his world from this point on. Their interpretation by someone else, like a therapist, will not prevail against the backdrop of the world at large; only his feelings about persisting problems that result from his maladjustments to unfortunate brushes with past events and his anticipated pleasant and compensating encounters with future ones will matter. So, treatment is undertaken not as marking the end of the problem and the encounter for him, but as only the beginning of an existential journey into his future. How he constructs and changes his world reflects how and what he is, and it is from this standpoint that his therapist will understand him through his actions, which reflect his construction of the world for himself. His adoption of the existing world-for-others that he cannot share will not be of any use to him and his therapist. She will understand his isolationism, his lack of communication, and his total dis-engagement from this strange world; thanks to Existential Psychiatry and to existential thinking, which Benda credits with opening "new horizons in the understanding of the [existential] predicament of modern man."5

# 6

# The patient analyzed

To be able to understand the whole, one often has to understand the parts that compose that whole in their functional relationships without committing the error of composition: attributing the properties of individuals to the whole that they compose. Hydrogen and Oxygen are gases that combine chemically in the ratio of 2:1 atoms to form a liquid, $H_2O$ or water, whose properties do not reflect the properties of its constituent gases. Trying to make sense of one without the other will invariably be unproductive and self-defeating. In this section, we will consider the meaning of normality as one end of the spectrum that extends between health and illness and as a necessary prelude to the abnormality that characterizes disease, illness, and disability. We will also contrast the humanistic and scientific attitudes toward the patient as the abnormal person defined in the preceding paragraphs, since the abnormal organism is only a part of that person.

## Normality

### The concept

Normality is a sometimes arbitrary concept that refers to the theoretically ideal state of conformity with what is commonly experienced and has been established as the acceptable standard or basic parameter in a particular discipline or manner of being of a people, depending on their culture, ethnicity, and now gender orientation. (Abnormality is any significant variation from this established norm that cannot be understood without a prior understanding of the normal). Following the derivation of the term from the Latin word *normalis*, meaning "made according to rule", the normal is the regular, the usual, or the most frequently occurring parameter in most of the population that is already considered not to be deviant, otherwise it would not be used as the standard. The definition is circular, because it starts off by excluding individuals considered to be supernormal or subnormal, using as its yardstick of normality only those whose selected parameters of grading fall within the most common distribution range for a specific population. The circularity is necessitated by the fact that there is no naturally normal individual who can be used as the universal standard of comparison, compelling us to regard normal persons as those who function adequately within their natural settings of age and sex, even with the proverbial lack of experience of youth, infirmities of old age, and postulated survival hazards associated with possession of the Y chromosome. We also have to decide if the normality in question is to be applied to the global functioning of the individual or to selected sectors of his existence, because there are always disparities in the levels of functioning of the many aspects of any single person's constitution—his organs, systems, and his whole person.

Normality cannot always be arbitrary, though, because the form and function of the human body was not arrived at by consensus, but by the natural process of genetic selection over which we did not bring our consensus to bear. Hence it is that when our consensus would have us abort evolutionary defence mechanisms that we mistakenly consider to be deleterious to our well-being, Darwinism, as we will see in the final chapter of this book, admonishes us not to suppress the moist cough or the fever that necessarily accompanies infections, and to be aware that someone may be normal in some respects and abnormal in other respects that still have to attain maturity; e.g., a child sees and hears normally before he can walk, run, or speak normally. How we experience the world, and how broad and seasoned that experience is, determines how our concepts of the normal world are formed. We might entertain concepts of how we would like the world to be, but reality is largely what is, it does not answer to our wishes. Therefore to realize our wishes for how we would like the world to be, we have to modify certain aspects of reality to provide a new and acceptably normal state to us from a previously uncongenial normal state of nature, such as tinkering with "bad" genes to replace their undesirable effects with desirable ones.

In the domain of health and disease, normality in life is not merely the absence of disease—disease is absent in death also—but the presence of good health and physical and mental well-being from which other states of personal value ensue. The notable feature of the definition of each of these states is its circularity in terms of the others; e.g., well-being for an individual refers to the good in that person's life, including medical wellness, good physical and mental health, a fully satisfactory social life, and contentment with his prevailing circumstances. Good health for the individual is believed to consist in the physiological state in which the bodily organs function as nature intended, or rather as they turned out by a process of natural selection, i.e., normally; and the healthy state is defined in terms of this normal state, which is also supposed to be the natural and teleological state in which every organ serves the purpose for which it has been adapted for the whole organism's survival; e.g., the heart pumping blood, the kidneys filtering waste, and the pancreas providing insulin for carbohydrate metabolism—many functions that help to maintain its homeostatic integrity. Therefore homeostasis is normality, and normality is homeostasis, health is a species of well-being, which is dependent on health. Homeostasis will be understood to mean "a relatively stable state of equilibrium or a tendency toward such a state between the different but interdependent elements or group of elements of an organism, population, or group."[1] But when we come to discuss non-linear dynamics we will have to modify this concept of reduction in transient, externally induced variability of a steady state in the maintenance of normal organismic function to make room for the increased, internally induced variability that is the norm in the healthy state of the same organism.

## Normal as standard

What happens normally happens ordinarily and as a rule in our experience; the extraordinary is the abnormal. It is what we expect to happen when certain established patterns and recognized basic conditions have been fulfilled, when the stage has been set, and no room has been left for unusual occurrences or exceptions. In this context, allowance is made for a slight deviation from the range of normal as a case of statistical or acceptable error, as long as it does not distort normalcy to a notable degree. We have to guard against getting caught in the compromising position of allowing yet a little more deviation, and still more, until our concept of normality is so far removed from what we started with that it becomes amorphous and meaningless. So the concept has to be based on the experience of what obtains standardly, not what could obtain potentially and non-standardly. The standard should always be our lodestar.

Only in mathematics, statistics, and artifactual designs do we have indisputable "normals" and "abnormals", because that is how we designed those disciplines and artifacts. Where we do not have the liberty of being the designers, we have to use various imperfect methods of establishing what will be pragmatic for the purpose of designating what shall be considered normal or not. According to pragmatism, what is true is what works in conformity with our beliefs and thereby affects our choices and their practical consequences. Whether we choose action $a$, with highly desirable result $x$, or we choose action $b$, with mediocre result $y$, will depend on the results that we expect from our choices, based on our beliefs about the ability of each choice to produce the desired result, barring a Hobson's kind of choice. If the expected result is the same for each choice, then there is no choice between $a$ and $b$, and causal relations do not matter. So it is that if Penicillin will result in complete cure of pneumococcal pneumonia while Streptomycin will only subdue it, we will choose to treat our patients with Penicillin and not Streptomycin; but if the two drugs are equally effective in producing a cure, our decision will be influenced by other conditions like the pharmacokinetics of the two dugs, mode and frequency of administration, possible side effects, etc. On the other hand, in the taxonomy of disease, diseases with different etiologies may be grouped together on the basis of their clinical courses and responses to treatment for pragmatic reasons; but once their exact etiologies become known, preventive measures aimed at their causation will also come into play to determine their pragmatic and conceptual classification based on their etiologies. The causal relation therefore serves as a partial and pragmatic explanation of why a complex of signs and symptoms constitutive of a clinical state styled pneumonia, the result of pathological changes due to pneumococcal infection in the lung, will be prevented and treated in a particular manner to ensure certain specific, anticipated responses and eventual outcomes.

Statistically, normality does not carry normative implications; it is only an evolutionary statement of fact about how our bodies came to be as they are and to work as they do by a process of non-teleological natural selection. The process itself started with a change in the genetic code of our remote ancestors, followed by successful reproduction of phenotypes that resulted from their progressively changing genotypes in response to environmental demands. Still, a condition can be statistically prevalent without being normal, e.g., dental caries or myopia, which are sources of disvalue to the subject. Nevertheless, value and disvalue are implied in relating normality to wellness or illness, because wellness places one at an advantage in life that one is deprived of by illness. However, when optimal conditions of health are under consideration, those individuals who fall within the distribution of one and two standard deviations are generally presumed to be the normal ones who will serve as the best examples of the desirable standard of health. At the extremes will be those who are either sub-normal or super-normal and constitute negative and positive exceptions to the norm.

## Normal as statistical

On the statistical scale, what is normal is what falls within certain agreed upon limits, because there is no automatic boundary between normality and abnormality. The limits have to be agreed upon for each specific purpose, although biochemical and other parameters sometimes dictate where the lines will be drawn when values outside those parameters are known to result in life threatening states, e.g., serum potassium levels below 2mMols/L and above 7mMols/L. Some ranges may be normal for certain ages but abnormal for other ages, e.g., total bilirubin of 200mMol/L (with direct fraction <15-20% of the total) in a healthy one week old infant of >38 weeks' gestation, and some ranges may hold irrespective of age, e.g., serum potassium levels. In the bell curve distribution, the range within three standard deviations, embracing 99% of cases, proves to be a very strict criterion of exclusion, but a loose one for inclusion within the normal range. On the other hand, a value falling within two standard deviations, embracing 95% of cases, and generally used as a standard criterion, appears to account adequately for the exclusion or inclusion of cases without undue dissonance between the two criteria. Within one standard deviation, however, the criterion for inclusion, embracing only 67% of cases, is very strict, but equally loose in excluding the 28% of cases that might otherwise qualify for inclusion within the 95th%ile range of normal, as shown:

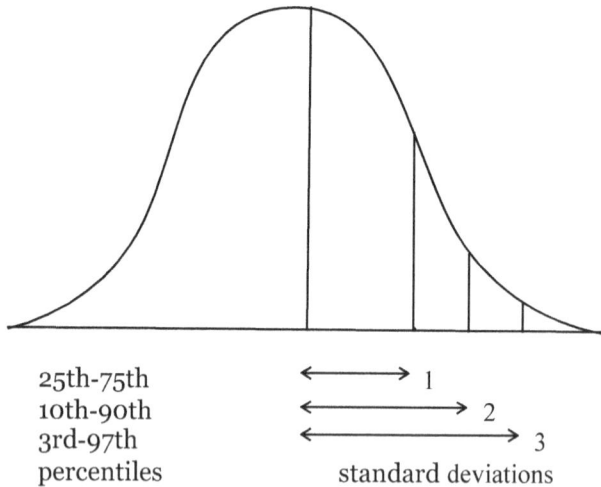

| 25th-75th | ←——→ 1 |
| 10th-90th | ←———→ 2 |
| 3rd-97th | ←————→ 3 |
| percentiles | standard deviations |

Alternatively, slightly different but corresponding ranges may be expressed in percentiles: 3rd - 97th%iles, 10th - 90th%iles, or 25th - 75th%iles respectively. Additionally, the measured parameters may be as varied as physical development (50th%ile for weight, 75th%ile for height—normal, or <3rd%ile for weight and height—abnormal); mental development (IQ or DQ 90 to 110—normal, 60 to 90—abnormal); biochemical values like cholesterol levels (>5.2 mmol/L—abnormal); or indicators like blood pressure (average of 120/80—normal for adult, but elevated for infant); vision (20/20—normal for adult, or 20/50—abnormal for adult, but normal for toddler); and hearing (25DB—normal, or 60DB—abnormal for all ages).

Normal ranges for peer groups (age, weight, height etc.) are established by compiling the aggregated performances of allegedly normal individuals in that peer group for use as criteria for the particular parameter under consideration—a circular criterion. On the other hand, the statistically normal heart, as described in standard reference books, is an ideal against which we measure the deviation in function and the adequacy of our corrective measures of all hearts in Clinical Medicine. In the case of disease, those who do not show clinical symptoms or signs of disease are selected to serve as "normals" for the study and categorization of that disease, without taking into consideration other factors that might influence the presence of the disease and its clinical manifestation or suppression. The assumption is that they do not appear to have the disease and thus constitute the perfect, normal standard with regard to that parameter. This is another circular criterion of normality that also entails an infinite regress, because standards$_1$ will be required for selecting those normal individuals, and more standards$_2$ for establishing the selection standards$_1$, and so on, *ad infinitum*.

## Normal as socially acceptable

A problem arises in a society of people with certain standards that they consider normal within the ambit of their locale and their historical, cultural, and scientific period, but which happen to be abnormal by the standards of other people in different circumstances. In that case we will end up with a population of abnormal persons who pass themselves off as normal by our standards; and if they were to establish parameters of normality for the rest of us, we would all end up as abnormal "normals". The implication is, therefore, that normal in these circumstances amounts to adaptability to one's particular social milieu; we are normal, because we can function adequately and survive in that environment; we have a pragmatic fit with the environment and its demands on us, and the abnormal individuals are merely the ones who are misfits in that environment, covertly or overtly, physically, mentally, culturally, educationally, politically, or economically.

Persons who appear to be normal may still be harbouring subclinical dysfunctions or a propensity for disintegration with slight changes in their circumstances, proving that they were not normal all along. In fact, no one of us is totally free of some abnormality of form or function, despite a healthy appearance. Contrarily, normality may relate to the satisfaction of a healthy appearance; the person looks normal, in spite of underlying problems of ill-health or processes due to age-related infirmities, or to deviations of anatomical form that do not result in deviant physiological function, like an extra lobe of the liver (Riedel's lobe), or a horseshoe kidney. It is also possible that what we regard as normal today may prove to be abnormal when our knowledge becomes more refined, as is the case with genetic and metabolic disorders whose victims may appear to be functioning normally, even though they have an abnormal constitution, which manifests only under certain conditions of stress. If they are not thus exposed, they will always be regarded as normal. For example, the victim of Sickle Cell trait will show symptoms and signs of disease only when the concentration of oxygen in his blood falls below a certain level that allows his erythrocytes to assume the abnormally functioning sickle form, and the victim of Fructosaemia will display his illness only when he ingests fructose containing foods. The suggestion here is that the concept, but not the existential state of normality, evolves with time and increasing knowledge, from which it appears that the concept is shrinking as knowledge of the molecular basis of disease expands to include more of those previously thought to be normal in the ranks of the abnormal, although the existential state remains the same—they have always been abnormal.

As an aside, the concept of normality suffers intentional distortion when it is used for the sake of promoting questionable practices driven by money seeking motives, such as comparing laboratory values that apply to a healthy population carrying in their blood small amounts of a chemical like lead to which they are constantly exposed with those who are not similarly exposed, and then claiming that the values shown by the exposed ones are abnormal and in need of therapeutic correction, even though they are not in the toxic range and are not causing any symptoms.

The above considerations imply that the normal is a functional range and not a specific, naturally occurring entity, in the same way as disease is not the name of an entity with an ontological status or a reified concept but rather a state of ill-being or functional deviation from wellness and integrity that adversely affects the subject and those around him with whom he interacts constantly.

# Holism

## The whole person

Holism, derived from the Greek word *holon* meaning "entity", (*holos*=whole) maintains that the components of a system are not enough to explain the properties of that system—the behaviour of the system as a whole is always more than can be determined from the mere sum of the behaviour of its parts; instead, it is the system that determines and explains the behaviour of its constituent parts. Holism can therefore be regarded as the antithesis of the reductionism to which biophysical medical practice reduces the person in its study of his underlying physiological, physical, and biochemical components in health and disease. The logical conclusion to this process of reduction is when the human body has been reduced to the atoms, neutrons, protons, and quarks in terms of which its form and function become ultimately explicable. At this point, the human body is inseparable from the rest of the material constituents of this world, which also ultimately consist of the same elements; the unique personal identity aspect of the patient has been retired from his being. And yet a person's humanity cannot be confined to these dogmatic limits of the measurements of science; it transcends those measurements to encompass his personal experiences in the particular social milieu that also determines his forced and freely chosen responses to events in it. So medicine should combine the personal (subjective) and the scientific (objective, mechanistic) perspectives in its treatment of all patients, with the personal superseding the scientific at all times without disabling it, since the normative practice of clinical medicine will always need the support of factual, value-free basic science. The doctor should also take into account the effects of social, psychological ("spiritual"), and biological factors on the patient in treating him and his illness, in addition to his disease and his biochemical and pathological parameters, as discussed in the section on biopsychosocial medicine. She should, however, guard against the temptation to incorporate in her holistic approach treatments that have no basis in rationality and utility.

Jim may weigh 70kg and measure 175cm, his visual acuity may be 20/20, and his IQ may be 103, but that still leaves out his personhood as a quality that cannot be expressed in quantitative terms. This personhood is central to his diagnosis and treatment as a person with a fasting blood glucose level of 21mMols/L who has to be subjected to certain therapeutic measures to reduce his blood glucose to an average of 5mMols/L, because persisting hyperglycemia will take its toll on all his nerves and blood vessels. Balancing the same figures against normal values in the laboratory without regard to their implications for the person on whose blood the tests are being done is driven by radically different motives. The laboratory technician is not concerned with the personality of the patient but with providing the correct numbers. On the other hand, toddler Jill may have Trisomy 21 with its attendant, irreversible measurable physical characteristics, including a low IQ, but the meaning of life for her will be found beyond the perimeter of these physical characteristics, in the whole of which they are only a part. It may not be possible to change her genetic and physical situation, but it is still possible to enrich her life with the appropriate type and amount of attention and care that will allow her to develop into a full fledged person. Too often the "normal" ones tend to relegate the "abnormal" ones to lower ranks on their own fabricated scale of human values, denying them personhood or denigrating their personhood.

Holism is, therefore, a concept that is closely related to the *art* of the practice of Medicine as reflected in the traditional methods of the legendary first physician Asclepius (Aesculapius), not those of the modern-day *science* of Medicine, because its area of application as art can't be subjected to objective measurement by scientific parameters. It is firmly grounded in the morality of the practice of Medicine, which takes particular note of the individual in his own context of multiple warm relationships without losing him in the universal generalization of cold scientific statistics and laws. Every person has emotional ties to family, community, and his health care personnel, which have a value that must be maintained during the elaboration and resolution of his medical problem in physicochemical terms. His role as a person in the evolution of his diagnostic and treatment programs can never be fully accounted for by generalizations of the type: symptoms *a*, *b*, *c* and signs *x*, *y*, *z* always suggest the diagnosis of Leukaemia; so let's treat the Leukaemia and the afflicted person will self-correct. The sequence goes like this:

(1)   blood tests consistent with specified standard findings increase the chances from possible to probable Leukaemia, without providing final proof of the diagnosis.

(2)   Bone marrow examination and genetic tests will, however, always provide conclusive evidence, confirming the diagnosis.

(3)   Treatment plans are specific for each type of leukaemia: chemotherapy, radiation therapy, bone marrow transplant, and adjuvant therapies, and all of them are adjusted to suit the individual biochemical parameters of the case under investigation.

(4) Therefore the course of treatment for this case of leukaemia will consist of such and such, based on the laboratory data.

If the long course of treatment administered to the patient results in remission or cure of the disease, then that constitutes a job well done. In the process, no regard has been paid to the person who is diseased and how much good or damage has been done to his ego.

Without the holistic approach, little attention will be paid to measures that promote patient well-being and avoid the adverse effects of disease and illness on him—a regrettable failing on the part of the doctor and a source of discredit to the practice of medicine.

## The object-human-being

In the midst of all this scientific maneuvering with his mechanical body, his total care has been lost to the cure of his body's disease. No attention has been given to his personal concerns like his fear of dying, the undesirable and terrifying side effects of his treatment, the threatened security of his job during his illness, and the jeopardized welfare of his family during his illness and also after his death, if he should die. These and many more concerns will influence his rate of recovery adversely, if they are not addressed at the same time as his disease. Therefore failure to include attention to these background aspects of the patient's life in his treatment constitutes a failure to treat the whole patient as something more than the sum of his constituent parts. Besides, during his treatment the patient loses his individuality and personhood to the many specialists and sub-specialists who concentrate on only those problems of interest to their specialties, to the exclusion of their owner. He is vivisected into as many parts as there are specialists; each one takes off with a piece of him to analyze, and each such part receives detached attention, contrary to the common sense view that no individual, object, or statement can be fully evaluated out of context, and no aspect has meaning outside the context of the whole of which it is a part; ask the people who are for ever complaining that their callous and sometimes insulting public remarks have been taken out of their context. The process is like removing parts of a car to examine and repair outside the car, and then putting them all back together to get the car running well again. It does not match up to what constitutes a person as an active agent and participant in the world that is being carved for him by his attendants. Thus there is a constant need for someone who can knit together all of his problems and keep him, as person, at the forefront of all these specialized but detached attentions to his parts in their relations with the laboratory and the pharmacy, and his personal physician should be that person.

Patients are often tagged in conversation among medical personnel as that case of pancreatitis or schizophrenia, or simply as that diabetic or schizophrenic, when they should be referred to as persons who are afflicted with pancreatitis, schizophrenia, or diabetes. This detached attitude ignores the person while attending to a part of that person, which can never replace him in forming cognitive and emotional relationships with those who treat him like a mere unfeeling thing. In any case, the pancreatitis, schizophrenia, or diabetes can never be understood apart from the whole person in whom it is occurring. The lived body or person versus the sick body parts that are examined and subjected to tests should be the object of attention, and it should be the common point of interest of the doctor as it is of the patient, so that the patient will not be treated like an object, and the doctor will not be regarded as a mechanic who has no empathy with the material object that she is fixing.

With this kind of empathic attitude in place, the doctor will not feel tempted to act as an infallible forecaster who can dare to tell a patient that he has three weeks to live and does not deserve any effort to have his disease process halted or reversed, forgetting that she does not control the complex physiological and pathological processes that are occurring in her patient's lived body, even though she may be able to control them to a limited extent in his biological body; forgetting also that her forecast of the course of the patient's illness is a probabilistic exercise that lacks the absolute certainty that is characteristic of deductive reasoning, which is not available to her in this instance. She is using her own and accumulated past experience for its presumed reliability in predicting that the future course of her patient's illness will follow past cumulative experience, in spite of what we know about random courses of events as represented in quantum mechanics, and also in spite of the circularity of using what still has to be proved as its own justification. Past courses of events are no guarantee of the future course of events at any time, and what is popularly styled the natural history of a disease process is not written in stone and does not have to follow a predetermined course. Even so called laws of nature are only the result of our systematization of the constancy in the relations of observed events; they are not absolutely imperative.

## Art versus applied science of Medicine

The preceding considerations blend seamlessly with that of Medicine as an art that is individualistic and warm, that recognizes differences in patients versus an applied science that is cold and general, even though its basis is scientific and its tenets, including diagnosis and treatment of disease, cannot be understood outside the ambit of science. The art of Medicine without science is anecdotal folk medicine, which is founded on tradition, not on science. So art and science in Medicine are, in the end, complementary. The science includes all the cold factual aspects of mechanics, experimentation, and clinical trials aimed at reversing pathology and effecting a cure. The art is instead focussed on humane care (rather than cure of the patient), compassion, understanding, sensitivity to the patient's needs in his existential predicament, and the desire to comfort on the part of the physician. In practising the humane art of Medicine founded on empathy, the doctor looks beyond the body to connect with the patient's mind where all his hopes, dreams, wishes, values, and fears reside, and she tries to understand how they will influence her patient's healing process as she strives to promote his autonomy and welfare in a sometimes selfish and exclusive society. She also tries to cultivate mutual trust between her patient and herself in their journey (short or long) together through the uncertain terrain of disease and illness, allaying his anxieties and boosting his genuine hopes without inspiring false hopes in him. Both practices are, however, victims of modern commercialization of Medicine by insurance companies and some doctors who operate their own diagnostic facilities to which they sometimes refer patients unnecessarily for diagnostic tests that do not add anything to the further delineation of their medical problems but only help to swell the incomes of the doctors concerned.

Disease processes sometimes take unexpected courses that defy prediction, as much as the drugs that are used in the treatment of disease may produce unexpected and sometimes deadly side effects, all of which should be humbling experiences for the physician. In a way, the holistic approach in Medicine is like the standard practice in Philosophy where the subject matter is the totality of existence; how things hang together in context, as opposed to how each state of affairs exists and functions on its own, out of context. The essence of human homeostasis is the total interaction of the person with his own psycho-physiological mechanisms and with his physical environment that includes other persons. It is in this social context that each person realizes his psycho-physiological integrity and becomes aware of the effect of disease and illness on his whole person.

Furthermore, the art of medical practice differs from the scientific basis of medicine as evidenced by the selective, atomistic interest shown by the medical scientist or pathologist studying the morbid histology of "sick" cells in a liver biopsy without making direct contact and empathic exchange with the sick patient whose liver cells he is studying, which is not his function. He is a scientist who deals coldly with the cold, objective facts presented by the specimens that he is analyzing and categorizing; his is not a discipline that requires emotions, compassion, or empathy with his specimens. On the other hand, the clinician who deals with the live patient, the subject of feelings and aspirations, and who thinks that she can also assume the cold and detached attitude of the scientist to her patient is misguided and needs to revise the meaning of her vocation, not her business enterprise, as medical practice has turned out to be. She might not think about the patient's emotional experiences, as long as she is helping to cure his disease and run her business, but to the patient it matters a lot that he is a subject with liver cancer who lives with unfulfilled responsibilities, wishes, hopes, and dreams in the midst of despair and fears of the unknown. He does not regard himself as one of the statistical five out of a hundred who will inevitably succumb to their disease in two months and does not, therefore, warrant too much fuss. Statistical probabilities have no meaning within the framework of his personal concerns; they do not take care of his emotional needs. The victim of lung cancer also is not, at this late stage, interested in statistics about its increased incidence in smokers compared to non-smokers; he wants to be cured of his affliction, or else to be helped to live long and comfortably enough to fulfill some or most of his outstanding wishes and obligations. That is what a physician friend was hoping for after receiving his diagnosis of metastatic lung cancer, telling me of one special project that he was planning to start and accomplish when he was in remission from his cancer. I knew from his condition that it would not materialize, but I refrained from damping his hopes, as that would have been unkind and it would have served no purpose. He died a few days later.

The upshot of these considerations is that while it is essential for the progress of diagnostic and therapeutic medicine to promote a reductionist approach through basic science, the patient, who is at the centre of this entire enterprise, should not be left behind or brought along stripped of his personality and reduced to the status of a mere mass of protoplasm on which science is to perform its wonders, nor should he be treated outside the context of his existence. The doctor should recognize these facts and cultivate the habit of communicating with her patient also as a subject and not assume the positivist approach of basic science which regards him as solely an object, a mass of protoplasm, a complex of physical and biochemical propositions that have to be proved true by experiment if they must mean anything in the assessment of that protoplasm's well-being. All such objects are subjected to batteries of routine tests to study their reactions: SMA12, Chem 20, X rays, etc. Their reactions will determine how they will be treated. The doctor should not treat her patient in this manner, because the goals of her discipline oblige her to listen to him first and then institute selected tests to substantiate her impressions of the patient's problem and how to resolve it. The aim of science is to explore knowledge and scientific truth primarily for its own sake, and secondarily for its applicability to the promotion of human protoplasmic health and welfare. It helps to elucidate some of the diagnostic and therapeutic problems of medicine while it tries to understand every aspect of the world for the general advancement of knowledge and its application to many other disciplines, but passing clinical judgments remains outside its sphere of operation, and the doctor needs to keep this fact in mind. Only she has the insight to assess her patient's clinical signs and symptoms and arrive at a rational diagnosis and plan of treatment of his illness.

Medical practice, on the other hand, deals in ethical values as they apply to individual subjects in search of relief from their various afflictions. Medicine's primary aim and purpose is the prevention of disease, healing of patients, and the restoration to homeostasis of the whole human person, or body and mind as the dualists see it. Hence, the interest of Medicine in Science relates only to the ability of science to provide information that will facilitate the prevention and treatment of disease. The doctor, as the presumed expert who knows

104

everything, is placed in the untenable position of having to provide all the answers by a combination of knowledge from basic science and clinical medicine. The patient, however, only expects assistance toward a total healing of his person from her, because she is the doctor, and her knowledge entails how to promote healing of people. He is not concerned with the doctor's store of scientific knowledge relating to the histology and biochemistry of his chondrocytes, however relevant that basic knowledge might be to his health problem. His daily life has been derailed by illness, and he wants to be able to return to his usual routines with the help of his doctor and without worrying about whether his chondrocytes will carry out their usual functions to enable him to carry out his usual functions. In as much as the patient does not regard himself as a biological specimen, the doctor should not regard her patient in that light but should always be mindful and respectful of his personhood.

The final question is, of course, when is the patient cured of his disease, or when is he considered to have reverted to normal from abnormal? Is it when he is permanently free of the symptoms and signs of that disease or of any trace of mental and physical dysfunction, or is it when he is functioning reasonably adequately in his circumstances, even on a short or longer term basis? The answer is that we can only venture so far as to call that state a cure which allows the patient to return to his previous level of satisfactory pragmatic functioning for a prolonged period.

# Reductionism

## Mechanistic Medicine

Reductionism, as we have already indicated, analyses the dysfunctional whole into its parts and treats those parts in the hope that the whole will be restored to its former functional state, since it regards the whole as merely the sum of its parts, versus more than the sum of its parts as holism maintains. In medical practice it is manifested in the scientific explication of the patient's illness in terms of disease and the behaviour of its causative factors (microbes, biochemical, chromosomal, and other imbalances) on the basis of certain tenets of basic science, wholly or in part, but not also on the influence of existential predicaments in the patient's life that demand empathy in place of laboratory tests and drugs. It regards the patient as no more than a complex of abnormal physiology (biochemistry, kinematics, optics, sound, etc.) and anatomy (atoms, genes, chromosomes, tissue components, etc.), and thus shares a common worldview with mechanism—that physical causes and mechanical principles can adequately account for all natural phenomena on the basis of quantitative natural laws that govern the succession of events known as effects on the occurrence of preceding events known as causes, and over which the individual does not have control. Reducing talk about persons, and therefore their nature, to discourse about quantitative physical mechanisms that govern their bodies precludes the possibility of spiritual explanations of their behaviour and the role of the qualitative exercise of human free will.

The criticism levelled against orthodox doctors is that they attack only the disease, leaving the patient behind; hence they are called allopathic, suggesting that, unlike homeopathic doctors, they act against the natural self-healing mechanisms of the body in their methods of treatment, when they should simply be helping to sustain the body in what it does well by itself. They, of course, see themselves as practicing scientific medicine based on basic principles of anatomy, physiology, and pathology, which are amenable to experimental confirmation and clinical documentation by any one, unlike those who practice alternative means of treatment, which they see as anecdotal and without scientific foundation. From this empiricist and materialist point of view, the body is regarded as an object with many parts that can be manipulated, modified, and replaced (with transplants) as circumstances dictate.

There is, of course, no doubt that modern medicine tends to be mechanistic in the sense that its scientific outlook rests on the theory that all phenomena are totally explicable on mechanical principles where efficient causes are followed by necessarily predictable effects in a law-like fashion, as in the case of a machine whose every movement is determined by its preceding state, e.g., physics of joint action, hearing, vision; dynamics of cardiovascular system, renal and pulmonary function; chemistry of metabolism and neuromuscular and synaptic transmission. This mechanistic structure enables physicians to prognosticate about the possible courses of physical diseases that inhabit patients, are pathologically demonstrable, and conform to a pattern. But it avoids speculation about the probable outcomes of illnesses whose courses are always uncertain, since they depend on the unpredictable, lived experiences and reactions of undetermined personalities that are never all alike.

The upshot of this perspective, as we have seen, is that all phenomena can be explained by physicochemical laws that govern all matter and its motion, thereby delineating the scope of operation of clinical medicine in its biomedical but not its biopsychosocial character. The elements of medical science are true and meaningful within the atomistic context of their reference range, e.g., a neonatal total bilirubin value of 200mMols in a term infant is not a pressing concern at age 36hrs, but calls for drastic action in a very low birth weight infant of the same age. On the other hand, the holistic perspective and all the concerns of the parents about the jaundice of their neonate and what it may mean to them extend far beyond the statistical facts and implications of bilirubin levels for certain gestational and postnatal ages. The clinical state of the neonate has meaning for them in the broader context of its position and role in the life of their family—a situation to which the doctor may be oblivious in her concern with numbers.

In its analysis of phenomena into their constituents and their relationship to what is happening in the patient's body, science does not try to bridge the gap to place itself in the anxiety-ridden position of the family, and nowhere does it suggest that its function is to care for the whole patient by correcting his deranged psycho-physiological status; it deals only in fixing parts of the whole. So it is up to the doctor to knit together all these aspects in her dealings with the patient and his proxies, because it is not possible to reduce the moral bond that should come into being between patient and doctor to mechanistic causal relationships where she acts on the patient as an object with only measurable parameters. If the patient feels that his humanity is being slighted by the doctor, he will not be receptive to her advice, and he will tend to thwart all her efforts to try to "repair" him when she should be promoting his healing, because he believes that she has not bothered to understand his narrative about his personal problem in her application of rules of thumb to him.

That said, there are times when prevention and treatment of disease and promotion of health are empirical, based solely on positive clinical experience without the overt backing of science; e.g., if a community has succumbed to E. coli infection from poorly monitored, contaminated water supply, the obvious preventive step is to monitor the water supply honestly and ensure its conformity with minimum acceptable standards of safety and potability. Communities where these measures have been overlooked have paid with their lives for the negligence. So even if people may not have known at the time of their infection that the culprit was E. coli, the exercise of responsible preventive measures would have been the same, and the outcome would have been the best for all concerned. The need for truthfulness in this limited regard is also the basis of the need for truthfulness in all scientific endeavours in their global application.

## Scientific Medicine

Medical Science has the specific function of providing causal explanations of medical phenomena, which enables the doctor to understand the functional derangements of her patient. It also provides her with the technology to diagnose and treat her patient's problem, access to the use of drugs that have been researched for safety and effectiveness, and the benefits of a hypothetical-deductive

107

method, which, even if it has no discernible role in her diagnostic procedures, yet enables her to prognosticate on the outcomes of her interventional endeavours. Medicine is based on and uses the methods of science, but it differs from science in its goals. As previously indicated, the sole function of science is to seek knowledge for its own sake and for its applicability, insisting only on truth in this quest; but the doctor cannot disregard the role of Medical Science in the total care of her patient, because without its special contribution to her understanding of disease, she would be reduced to the status of the earliest physicians who depended only on the historical narrative and appearance of a patient to make a diagnosis. After this superficial diagnostic assessment, the ancient physician prescribed potions of home brew and then waited for the unpredictable results of his treatment. He had no way of proceeding from historical narrative to clinical findings, and, with the addition of supporting investigations, to therapeutic measures and prognosis. He lacked the ability to predict and amend the clinical courses of diseases to produce desired outcomes by resorting to one or more scientifically backed modalities of treatment to replace the less successful unscientific ones.

The modern doctor has to be a promoter of healing and part scientist if she is going to practice her vocation with understanding, compassion, empathy, and respect for the patient's personal values. These moral components always have their therapeutic effects on the mental outlook of the patient, but they cannot reverse pathology. Only specific, scientific, therapeutic modalities have the potential to reverse specific pathological processes. Since there is no equivalence between medicine and science, so that medicine can be reduced to science and still retain its characteristic contextual character, there remains only the possibility of using science to explain the scientific basis of medicine, with the addition of the personal element to produce a whole that is more than the sum of its parts and that cannot be reduced to its parts without losing its essence or its character in the process.

The goal of medicine is the improvement or change in the total conditions of health of a patient that add up to a life that he can live with satisfaction. It may entail prolongation of the lives of individuals and communities by the prevention and cure of disease and the maintenance of good health, and it will entail bodily and psychosocial healing through the relief of suffering and pain; hence the designation of bio-psycho-social to its method, meant to encompass all aspects of the living person and to encourage understanding of all levels of his existence. Problems relating to these goals and the doctor-patient bond cannot be reduced to problems in genetics, biochemistry, or pathophysiology, which answer only to the scientific truths on which the general laws that govern these disciplines are based. There are patients who cannot be cured, and they will require beneficent care and support so that they can tolerate their conditions with their self-esteem or dignity intact, and finally die with the same dignity unblemished. The doctor who understands that life-as-lived by the patient is different from the same life-as-seen by the scientific cure-oriented doctor will readily come to appreciate that this kind of patient is not looking for a miracle from her; but he is also not looking to be abandoned as incurable and, therefore, not worth the doctor's precious time that can be more profitably spent on patients for whom there is a hope of cure. Not all sickness requires medicine, because not all sickness can be alleviated with medicine, as in the case of Mr. M. The sick body may be necessary but not sufficient to justify the need for medication. Other modalities of therapy like caring and compassion fill the role of medication just as well.

# Systemism

Simply stated, a system is a unity of reciprocally interrelated elements in constant interaction, unlike an aggregate in which such bonds don't exist among its self-sufficient elements. The human agent (body and mind), the human community, and the environment in which all have their being should be viewed from that perspective: "everything is either a system or a component of a system and every system has peculiar (emergent) properties that its components lack."[2]

## Structure of systems

At this juncture, it may be useful to refer briefly to the place of systemism in knitting together the holistic and reductionist approaches to persons in health, disease, and illness. Systemism, like holism and clinical medicine as an art, assumes a panoramic view of the patient. It goes beyond quantitative limitations to embrace qualitative analyses of the dynamic interactions among component systems of the person that reductionism and medical science make their foci of isolated mechanistic attention. We have seen how they treat each aspect of the patient's disease with its individual risk factors and not the whole patient and all the risks to which he may be exposed in his dynamic interaction with his environment. We have also seen how they will restore his laboratory and other parameters to "normal", as it is conventionally defined, without regard for the integrity of his entire personhood in this exclusively mechanistic endeavour.

When we consider that the human organism in its infinite variety of shapes and character developed from a single (fusion) cell, it is not surprising that the many unique systems of form and function that emerged from this developmental process at different levels of organization should be considered as a unity of physical parts on one hand, and physical and psychic aspects of this single organism on the other. They exist as interrelated and integrated systems within systems in the same complex, interactive environment that exerts its diverse influences on them. Therefore, to understand the whole person, one must understand his parts in their own complexities and ways of being within the system; and to understand the parts all the way down to the subatomic level, one must understand their roles in the integrity of the whole in its unique way of being; none of these components can be understood in isolation.

That is why some philosophers regard systemism as a synthesis of the approaches of reductionism and holism in overcoming their respective deficiencies—reductionism by underrating and often overlooking the natural bonds that exist among people, focusing on subsystems independently within the system, e.g., mitochondria in cells, cells in tissues, tissues in organs, organs in systems, systems in the body; holism by minimizing the individual and often totally engulfing him in the being of the whole system. Systemism emerges as goal oriented or teleological functionalism, as the referent of a simple or complex transcendent collection and organization of parts, which are smaller system units made of even smaller system units, all of which are essential for the functioning of the larger system, and without whose complete, cooperative participation the larger system is bound to malfunction. The subsystems influence one another's function by a relationship of constraints and facilitations, while they also function as hierarchical steps in the dynamic equilibrium of the whole or larger system, a state styled as homeostasis. On the holistic perspective, the system is a complete functioning, hierarchical unit with higher levels regulating lower ones, while on the reductionist perspective it is the sub-systems within the system that determine the behaviour of the system. The functionalist approach combines these two viewpoints to relate the system with its transcendent properties to other systems by overriding the mutual constraints of the other two perspectives and emerging with some properties that both of them lacked, even though the system could not have come into being without those parts.

In open or controlled biological systems, which exchange matter and energy with their environment, changes in any of the components are potentiated or inhibited to the extent that they contribute to or diminish the adaptability, growth, and survival of the whole system at a higher level of organization in the environment where it exerts reciprocal exchange and influence. As an example, let's look at the kidney from the cellular functional point of view, with each component cell of the kidney having a specialized function in the service of the whole organ and its co-operation with other organs in maintaining the integrity of the person's physiological functioning. Glomerular capillary cells filter urine from blood, proximal and distal convoluted tubules absorb water and solutes differentially, collecting tubules store water, and interstitial cells execute hormonal and supportive functions. The successful performance of all these functions and their regulation through feedforward and feedback loops is modulated at different sites that relate to the elimination of toxins and their effect on the brain and other organs. Furthermore, these integrated systems serve to preserve water when brain cells become aware of body water shortage, and to eliminate it when excessive intake impels the brain to call for its increased excretion. They also facilitate secretion of erythropoietin to stimulate iron metabolism and the formation of more hemoglobin to make up for losses sustained in the circulatory and metabolic systems.

Analysis of cellular function at the system levels of any organs in the human body will reveal the same kind of organization and integration moving upward from the cellular level to the organ level and on to the constitution and service of the living body and person.

## Organization of human body systems

In a reciprocal manner and from the same functional point of view, the descending hierarchy from the human body and its parts constitutes a network of causally related sub-systems in the service of the whole system. Hence the organization of the human organism into cardiovascular, urinary, respiratory and other systems with their uniquely functioning elements (heart valves and chordae tendinae, nephrons and ureters, bronchi and alveoli), which are ultimately reducible to cells with specialized functions, as in the example quoted above. The cells share the same basic structures: mitochondria and other organelles, chromosomes and their genes, all of which are in turn reducible to the ultimate constituents of matter, including the proton with its system of electrons, hadrons, and other particles. What distinguishes these common elements are the activators of the genes of each system, which are programmed to yield a product that is characteristic of the system of which each is a part and which is necessarily different from the products of other systems with which it is coordinated in the service of the organism.

The system exists solely for the survival of the organism, as evidenced by the fact that a single uncompensated derangement in this hierarchy is enough to produce a vastly different end product from the normal; e.g., three X chromosomes in place of the regular two at the 21 position yield a person with Down Syndrome and its disadvantages in the struggle for survival of the individual and the species. Other deviations produce defective heart valves, hypoplastic kidneys, adenomatoid lobes of lungs, etc. On the other hand preventive measures against disease in one area pay dividends in the simultaneous prevention of disease in other areas, e.g., dietary control of obesity for management of diabetes mellitus benefits the patient with hyperlipidemia and helps to prevent coronary artery disease and thrombosis with myocardial infarction, stroke, and gangrene that results in amputation. Also, smoking cessation, intended to minimize cardiac insults, helps against hypertension, chronic obstructive pulmonary disease, and lung cancer.

A systems approach will not permit the interpretation of the fractal variability of normally chaotic physiological states as abnormal, nor will it seek stability where dynamism is the normal state. Systemism recognizes that static homeostasis does not contribute to growth and development and is not the normal state of the organism; it rather makes for decay with its constant positive and negative feedback cycles that tend toward high entropy. (Fractals are discussed in detail in chapter 13).

Entropy is a characteristic of all energy consuming systems that do any amount of work. It is a measure of the disorder that obtains in a system and can either increase or remain stable if the system is isolated or closed to the flow of matter and energy. Any decline in energy of the system results in an increase in entropy and the tendency to system decay. A system that hopes to sustain itself must keep its energy from rising to a maximum, otherwise it will perish. It must maintain a constant flow of matter and energy at all times through its sub-systems to keep its entropy levels below the maximum, because increasing and high entropy levels induce greater disorder and inability to do any work, thereby failing to sustain the system and propelling it to dissolution or death. In this regard, integrity of the brain as coordinator is essential for opposing entropy in the system The human living system, in its ever increasing organization and order detailed above, maintains an entropy level below the lethal by being open and allowing inflow of energy in the form of food, which facilitates the decrease of entropy as a hedge against energy expenditure from limited internal sources and a tendency toward high entropy levels and death. The integrity of all of the living organism's functional systems (its homeostasis) depends on the brain as the centre of their integration; but it has been shown that artificially sustained bodies with dead brains can also carry out many homeostatic activities for prolonged periods; hence the use of their parts in transplantation. So, is brain death death of the organism?

# 7

# Causality

Explanation for the occurrence of events and objects in philosophy, science, and history embraces more than causal, deductive (nomological or law-like), inductive, and statistical explanations that are discussed at some length in this book. There are also functional explanations such as those that refer to the various beneficial functions carried out by our organized body parts: kidneys, heart, eyes, skin, bones, etc. Then there are narrative explanations where events are recounted in an intelligible manner so that the information that they transmit can be clearly understood. Behind all these different methods of explanation, however, is a current of causality that binds them.

## Principle of cause and effect.

### Causality and causation

Causal explanation of disease entails identifying predisposing factors (necessary conditions) that are related to the patient's life-style, as Bradford Hill points out, and triggering factors (sufficient conditions) that take advantage of the groundwork laid by predisposing factors to precipitate the diseased state. On this basis, we are able to designate one or more pathological factors as causes or co-causes of specific diseases. Causality is one of the central concepts in life that allows us to connect like causes of events with their like effects, while causation merely states the truism that every event has a cause by virtue of our postulation of causality, although the many effects of the cause or the many causes of the effect may not be consistently the same in each case. We go through life glibly making assumptions about one event physically causing another without stopping to ask what we mean and how we know that this is the case. We often say event $C$ is the cause of event $E$, meaning that in our estimation $C$ is the most proximate and relevant of the many known factors apparently contributing to effect $E$, and leaving out of consideration unknown and apparently irrelevant factors, however much they may influence the assumed cause-effect relationship. If billiard ball *A* strikes ball *B*, and ball *B* subsequently moves, we say ball *A* caused movement of ball *B*. We interpret antecedence as causation and thereby generate the mental idea of causation as a general principle to explain the observed succession of events. In so doing we leap far beyond the bounds of an observed, logically contingent sequential succession of events to the postulation of a presumed, logically necessary causal relationship between them. The implication is that non-movement of *B* after contact with *A* would be a contradiction of logical laws, which it is not.

Technically, therefore, causality is the relation between events, processes, or entities in the same time series such that when one occurs as cause or sufficient condition, the other necessarily follows as effect (invariable antecedence); when the effect occurs, the cause as necessary condition must have preceded it. We assume that one has the efficacy to produce or alter the other if the appropriate conditions obtain at all levels, down to the molecular level where physico-chemical reactions occur. Without one, the other could not occur in spite of the presence of those conditions; i.e., event *C* is always followed by event *E*, and event *E* does not occur without event *C* preceding it, whatever else obtains. It is also the case that event *E* (syphilis) does not occur in the absence of event *C* (infection with Treponema pallidum), although event *C* may fail to cause event *E* for several reasons, e.g. prompt therapeutic intervention, even if all the other conditions that favour the establishment of syphilitic infection are present. They alone will not cause the disease. A third scenario is where *E* occurs only after *C* in specified circumstances; e.g., acute glomerulonephritis after group A Streptococcal or other infections, or as part of primary renal diseases or systemic diseases, e.g., Berger disease or vasculitis. All these scenarios provide us with the rational means of explaining the causes of diseases, predicting their outcomes, and preventing them, because they derive from coherent relationships among events. The class of *C* (type) entails individual instances $C_1, C_2, C_3$...(tokens), and the class of *E* (type) entails individual instances $E_1, E_2, E_3$...(tokens), thus providing the basis for statistical correlation and justifying token event $C_1$ as the cause of token event (effect) $E_1$ by virtue of the causal relation of class C to class E. (See page 125). Also $C_2$ as the cause of $E_2$, ...etc.

This sequence of logical entailment has been extended to cover the relation between a thing and itself when it is dependent upon nothing else for its existence (self-causality), and between an idea and the experience whose expectation the idea arouses, because of the customary association of the two in this sequence. In the medical domain, we extend this principle of causality to the postulation of the causes of diseases, their management, and their expected outcomes by adding the principle of induction. Induction permits us to predict

114

outcomes from the observed regularity of successive events of a certain kind by projecting only their projectable features, and by positing covert cause and effect mechanisms underlying the overt causes and effects that correspond with them. Experience teaches us to anticipate $E$ after $C$. Hence we scientifically and justifiably blame the pneumococcus, $C_1$, for causing the patient's pneumonia, $E_1$, regardless of other antecedent factors, because of the observed causal connection between pneumococci and pathological changes in lungs that produce the disease styled pneumonia. We unscientifically and unjustifiably blame the draught ($C_2$) for causing his coryza ($E_2$) even though there is no underlying evidence for a causal connection between draughts and coryza. We also rightly infuse crystalloid solutions into patients in shock to restore their circulating fluid volumes on the basis of sound universal evidence that overt or covert loss of circulating fluid volume ($C_3$) caused their state of shock ($E_3$).

To be able to postulate the causal reason *why* (not only how) in a particular situation $C$ causes $E$, and not $X$ or $Z$, we should be able to identify the universal nomological mechanism of invariant relationship that confers conformity on this process, so that we can also be able to postulate that if one brings about condition(s) $C$, then condition(s) $E$ will follow in all situations, and so will the chain of $F$... to...$Z$. If $C$ is not present, then $E$ *and* $F$...to...$Z$ do not occur. Five thousand single cases of $E$ following $C$ invariably in similar sets of circumstances are knitted into a general statement or law stating that whenever event $C$ occurs in those circumstances, event $E$ can be expected to follow, because it has followed regularly up to now, and not by mere chance. Hence a causal law can be enunciated that $C$ causes $E$ in circumstances $Z$, and future plans can be made based on this assumption, even if we never see or experience $C$ producing $E$ at both the macro- and micro-levels. Of course, the whole process relies on the bold assumption that induction is always valid.

So far we can only claim belief from frequent observation and not certainty or logical necessity for this habitual relationship; that means our claim for causation as the physical idea of one event causing another can be empirically justified on observed occurrences only, but causality as a mental idea entailing the attribution of effect to cause still eludes us, because we have no means of perceiving it, although we have enough evidence to be able to postulate it without proving it in the positivistic sense. We reason backwards from habitual effects to their probable causes, even though we know that the interrelationship among causes on one side and effects on the other is subject to a chaos effect, in which a change in any one of the multiple antecedents to the final effect can abort, distort, abbreviate, or delay its occurrence. Ordinarily, whenever Varicella infection occurs, a typical rash also occurs. So, when Jim succumbs to infection with the Varicella virus, we expect that he will develop the typical rash as a result of his illness. Hence we postulate a quasi-deductive nomological explanation of the cause and effect relationship between the virus and his rash, which falls short of using demonstrable etiological evidence to deduce the causal occurrence of the rash (effect) from colonization with the virus (antecedent condition). This maneuver helps to exclude other regular but accidental relations from claiming causality. On the obverse side of this relationship are some events like quantum events that are believed to occur spontaneously, without known cause, while others, like those of volition, are brought about by the will of the agent as voluntary but undetermined events.

Generally, therefore, our stance is dictated by the application of these laws that we have extrapolated from observations of multiple and constant conjunctions of $C$ and $E$ in mechanical situations that have thus far proved to be unlimited by time or place. We have learned to expect and even predict this conjunction between $C$ and $E$, and to rely on its universality within that particular sphere of operation where it has rarely been falsified, because we combine the known facts relating to $C$ with the universal laws by which we can logically deduce the occurrence of $E$ from $C$. We then extend this expectation to the case of the non-mechanical human agent in whom the effects of multiple operative causal factors are not always easily predictable. Our minds mistakenly establish a necessary causal connection between $C$ and $E$ from their constant, contingent, sequential physical connection, even though we know that inductive arguments of this type do not have to measure up to the standard of deductive arguments where the conclusion is contained in the premises and the proof is provided by hypothetical-deductive arguments, as we shall see later.

## Actual and possible worlds

Philosophers postulate logically possible worlds or "world[s] that can be described without contradiction"[1] where the truth of 2+2=4 could be, and probably is, that 2+2=5, and where our notion of 2+2=4 in this world does in fact not hold. However, these worlds are inaccessible to our empirical knowledge and therefore otiose and irrelevant from the standpoint of our world. If they are worlds in which truths are necessary, those truths are so by convention and meaning, and not by virtue of the intrinsic natures of those worlds (including our own). Therefore conventions and meanings that hold in this our world need not necessarily hold in other worlds and vice versa. So, do we really need these other worlds to assert necessity, or will our parochial domain suffice for that purpose? The convention 2+2=4 could easily be replaced by one that renders 2+2 equal 5 logically necessary or *a priori* in this world, thus making 2+2=4 internally contradictory, while the old convention 2+2=4 now applies to other logically possible worlds without contradiction in those worlds, because they have also changed their conventions. Does this render the concept of possible worlds a confusing exercise in futility? Not quite. Its supporters argue for its utility, stating that the proposition "if C did not occur, E would not have occurred" is true if such a world is more like our real world than any such world where E does occur.

In our actual world, anticipating the bitter taste of medicine *x* when we see it, because of previous experience, or anticipating an illness with bloody diarrhea, microangiopathic hemolytic anemia, and renal failure as a result of ingesting food contaminated with E. coli O157:H7 are standard mental events. But the mental connection is not a necessary or a logical one in the sense of the cause entailing the effect. It is equally possible that the particular effect may not follow the particular cause, because each one is only contingent and therefore logically distinct from the other; they do not share a necessary connection. The medicine may have been sweetened enough to disguise its bitter taste, or the defences of the body may interrupt the causal chain by preventing the anticipated but dreaded complications of the E. coli infection. From another point of view of our real life experiences, the causal connection is probably beyond doubt in many situations where events occur from apparently known causes, such as when fractures occur after excessive traumatic force on healthy bones, and where we produce effects by known means all the time, such as inducing convulsions with the passage of an electrical current through the brain. But there are also situations in which this constant conjunction does not constitute a causal sequence, such as the succession of night and day where night is not the cause of day, or day the cause of  night, but both have a common cause in the rotation of the earth on its axis; and the occurrence of fever before the spots of Roseola infantum where fever does not cause the spots, but both fever and spots are sequentially caused by infection with HHV$_6$ virus that causes these disease manifestations.

# Koch's postulates

In our further analysis of cases where the relationship between cause and effect is presumed, let us begin with a set of postulates that is accepted readily in medical circles, viz., Henle-Koch postulates[2] (Dr. Jacob Henle and his student Dr. Robert Koch, 1840):

A disease is an infectious disease, and a micro-organism is the cause of the disease if the micro-organism

(1) is present in all cases of the disease and is isolated from diseased but not from healthy hosts;

(2) is cultivated in the laboratory in pure culture;

(3) causes the same disease in healthy, susceptible hosts after its inoculation;

(4) is isolated in pure culture from these susceptible hosts without change in characteristics;

(5) is not present in any other disease as a fortuitous and nonpathogenic agent.

Henle and Koch proposed these postulates on identification of causative agents of infectious diseases as a sequel to the postulates used by Friedrich Loeffler to confirm the exotoxin of Corynebacteria diphtheriae as the cause of diphtheria. Initially about tuberculosis, the postulates have evolved through time to form the basis for causal concepts in many other diseases, e.g., Severe Acute Respiratory Syndrome (SARS) and its causation by the SARS-associated coronavirus: the virus has been isolated from diseased hosts, it has been cultivated in host cells, and it is filterable. The other modified Koch's criteria for viral diseases have also been tested and confirmed: production of a disease like SARS in healthy hosts, re-isolation of the virus, and detection of a specific immune reaction to the virus in these subjects. Nevertheless, the postulates are not applicable to all infectious diseases, as evidenced by the case of HIV Aids. Some researchers in this field claim that the correlation between HIV antibody in infected persons and AIDS does not amount to causation, because of the low activity of the virus and the constellation of AIDS-associated diseases resulting from the many variants of the virus caused by its imprecise replication. Similar concerns have been raised in the case of typhoid fever and leprosy, where it is not possible to produce disease in experimental hosts or to produce pure cultures of the implicated micro-organisms; in diseases due to parasites, because they cannot be cultivated; in diseases with multiple causative factors but similar symptoms, and where the presumed agents cannot be pinned down and subjected to the same rigorous procedure of identification before and after cultivation; and in the case of pathogens that cause multiple diseases, e.g., Streptococcus pyogenes.

# Hill's criteria

A significant addition to the preceding concept was made by Hill with his emphasis on association between the incidence of disease and environmental factors versus indisputable "hard-and-fast rules of evidence"[3] for causation. His viewpoint helps to give statistical validity to connections between events regarded as causes and effects by combining mechanistic and probabilistic evidence for

118

causation of disease. It takes into account the fact that strong correlation provides good evidence for causal relation without proving it, and the more strongly correlated the cases, the better the evidence, especially if they fit well into our existing domain of knowledge; e.g., the strong and invariable correlation between striking a dry match-stick on an appropriately treated dry surface in the presence of oxygen and thereby producing a flame is good evidence for a causal relation between the two events. It illustrates the fact that an event is explained by showing what factors are statistically relevant to its occurrence—factors that support the causal belief, as we also saw with the case of fractures. Hill states that

(1) Statistically strong associations are more likely to be causal than weak ones; e.g., between active smoking and lung cancer.

(2) Similar results obtained from several studies done in different settings by different methods, e.g., second hand cigarette smoke, also fulfill this condition.

(3) The cause must necessarily always precede the effect— smoking for a prolonged period before developing cancer.

(4) Increasing dose or amount of exposure, e.g., more cigarettes smoked per day, must lead to increasing risk and frequency of disease.

(5) The association should not be in conflict with current knowledge about the disease, i.e., the pathogenesis of the disease can be traced from the initial stages of respiratory mucosal irritation to cellular neoplastic change.

(6) There should be some biologically acceptable or relevant reason for the cause, which is the toxic chemicals in tobacco smoke, to produce a certain effect.

(7) Alternate explanations in the form of multiple hypotheses, e.g., cancer arising de novo, should receive consideration as possible causes of other conditions.

(8) Experimental introduction or removal of a disease agent should lead to a change in the effect, e.g., restriction of smoking should and has produced a drop in the incidence and prevalence of lung cancer.

(9) Previous experiences can be used as analogies to make causal inferences, e.g., thalidomide is statistically, significantly, and antecedently associated with congenital anomalies, so is drug $x$ with its as yet undefined but similar effects. Therefore both drugs can be labelled causes of those anomalies.

# Hume on cause and effect

In the beginning, though, it was David Hume's seminal analysis of cause and effect that set the process in motion and is still worth re-stating in its original form:

(1)  The cause and effect must be contiguous in space and time.

(2)  The cause must be prior to the effect.

(3)  There must be a constant union betwixt the cause and effect. It is chiefly this quality that constitutes the relation.

(4)  The same cause always produces the same effect and the same effect never arises but from the same cause. . . .

(5)  Where several different objects produce the same effect, it must be by means of some quality which we discover to be common amongst them . . . like effects imply like causes, . . .

(6)  The difference in effects of two resembling objects (causes) must proceed from that particular in which they differ. . . . 4

Hence, we can say that if the cause is present, the effect is invariably present, and if it is absent, then so is the effect. But Hume also maintains that all empirical enquiry presupposes and relies upon the causal principle; therefore appealing to empirical enquiry in support of causal principle is viciously circular. However, he appears to exclude simultaneous cause and effect where, for instance, the steady state of my desk as effect is simultaneous with its causes, i.e., the support provided by the floor on which it stands and the force of gravity, which stabilizes it, and the books resting on it in yet another simultaneous cause and effect relationship. A parallel case could be made for a person moving a car (effect) at the same time as he is continuously pushing it (cause), if we ignore the initial push (cause) that started motion (effect). What has never been demonstrated, though, is a case of effect preceding its cause.

According to Koch, the chain of cause and effect is clearly laid out by the consistency of the sequential steps and the explicability of every case of observed effect by the preceding presumed cause. Where the chain can be proved to have been interrupted, there is no causal relationship between postulated cause and observed effect, only a chance one calling for a search for other causes. Medical treatment uses the principle of causality to interrupt the sequence of events in the genesis of disease at a point preceding the onset of dysfunction from the disease, i.e., as soon as its presence can be ascertained, or while attempts are being made to confirm its presence, categorize it, and institute methods of eliminating it. This happens in the treatment of pulmonary tuberculosis from the evidence provided by Xray and tuberculin test before or without the confirmatory evidence of growing the bacillus from sputum culture. Interrupting the causal chain in this manner is made possible by scientific medical research, which discovers the drugs and other modalities of doing this, stressing once more the important role of basic science in clinical medicine.

According to Hume, the necessary, but not sufficient conditions for causal connection are the spatial and temporal relations between objects. The effect on the object should succeed its presumed cause directly and immediately to exclude the interposition of other contributory factors as co-causes in the conjunction of cause and effect. But we know that there may be many necessary remote past and present causes producing observed effects, which manifest in a constellation of symptoms and signs that we label as a certain disease, and that the proximate cause may be only the trigger or sufficient cause of the effect. We also know that the manifestations of the disease are as multifarious as the patients, their individual constitutions, and their diverse reactions to the same causes, and that we sometimes recognize a disease (effect) before we have discovered its cause; e.g., Kawasaki Disease.

# Logical relationships

Logically, if a bout of sneezing is always preceded by the colonization of the nasal mucosa by a rhinovirus, then the presence of the virus is a necessary condition for coryza and sneezing, because coryza will not occur in its absence. But the virus is not a sufficient condition, because some people have it without developing coryza and sneezing, and sneezing may be a sequel to irritation of the nasal mucosa by other substances, like a whiff of pepper or pollen. So attributing sneezing solely to viral colonization of the nasal mucosa in this instance is an exemplification of a fallacy called affirming the consequent as illustrated in the type of argument of the form:

if p, then q;
q,
therefore p.

q might be due to x, y, or z, and not due to p, as asserted in the premise "if p, then q"; therefore the conclusion that p (is true) in this context is false (not-p).

So we conclude that empirically the immediate cause of sneezing is irritation of the nasal mucosa from a variety of factors, but we also recognize the fact that other factors may intervene to offset the effects of those factors that we consider to cause the sneezing. At no time does the question of necessity in the relationship of these events arise, because in nature there appear to be only regular sequences of events and experiences whose factual relations may not be necessary but contingent. The only relations of necessity are found in Mathematics where definition makes them so; e.g., 2+2=4, a triangle is a three sided figure whose internal angles add up to 180°.

121

Furthermore, we encounter causal factors that we can control in spite of underlying hereditary disadvantages, e.g., control of trans-fats and cholesterol in one who is genetically predisposed to arteriosclerotic heart disease prevents the ready occurrence of heart attacks.

We operate on the principle that nothing comes from nothing or nothing happens without cause or sufficient reason, and what is unexplained is not readily believed and accepted. So for pragmatic reasons we have found it useful to interpret regularity of succession as a causal sequence; it helps us to explain disease and its causes, and to plan treatment and anticipate outcomes by enlisting causality, depending on the extent of our interest in a particular causal sequence and its pragmatic significance. Meconium ileus is invariably followed by other manifestations of Cystic Fibrosis as the newborn infant grows older, so we have learnt to look for clues to this disease in these children. Sometimes we simply convert *post hoc* to *propter hoc* on the assumption that there must exist a basis for the facts of one event to causally explain and justify those of another event completely in a particular context, even if we are may not be aware of it at the time. We conclude that there must be a reason why $x$ is always followed by $y$ and not by $z$, even if the empirical sequence $x$ to $z$ is not *a priori* or logically necessary, although it can still be logically or nomologically possible. Perhaps the reason resides in the fact that we read meanings into our perceptions, and we interpret them like Alexander Flemming who attributed his serendipitous observation of the destruction of bacteria in the vicinity of cultures of Staphylococcus aureus to the mould Penicillium notatum that had contaminated them. From this observation followed the extraction of Penicillin from mould by Ernst Chain and Howard Florey and its subsequent use in the treatment of bacterial infections.

The explanation that we proffer for causal relationship is that events happen in conformity with the laws of physics, chemistry, etc., and they fit snugly into our existing system of knowledge, beliefs, and expectations. We elevate the statistical correlations and probabilities garnered from past experience to the level of causal laws, because we cannot offer explanations for the occurrences of events *ad infinitum*, without halting the regress by providing an ultimate explanation or reason. Event $A$, which is caused by event $B$, which is caused by event $C$, which is caused by event $D$, and so on, cannot go on for ever; there must be an ultimate causal event or set of events on which the entire series rests. What matters most in our particular universe of discourse depends on where our understanding and explanation of events stops for pragmatic reasons. Furthermore, causal relations represent potential and logical predictability, and causal predictability is in turn limited to postulating that if p, then expect q, in accordance with the relevant laws that apply to that situation, without implying that the relation must be so of necessity. The following assertion illustrates the point succinctly: (x)(px[?]qx), which means that in any place, at any time, if it is true that p, then it is also true that q. This contingent relation is due to the existence of many more antecedents to events that we call effects than those that we are aware of. So we choose the ones that we consider most significant for our purposes; e.g., the environmentalist looks for causes of illness in the environment, the pathologist looks at micro-organisms, and the psychiatrist looks at mental phenomena. The anatomical pathologist stops at the macro-level of gross pathology, but the histopathologist refines the search for causes to the microscopic level, and even further still to the ultra-microscopic level. The geneticist goes much further to the level of chromosomes and genes. With these successive refinements, the essence of the ultimate cause and description of disease is shifting all the time.

# Induction / deduction

We are always dealing with multiple causes and effects. For instance, the patient's immune status may be weakened by his genetic constitution (necessary condition), and he may be managing just fine until he picks up a virus (sufficient condition), and then he comes down with a disease that would ordinarily not have disabled him. The converse is also true: if someone succumbs more readily to a virus that is not considered to be virulent, we look for the cause of this effect in his weakened immune status. Such reasoning is guided by repeated experience with the external relationships among events. Internal logical relationships of the constituents of those individual events do not appear to us to have any role in this causal process.

The basis for our ability to retrodict to initial causes (causes that obtained at the onset of events culminating in and explained by the present situation) and thereby predict from the combination of past and present events to future events is also made possible by taking causality for granted. If we do not take causality for granted, we are caught in the trap of being unable to predict future events from past events, and our experiences become useless, because we cannot infer anything from them when we impose the strict criteria required for drawing deductive conclusions on procedures used for drawing inductive conclusions. We do not have any justification that what has held good in our past will necessarily hold good in our future, because it is not possible for us to perceive our future—that our future futures will resemble our past futures, or else that the future of today will turn out as we expect, just as the anticipated (past) future of ten, five hundred, and fifty thousand years ago turned out as expected in those past years; that *C* will cause *E* next week and next year in the same way that it does today, and as it did many years ago. We boldly assume this relationship by wrapping the probability concept within the idea of cause, forgetting that all we can go by is chance and probability as a result of our deductions from the hypothesis of causal connections, although we have no proof that it will hold; that still remains to be proved empirically.

Our scientific conclusions, based on an accumulation of facts or premises as they are, are always one step removed from the premises that justify them, because they are not contained in the premises, as we shall see in the discussion of Logic and Scientific Method. We claim justification for this move from using universal laws as the major premises of our arguments for arriving at our conclusions: All metals expand when heated is a law of Physics based on the fact that all the metals that have been heated over a period of $n$ years to date have expanded; therefore this piece of metal will expand when it is heated. But what is the justification for these laws, if not the uniformity of nature, which derives from induction and the principles of causation and causality, all of which we are supposed to be trying to prove? Forgetting that scientific laws state only probabilities, not universal connections, or logical necessities, we form quasi law-like generalizations embracing future conditional and counterfactual statements from an accumulation of individual causal sequences: e.g., if Jim is infected with a certain virulent strain of Streptococcus pyogenes, he will succumb to flesh eating disease; if his present infection were not one with that virulent streptococcus, he would not have developed flesh eating disease. Our type of conclusion, not being the deductive type in which the conclusion is already implied and entailed in the premises as in

All $As$ are $Bs$,
$C$ is an $A$;
therefore $C$ is a $B$,

is, however, necessitated by the statistical significance of its regular occurrence which approaches, but does not amount to certainty or 100%. It is only made virtually certain or highly probable by the frequency, extent, and variety of situations in which it has occurred, thereby rendering the chance of failure unlikely without mandating its exclusion; e.g.,

all the $As$ observed so far have been $Bs$,
this (something) is an $A$;
therefore it most likely is a $B$.

In the deductive process, one case of a $C$ that is not an $A$ will destroy the magnanimous edifice that has been constructed on this foundation, together with all our theories of causation and prediction that are related to it or depend on it. Karl Popper maintains, therefore, that to base Science on such a shaky foundation is to

124

expose it to vulnerability by only a single exception to the rule. In the case of induction, the aberrant *A* that is not a *B* only reduces the probability of more *As* being *Bs*, but many more *As* that are *Bs* can make up for the shortfall. In Science very good reasons and explanations must be advanced for an aberration, if the viability of a theory and its application are to be retained, because Science needs desperately; without it there can be no scientific hypotheses, no scientific prediction and progress, and therefore no Science. All scientific laws are contingent, because they are based on repeated observations of individual ontological phenomena and their conceptual classification into general systems from which predictions are made; but their denials are true without contradicting the laws of logic, which are necessarily true by definition. Also, scientific and so-called laws of nature are empirically derived and cannot be used to prove inductive inference without incurring the circularity of using a conclusion derived from accumulated data to prove the validity of the same and more data (see 'induction' on pages 217-218 and 220).

# Type-token relationships

The type-token distinction categorizes causality into statistical and specific varieties, as in the following example:

1. John succumbed to lung cancer as a result of chain smoking for many years. If the two are causally related, the single case of John constitutes specific or token causation, and it is true only if John was a smoker and untrue if he was not, because he could still have died from lung cancer without being a long time chain smoker.

2. Chain smoking over many years causes lung cancer. Bill's case of lung cancer is one of statistical probability or type causation, because long time chain smokers, as a group, have a much higher probability of getting lung cancer and dying from it than non-smokers.

An unending concatenation of token instances of individual deaths from lung cancer temporally related to smoking strengthens the statistical or type causal relationship between smoking and lung cancer, thus making possible the quasi-nomological relationship between the two events. We hereby exclude the possibility of a chance relationship such as that of the single token events of smokers John and James dying from lung cancer being claimed as a causal one. On the other hand, the occurrence of a single case of unchecked, profuse hemorrhage followed by the death of the subject is sufficient to establish convincing evidence of a cause and effect relationship between the two events of unchecked profuse hemorrhage and death of the victim without demanding a succession of similar incidents to prove its quasi-nomological status: statistics aside, people who bleed profusely to the point of exsanguination will die.

These are cases in which effects can be said to necessitate their causes, because without the occurrence of those specific preceding events, their specific sequels would not have occurred, and there is no way of altering that sequence of events now, although at the time of the occurrence of those causes events might have been made to follow a different course. On the other hand, causes do not necessitate their effects, because anything is possible; unexpected results may contingently follow the best laid schemes, although the principle of sufficient reason would have us believe that causes explain their effects in the sense of determining them. But we know that the world is not deterministic; it does permit freedom of choice by the agent, and it is also subject to chaos effect; therefore there is no symmetry between causes and effects, with one necessitating the other in a predetermined or fixed way. All relations are contingent, because the objects and events involved do not have intrinsic relationships that necessarily and intimately bind them to render them necessary.

So, to use a borrowed example, if causation can entail relations between property tokens (token=specific occurrence of type of event), and if a mental token $(m_1)$, like the thought of raising my arm, causes physical token $(p_2)$, which consists in my arm going up, and physical token $(p_1)$, like someone raising my arm, also causes $p_2$, then $m_1=p_1$ described in mental and physical terms. That means the sense of $m_1$, which is mental, equals the sense of $p_1$, which is physical, because they have the same referent, $p_2$. But this is not possible, and so the general theory of causation can't be based on token causation, because token causation lacks universality; its validity depends on type causation, which transcends particularity. So we bridge the gap between them by using type causation as nomological principle to explain token causation and cover all causal possibilities logically. In justifying the treatment of disease as effect on the basis of its microbial and other etiology, we search for pathogenic mechanisms that will enable us to jump this gap, because without them there cannot be a correlation or reasoned explanation of observed cause and effect. Cause must explain effect completely as per mechanism postulated, otherwise the link breaks down.

## Mill's methods

To round off the discussion on causality, I will mention the methods of John Mill[5] who also enunciated his tenets for correlating events where a causal connection between events can be established only when we already have an idea of some link between or among them:

(1) Difference - z (invariably a multi-constituent factor composed of $z_1$, $z_2$, $z_3$,...) is the cause of event $E$, if $E$ occurs only when $w$, $x$, $y$, and $z$ occur together, and never when only $w$, $x$, and $y$ occur together without $z$. The additional factor z may act as a triggering, necessary, or sufficient cause of $E$. This method is used when controlled trials are set up. The subjects and experimenters are blinded so as to exclude biases in regard to some of the variables (e.g., factor $z$, a new drug) whose causal effect is under investigation, and the subjects are matched as closely as possible for all the other variables $w$, $x$, and $y$. Any difference in results at the end of the trial is presumed to be due to the only variable being added or subtracted during its course, i.e., the drug. The method is easier to execute with animal experimental subjects than with humans for obvious ethical reasons.

(2) Agreement - z is the cause of event $E$, if $E$ occurs when $w$, $x$, $y$, and $z$ occur together and when a, b, c, and z occur together, z being their common denominator, but not when w, x, and $y$, or $a$, $b$, and $c$ occur alone without $z$; e.g., John has bloody diarrhea and abdominal cramps, and subsequently develops hemolytic uremic syndrome. E. coli O157:H7 is cultured from his stool, and verotoxin is also isolated. Mike has similar symptoms, but does not develop the syndrome. His stool culture yields entero-invasive E. coli or Salmonella species, or even E. coli O157:H7, but no verotoxin. On the other hand, Harry does not have diarrhea but develops the syndrome, and shows the toxin in his blood. The occurrence of several similar sets of cases makes it more and more likely that the verotoxin as found in E. coli O157:H7 and other bacteria is the cause of the syndrome, which constitutes complete causation.

(3) Concomitant Variation - z is statistically correlated with $E$, suggesting that it may be causally related to $E$, either directly or by way of a common cause, if variations in z are accompanied by variations in $E$ as quantitative aspects of causation; e.g., the severity of the hepatotoxicity of acute acetaminophen poisoning varies directly with the amount of acetaminophen ingested by a subject at one sitting; hence the utility of the Rumack-Matthew nomogram for correlating drug concentration in plasma, precise time after ingestion (past four hours), and hepatotoxic effects.

(4) Joint Method of Agreement and Difference - z is the cause of $E$, if z is present in all cases where $E$ occurs, and absent in those cases where $E$ does not occur, or else $E$ is the cause of z. The method is a combination of the first two methods, constituting complete and necessary causation.

(5) Method of residues attributes antipyretic effects to aspirin if aspirin as the sole remaining member of any combination of drugs lowers fever when each of the other constituents fails in that role.

Joe Lau and Jonathan Chen[6] have added other forms that the causal relationship can assume:

(1)  Loop causation where $C$ causes $E$ and $E$ causes more $C$, which in turn causes more $E$, and more $E$ causes more $C$, and so on; e.g., depression causes low self-esteem, which causes depression, which causes low self-esteem, which causes more depression, which causes more low self-esteem, and so on and on.

(2)  Minor causation where c is a minor contributor to $E$ in the presence of $C$; e.g., stress is a minor contributor to heart attack in the presence of arteriosclerotic heart disease.

(3)  Common effect where $C_1$ and $C_2$ cause $E$, as in the toxic systemic effect of acute iron overdose and its gastric irritation causing emesis.

(4)  Common cause where $C$ causes $E_1$ and $E_2$, as in the HHV$_6$ virus causing fever before rash, not fever preceding and causing the rash; or group A streptococcus causing Rheumatic fever and Chorea.

(5)  Side effect where c appears to cause $E$, but no underlying causal mechanism exists between c and $E$, as in the placebo effect of a drug.

Lastly, let's examine counterfactual (contrary-to-fact) causation that was introduced on page 124: if C had not occurred, E would not have occurred; alternatively, if C (lead ingestion) had not occurred, the chances or probability of the occurrence of E (lead poisoning) would have been zero. The counterfactual is supposed to occur on type and token levels of events; it indicates that E was not realized because C was not realized. But we can see right away that E could still have occurred without the occurrence of C for reasons that may or may not be apparent to us, e.g., inhaling lead contaminated dust, because we have no way of telling now that only C is responsible for the occurrence of E; the occurrence of E could be a sequel to the occurrence of any one of $C_1$, $C_2$, $C_3$....$C_n$ that may be unknown to us, or a concatenation of them, as happens in all illnesses; e.g., if only I had treated the patient with cefotaxime instead of erythromycin, he would not have succumbed to his infection. But I don't know that, and I can't prove it now; I can only surmise.

Nevertheless, the counterfactualist maintains that without the precedence of any one or more of the antecedents, the result or consequent would not have occurred as it did—if I didn't have teeth, I wouldn't be having this toothache. As my grand-daughter said to me: if it were not for you, I would not be here.

No doubt, the philosophical problem of causality will continue to vex our minds for many years to come while we continue to witness the interplay of physical and mental causes and effects all the time, even if it is only in the pragmatic sense.

128

# 8

# Challenges to the person's survival

Even before birth, the human organism's survival is already threatened by multiple hazards of maldevelopment, maladaptation, and aborted viability. The threats multiply with the advent of birth and after, as the constant, looming presence of illness, injury, and disabilities reduce its chances of survival. These conditions define a person's being-in-the-world so much so that his being-in-the-world as a diseased, ill, injured, or disabled person may appear to be only a variation on a theme, and yet his diseased body imposes radical change in his existential balance and his goals, including efforts to avert death. Disease and illness render the person dysfunctional, strip him of the values that he attaches to his life, and prevent him from achieving his goals by degrading his general performance to a pitiable level below that of his fellow competitors in the struggle for survival. In this age of limited and mismanaged healthcare resources, it is important to delineate diseases with a view to determining which ones are more likely to incapacitate people, render them dependent, and thus deserve expenditure of money and time to be controlled.

The ensuing discussion will proceed on the assumption that it is possible to define the normal state of a person's health without begging the question of how the standard of normality was selected without assuming that some persons are diseased—which is what we are trying to prove—and should be excluded from serving as subjects for establishing that standard. The evolution of the concept of disease begins with disease as a mysterious metaphysical, foreign entity that invades the human organism, jeopardizing its survival. It fans out into disease as a state of uncorrected disturbance in homeostasis that results from a process of disadvantageous cellular and tissue change *per se,* in turn caused by microbial and other external or internal agents. It ends in disease as a complex process of deviant and unstable physical and psychological manifestations of the person's total, unique, responsive adjustment to the many different kinds of detrimental environmental stimuli impinging on him and tending to this derailment of his biological functions within his social milieu. These functions will be understood as the causally determined roles of the various structures that comprise the organism whose operation they determine, not the presumed goal-directed roles that are often arbitrarily assigned to them in the service and survival of that organism. Their failure is some of what constitutes disease.

When we consider the difficulties posed by the biomedical definition of disease and where to place other abnormalities of form and function of the person that do not emanate from invasion of his body by microbes, e.g., injuries (including poisoning and suffocation) pain, chronic fatigue, vague feelings of indisposition, malnutrition, handicaps, and symptoms and signs that are so disparate that they cannot be grouped together to constitute distinct diseases, we will appreciate the need for the promotion of the category of "malady" to account for all these functional abnormalities that disadvantage the person from adapting successfully to his environment. According to the promoters of this view, malady refers to having something wrong with oneself in a way that makes this dysfunction a self-sustaining, assimilated, and inseparable part of one's person (regardless of its origin), impeding one's enjoyment of life until that thing is eliminated from one's life. The term embraces nominal entities that belong to different species that form a heterogeneous collection but are brought together under one genus by their incompatibility with a life that is value-filled and flourishing, as also by their ability to harm the person beyond simply rendering him unable to perform his life tasks normally, i.e., in accordance with species-design.

# Disease and Illness

These two concepts will be considered together because of their close connection, since talk about one always spills over into talk about the other, although different disciplines within the field of Medicine will define disease differently from the perspective of their specialties—pathology, psychiatry, paediatrics, biochemistry, radiology, etc. Meanwhile, society mislabels conditions as diseases in keeping with its own interests, e.g., drapetomania and homosexuality, claiming that they need to be treated like diseases caused by microbes that should be eliminated from a person's system.

Notwithstanding their close association as negative states of health, the two conditions call for the need to maintain a logical distinction between the referents of their appellations: "disease" and "illness" and the concepts that they represent to avoid confusion in meaning and in the management of patients who come complaining about one or the other condition. The term disease implies the need for objective biophysical investigation and remediation to relieve the person thus afflicted of the subjective state of illness, discomfort, or suffering imposed on him by the disease. It also has implications in the social sphere by placing him in the category of sick people, apart from those who are not similarly afflicted, and demanding sympathy and all possible assistance for him, e.g., exemption from duty or exculpation from moral and legal culpability in the case of one who is mentally afflicted. The person who complains of feeling ill may elicit our sympathy even though he does not often present us with a handle on the subjective condition of his dysfunctional body, unless he also has an underlying physical disease state that we can investigate to establish a connection between it and his experience of illness in his lived body as a result of the disease.

To anticipate a later discussion, this attitude attaches a character of value to the term 'disease'. We have imbued the term 'disease' with a value connotation, because it represents an undesirable clinical state of affairs that we would like to prevent or be rid of more than many other undesirable states, like wrinkles and senility, that are not diseases. Disease makes us feel unwell and disinclined or unable to pursue our usual functions, deferring their performance to others while we feel inadequate and ill-adjusted as we begin to lose our self-image as a result of what has been termed our "wounded humanity".

The need to delineate the two concepts "disease" and "illness" receives perplexing treatment in the opinion of James Birch as expressed in his paper, *A misconception concerning the meaning of 'disease'*[1]. To use his example, when my car has broken down (has an illness), I go to a mechanic to find out the cause of the breakdown (the disease or non-disease that is causing the illness), not to ask if it has broken down (has an illness), because I know that. But then Birch loses the reader when he says the man whose foot has turned backwards "*knows* he is ill because he knows the conventions of what is or is not disease"; but he still goes to the doctor "to know *what* is the matter" (what is the disease, if any, that is causing his foot to turn backwards, even though he knows the answer to that question from his knowledge of the conventions of what is or is not disease). He does not go to the doctor "to ask *if* he has an illness", as Birch rightly maintains, because he knows that. So why does he go to the doctor? He goes to have his personal problem that he knows fixed, and to have his well-being restored, not to know what the matter is or to ask if he has an illness, which he already knows.

On the conceptual level, the philosophical relationship between disease and illness is demonstrated in the following statements: disease is a necessary condition for illness if, and only if, (iff) when there is no disease there is no illness. As a necessary condition, the presence of disease guarantees that illness is or will be present at some time. But we know that people can complain of feeling ill, and even look ill, when they do not harbour any disease but are exhausted from overwork or battling with social or psychological problems, and people can unknowingly harbour disease and yet not complain of feeling ill. So the case for a necessary connection between disease and illness is rendered false by these two counter arguments. Disease is a sufficient condition for illness iff when disease is present illness is also present. The condition does not exclude other sufficient conditions, because illness can be precipitated by any one of many other conditions besides disease, as we just noted. Occult disease, which is not accompanied by illness, but may result in illness in some cases if it is not treated, also qualifies as a necessary condition for the occurrence of illness in the near or remote future, depending on when one or more of a multiplicity of sufficient conditions will trigger the manifestations of the illness.

When a patient falls ill with diabetes he does not come to the doctor complaining of disease or a scientifically delineable anomaly of the histology and biochemistry of his pancreatic β cells, he comes because he is feeling unwell, distressed or in pain, and failing to perform his daily tasks as he usually does. He seeks out the doctor to obtain help that will enable him to return to his usual (normal) functional state. The doctor listens to the account of his experiences and subsequently diagnoses his problem as a derangement in the form and function of his pancreatic β cells, and then she decides to prescribe Metformin or Insulin, communicate dietary, exercise, and other health style guidelines, and she also educates the patient about the disease that is making him feel unwell. She attaches a diagnostic label to the patient's conglomerate of signs and symptoms for ease of communication and re-identification of the problem by those who will treat the patient's disease and illness, and for those who will identify a similar health problem in other persons, as well as for its focused study.

Furthermore, the patient's cultural milieu does not delineate the diagnostic and treatment guidelines of his disease, even though his illness occurs in it and different cultures have differing attitudes toward the same condition *vis à vis* its disease status. The doctor certainly does not manage his illness on the basis of her common or non-professional knowledge of human behaviour, as proponents of the strictly socio-cultural characterization of disease would have us believe. She gets her cue from the patient but does not let her diagnosis and treatment be determined only by what the patient relates about his feelings. Some patients, like Mr. M, are spiritually at peace with their bodily disease and do not wish to have their lives prolonged. But to assist those who are looking for a cure, the doctor resorts to her clinical and technical knowledge to guide her in the quest for the objective causes of the patient's subjective illness, for which there is never just one cause but several physical and psychosocial causes and triggers determining at what point in time the patient will succumb to their onslaught. The interaction between patient and doctor reflects the interface between subjective and objective outlooks on disease and illness that dominate their respective thinking about it, until the doctor experiences the same disease and illness and comes to assume a subjective viewpoint that may then become congruous with that of some of her patients specifically and of all patients generally in a compassionate sense.

Some Philosophers believe that, because the experience of illness in different cultures carries different meanings, sickness has many meanings attributed to it by virtue of its origin from these varied experiences, otherwise it would amount to a case of a simple matter of pathological cause and clinical effect that is favoured by the biomedical concept of disease. From this belief, they rightly conclude that diagnosis and rational treatment of disease should not neglect the influence of the social, cultural, political, economic, and psychological factors involved in the genesis and clinical expression of disease as represented by the patient's verbal expression of his particular experience of illness and loss of function in the context of his social and cultural group. Accordingly, different social factors also become determinants of diseases and sicknesses suffered by all groups of people and of the relief that they will receive from their suffering, simply because attitudes to health and disease differ among the groups. A portly body build among some indigenous peoples is accepted as a sign of well-being while the slim shape is considered ideal in others; hence the ease with which some people slip into obesity or anorexia nervosa, depending on their cultural affiliations.

# Definition of Illness

At the risk of sounding redundant, because one cannot discuss disease, illness, and wellness apart from one another, let us attempt to isolate illness *per se* and consider it as a state of discomfort of the individual and disarray of his life's events as a result of bodily and psychic dysfunction induced by disease and other adverse factors in his environment. In the case of physical illness, the clue is the facies of the patient and his demeanour resulting from the quantifiable tissue disruption that has caused his illness; but in the case of someone with so-called mental illness, his illness cannot be linked to and defined by any gross pathological changes in his tissues and common biochemical parameters, unless he is afflicted with an organic brain syndrome. The diagnosis of his condition relies on observation of his behaviour and verbal expressions as indications of his suffering, and his response to treatment is reflected by these same parameters. The handle on his illness is purely phenomenological, and that is why it can be more easily feigned by the malingerer than can organic disease where a vast array of technical tools is available to assess and define the disease that is causing his complaints of illness.

133

The person who feels ill or impaired in health and well-being is usually afflicted with a pathophysiologic disease of the body, in which case his illness is the clinical result of the interaction of his body tissues and mind with external agents like micro-organisms and toxins. Alternatively, the ill person may be mentally afflicted with a socio-cultural disease of living, which has no pathophysiological basis but results from the stressful pressures of modern civilization, inducing a negative effect on his well-being. Physical disease by itself cannot serve as an index of illness, since it does not have a linear relationship with it, and one can exist without the other. In fact, some patients do not consider themselves to be diseased or ill until they receive a technical diagnosis, while in other cases objectively ill persons may claim to be healthy. Quite often too, patients complain of still feeling ill when the doctor has cured their disease and is happy about her accomplishments but has forgotten about other factors that have contributed to her patient's illness, while others may complain of illness but have no objective features and laboratory data to define their complaints, e.g., cases of chronic fatigue syndrome.

## Definitions of disease

Is it true that "there are no diseases, there are only sick people", in the same way as Szasz says that there are no mental diseases, only behaviours of which psychiatrists disapprove and call mental illness? Disease is a term that denotes a variety of conditions that can't be confined by formal criteria or be assigned common necessary and sufficient conditions and their exceptions to qualify for the designation. In Wittgensteinian terms, these conditions only share a family resemblance name. It thus seems impossible to provide a univocal and *a priori* definition of 'disease'. Fundamentally, the concepts of disease and illness are parasitic on the concept of person, although the concept of disease is commonly reified so that it is often regarded as representing a morbid entity that possesses people and their lives in an adverse manner. This latter is the realist concept of disease, which grounds it in nature as a distinct, identifiable, ontological particular or natural kind that causes illness and must be identified and eliminated to cure the patient. If a patient is suffering or distressed, something inapparent that has possessed him, a disease entity, must be causing the suffering. If the character of some diseases changes over time as more and more facts come to be known about them, and also as some of them evolve with time, it is this mysterious entity that is changing and evolving; e.g., General Paralysis of the Insane was previously thought to be a purely psychiatric disorder, not a physical disease, but is now known to be the neurological manifestation of tertiary syphilis, and diseases like tuberculosis that were known only by their clinical manifestations are now also known through their etiological agents as defined by Koch, making their management much easier than before; treatment is now aimed at their causative organisms than at their clinical presentations. Surely these changes are not those of entities but of concepts about causes of the processes whose manifestations are being elucidated. On the other hand, as the realist will point out, the evolution of the virulence of tuberculosis to its present level of extreme concern with the development of resistant strains of mycobacteria is the result of physical change in the causative microbial organism. The concepts used to characterize its actions remain unchanged, as does the concept of this and all disease as a non-entity without an ontological character.

Contrary to the realist view, the nominalist view reduces disease to the name of composites of abnormally caused and disadvantageous phenomena consisting of symptoms and signs manifested in a particular species. These phenomena do not have an independent existence apart from the subject in whom they manifest, and the diagnostic label that names them is only a marker of convenience. It guides treatment without designating an entity named disease as the cause of the illness and can be changed when the character of these phenomena changes. In keeping with this perspective, Charles George defines disease as "a *word* used by observers to describe a *process* that occurs when one or more *external factors* interact with a living organism to produce physical and/or mental changes within the organism that the observers consider *disadvantage* the organism as compared with its former state."[2] (*my emphases*—disease is a name applied to a process; not a thing. External factors may be things. The result of their interactions with the organism places the person at a disadvantage in life by disrupting his being in the world with experience of pain and suffering).

To the conceptualist, disease is an idea conjured up in the observer's mind by a particular group of phenomena. This definition contrasts with that of the concept of genetic disease, a realist position which implies an ontological and tangible existence for the causes of disease by these physically identifiable internal elements of our constitution (genes) that are as particulate as and on par with the bacteria, viruses, fungi, and external elements that cause our other diseases; e.g. genes responsible for Huntington disease, Sickle Cell Disease, and schizophrenia, *vis à vis* the viruses that cause herpes stomatitis and the bacteria that cause tuberculosis. Attention has been drawn by other authors on the subject to the need to distinguish between real diseases and human behaviours or characteristics that we just happen to find disturbing e.g., ADHD, a label that teachers love to pin on little boys who do not conform—the ones whose behaviour lies at the periphery of the bell curve and challenges their lack of skills in dealing with exuberant demeanours, and for whom I have refused to prescribe calming medication; hyper-extensible joints, which are often regarded as a curiosity but may be harbingers of possible complications like spondylolysis in gymnasts.

At the root of the problem of the nature of disease is the ontological question of whether, after the death of the patient, disease survives as something that transcends the pathology seen in his tissues, or it terminates with the cessation of physiological function in those tissues that marks the end of his life. The answer is that as an existential concept, disease survives the death of the individual as a pointer to other people in the same circumstances; but as an actual existent it ends with his death, even though it leaves behind histopathological features by which it can be identified after the fact.

So far, we have considered only some of the conceptual definitions of disease to which we can add the following that have been knitted from suggestions by other persons who have also tried to define this concept: Disease is a deviation from or a disruption in the normal anatomical and physiological integrity of any part, organ, or system of the body or person in response to harmful events, which results in symptoms experienced by the person as illness or feeling unwell; a psycho-physical change in the body that entails impairment of normal functional ability, and that has a specific etiology and physical signs and symptoms that can mostly be measured in the laboratory; a dynamic reaction of multiple etiologies that places individuals at increased risk of the adverse consequences of physical or psychological impairment resulting in restrictions of activities and limitations of roles. It is believed to appear when the person's homeostatic mechanisms fail in their functions, although they may also fail for other reasons than his being diseased.

The common threads running through these definitions are those of harmful etiological events causing deviations, disruptions and inversions in a patient's normal states, and impairments of his physical and psychological functions and integrity. These all result in the disintegration of the person's ongoing internal organization and his interaction with the rest of the members of the community in which he normally has a specific but now unfulfilled social, political, and economic role. The result is the advent of untoward symptoms that call for radical correction or, alternatively, amelioration of their many causes. The condition is easily recognizable to every doctor as one of a certain specific type that requires specific management, even if its conceptual definition is still shrouded in controversy. In this situation the doctor has a practical advocacy role to play in providing the kind of atmosphere within which the patient can realize most of the new goals that he has now acquired in the aftermath of this new existential predicament that he now compelled to relish as he tries to

136

to find new ways of adapting physically, psychologically and socially to drastic changes in his personal and community status. The doctor bears this responsibility as the one to whom the patient has entrusted himself, his values, his health, and his now dubious future life beyond her theories and conjectures about conceptual import of the term 'disease' as the univocal designation of widely divergent conditions.

Another definition of disease that does nothing to elucidate the problem is the circular one that defines it as the absence of good health, health being defined as optimal bodily and mental balance and integrity that results in homeostatic equilibrium between a person's abilities and goals in the absence of disease and suffering. This definition is in contradistinction to defining health as a reflection of how well the person feels or how much the function of his bodily organs conforms to the statistical norm, typically within the two standard deviations limit, implying that people whose functions are above the statistical norm are diseased or unhealthy, although we know that a super athlete and a genius are not always unhealthy. Thus health becomes the intrinsic good, the personal feeling of wellness, the value that transcends statistical considerations, and the state whose absence spells disease. But the use of personal feelings of wellness to assess health overlooks the ease with which they can be misrepresented by tying them to happiness and social well-being, so that the ones who have neither of these latter qualities are not healthy, which is untrue. We know that any one can be healthy even in the absence of these qualities that carry different meanings to different people, depending on their station in life. Health is also not a negative concept defined in terms of the mere absence of disease, as shown by its retrospective appreciation after it has been lost, but it refers to the total successful dynamic operation of the person in his milieu, defining disease as its opposite.

Two of the notable examples of the definition of disease and health that we will address, those of Boorse and Lennart Nordenfelt, should be assessed against the backdrop of the preceding comments. In his biostatistical theory, which uses the functional parameters of species design for species survival and perpetuation, mainly sex and age, Boorse says, "a disease is a type of *internal state* which is either an *impairment* of [statistically] normal functional ability, i.e., a reduction of one or more functional abilities below typical efficiency, or a *limitation* of functional ability caused by environmental agents."[3] (*my emphases*)—disease is a measurable and evaluable state, not a thing; environmental agents may be things whose interactions with the organism disadvantage it by impairing or limiting the person's functional ability, as George also observed. Nevertheless, we have no guarantee that biological dysfunction is a sufficient condition for disease, although it may be necessary, since some deviations from the statistical norm of particular members of the species are not diseases but natural features of development, e.g., edentulous gums of infants, or anatomical variations like accessory navicular bones. See diagram on page 157 for the role of statistics in the definition of disease.

Boorse assumes that the species-typical efficiency of an organism is a function of the duration of its survival; but stability of a feature over a long period does not necessarily imply or guarantee that it will have an advantage over a recently acquired one. Normal form and function is arbitrary, since every person is different from every other in form, function, and response to derangement of these parameters, and this lack of strict uniformity does not permit the singling out of any one or any group as the typical one; and that deprives deviation from the norm of an edge over a value-based conception of disease. On Boorse's definition, health is necessarily freedom from disease, which entails statistically normal function and the performance of all typical physiological functions with efficiency. His definition is supposed to be empiricist-reductionist in the sense that it is based on observation of phenomena without their concomitant evaluation, dealing with purely statistical measurements of normality, their significance for delineating a state of health, and the importance of their deviation from the norm to a lower standard. As we have already noted, this approach is subject to the criticism that, besides assuming the concept of normality like the other definitions mentioned below, it renders champion high-jumpers and courageous soldiers diseased, because their performances lie outside the statistical norm for their species-typical efficiency, although it is in the opposite direction to the stated deviation from the norm. It is, however, consistent with the Darwinian concept of ascription of function to an organ in relation to its evolution by the process of non-teleological blind natural selection and its contribution to the survival of the species through the survival and propagation of its genes. As per Cartesian dichotomy, the human heart is a mechanical pump; it is normal and functions so when it pumps blood to maintain the circulation, provided that the environment is conducive to its proper functioning. It does not have teleological or "propensity" functions that are geared to the future developmental success of the organism that transcend environmental influences when it is carrying out its normal functions; it simply pumps blood or fails to do so, period. We add value to this function.

Boorse's position is also an example of objectivist realism in that it stems from regarding disease as a real, causally induced malfunction of bare, valueless, measurable physical parameters (anatomical and physiological) of the body with undesirable consequences that do not depend on the value-laden conceptualizations of what human beings are. It contrasts with the contentions of constructivists who would like to see disease as a purely evaluative concept of the particular state of

138

the human organism in which it exists at that time as a result of some detrimental biological process, They do not subscribe to the idea of a valueless world, even though some people believe that value is always a mere descriptive property that is added to a valueless world. They will, for instance, claim that any condition that runs counter to our values as a society, like pedophilia, homosexuality, or even drapetomania, qualifies as a disease for which we then look for bodily malfunctions to justify the objectivist standpoint of disease. (See pages 58-59).

Nordenfelt defines disease as "a *subnormal functioning* of a bodily or mental part of the human being"[4] (*my emphasis*—disease is a functioning or process; not a thing), and health as "a mental and bodily state, given standard circumstances, which is such that A has the second-order ability to realize all his or her vital goals, i.e. the states of affairs which are necessary and together sufficient for A's minimal happiness in the long run"[5], except that people who are easily contented with less than noble ambitions will claim to be healthy in undeserving circumstances, and those who fail to attain their goals need not necessarily be diseased. Nordenfelt's definition is value laden in its emphasis on the achievement of vital goals of survival, reproduction, and perpetuation of its species, amongst others, by an organism that is constantly exposed to a variety of noxious agents that place it at a biological disadvantage and threaten its existence at every turn. John Scadding concurs with the views of Nordenfelt and Boorse in the statement: "disease refers to the sum of the *abnormal events* shown by a group of living organisms in association with a specified characteristic or set of characteristics by which they differ from the *normal for their species* in such a way as to place them at a *biological disadvantage*."[6] (*my emphases*— diseases are abnormal, biologically disadvantaging events for the human species; not things). Scadding's definition is not of an ontological existent named disease that attaches itself to human organisms to make them sick, it is rather a nominalist elucidation of the use of the term 'disease'. It concerns itself with the concept "disease" to the exclusion of the object 'disease'.

The above discussion is focused on disease in its overt presentation; but overt disease is different from occult or asymptomatic disease that is defined by and can be revealed only through laboratory or other technological means, as in the case of hypertension detected incidentally during the routine physical examination of an otherwise healthy individual with the use of the sphygmomanometer or other device. Actuarial calculations consider this a disease and a risk factor not to be overlooked.

Medicine recognizes illnesses like hypertension and Cushing's disease that are the outcome of systems in a poorly regulated, but stable state that is, however, suboptimal. It also recognizes disease in unaffected and healthy looking members of an index family revealed by genetic screening, occult cancer of the prostate or cervix detected by tissue biopsy and microscopic examination during screening, early colon cancer detected by colonoscopy screening, and incipient myocardial infarction detected by electrocardiogram and serum enzyme studies in high risk persons. These conditions suggest that disease cannot be sensibly defined in negative terms as the absence of good health, even if we are personally acquainted with good health, because we cannot handle absences mentally or recognize them as answering to the particular description that differentiates them from all other absences; all absences are the same. Attempting to define any entity in negative terms is impossible, because absences, lacking in distinctive features of recognition, leave us with nothing to hold on to. Thinking of absence automatically triggers first thinking of the presence of the absent object and then trying to imagine its absence; an impossible feat. Also, absences do not call for restorative efforts like a diseased person who needs restoration to good health. One cannot restore what is missing from nothing, because one cannot take anything away from nothing, in the first place; so there is nothing to restore. Besides, some persons who lack the well-being associated with good health are not necessarily diseased, even though well-being implies a state of good health.

Other situations in which we see illness without disease in patients include complaints of feeling ill and feigning signs like paralysis in hysterical persons, and actual experiences of chest pain and feeling ill without overt physical signs in patients with Da Costa's syndrome. Here the subject's self-evaluation of his dis-ease or lack of ease does not appear to be supported by the facts on which he is erroneously basing his presumed disease. Sometimes illness may only be a reflection of a multiplicity of untoward social effects that adversely affect the ability of the patient to function normally, or of the limits of his coping skills, e.g., how well he can tolerate the weight of disease without suffering disruption of his social life or succumbing to the stress imposed by the illness that he may be denying. Alternatively, it may reflect how long other people can tolerate the obvious symptoms of his illness before they become very concerned about them and him and feel compelled to draw his attention to the need for correction of his illness. Hence the claim by some people that there are no diseases, only sick people with clinical presentations causally related to certain adverse factors that impinge on them—these simple objectivists choose to define disease as simply abnormal function of personal physiology or psychology. It is indeed impossible to give any meaning to diseases like pneumonia, appendicitis, tonsillitis, etc., without the persons whose diseases they are and through whom they came to be designated. As Michel

Foucault has stated, "the solid, visible body [is] the only way . . . in which one spatializes disease."[7] No disease can be conceptualized as a thing existing without a subject of instantiation (not object); if it had objectivity, it would be something, an independent existent.

The World Health Organization has provided its definition of types of disease in "Disease Types" that were first introduced by the Commission on Macroeconomics and Health

> using these disease Types as a proxy to identify the burden of disease as it relates to the income of the population it affects is an important element to enable the monitoring of resource flows.
> The definition of diseases into Types mixes a number of concepts together including the wealth of a country between rich and poor; the state of its development between developed and developing and most importantly a measure of the burden of diseases by the incidence of the disease within the population. The definitions themselves are combined such that:
> Type I diseases: are incident in both rich and poor countries, with large numbers of vulnerable populations in each.
> Type II diseases: are incident in both rich and poor countries, but with a substantial proportion of the cases in poor countries.
> Type III diseases: are those that are overwhelmingly or exclusively incident in developing countries.
> examples: cancer is a typical Type I disease, TB and AIDs being typical of Type II with Type III diseases broadly corresponding to the infectious tropical diseases such as leishmaniasis or malaria.[8]

# Disease and wellness

Among the definitions of health that we recognize are those that characterize it as well-being, which is a variable state of flourishing, being happy, functioning adequately, and in control of practically all dimensions of value in one's life. The boundary line between being well and being unwell is mostly nebulous and will continue to be so, because the functional nature of these conditions doesn't render them subject to quantitative parameters and mutual exclusiveness that allow us to draw lines of demarcation between them; they defy the principles of excluded middle (either p or not-p is true; there can't be a middle truth) and non-contradiction (never p and not-p at the same time). Therefore it is not always the case that anyone is either well or unwell, but not both simultaneously; health is much more than the absence of disease. A feeling of wellness does not necessarily exclude the presence of disease, nor does a feeling of illness entail the presence of disease. One may also be physically well, but mentally ill or vice versa, while disease in one organ (e.g., pneumonia) can affect other organ systems causing general illness, or else limit its effects to its circumscribed function (e.g., deafness).

Wellness is expressed in homeostasis, the lack of which is a clear indication that the person is unwell and undergoing a threat to his existence, and homeostasis is assessed by biochemical parameters that often have a wide range and fuzzy boundaries. Kazem Sadegh-Zadeh calls illness and wellness "particular fuzzy states of health"[9] or stages along the scale of health that defy bivalent characterization, but overlap to varying degrees, and thereby admit of gradations, in conformity with the fuzzy boundary between them. We do not believe that this is the case with disease; it is either present ontologically or absent, although the doctor may be unsure about its functional and epistemic categorization, and although it may manifest differently, to varying degrees in different people or in the same person in its different stages. In the latter case it displays intrinsic fuzziness only that is shown by the commonality of symptoms (fever, aches, pains, etc.) and signs (lymphadenopathy, hepatomegaly, tachycardia, etc.) in diseases with different etiologies, pathogeneses, and clinical presentations, which fall under distinct ontological classifications. Sadegh-Zadeh's arguments lead him to conclude that illness and wellness admit of contradictory properties like the impossible objects of Alexius von Meinong (see page 254), from which he concludes that the different branches of Medicine have logics that differ from traditional logic and from one another; e.g., anatomy, by its nature, is not subject to probabilities, which are the hallmark of diagnostic medicine, because variations in anatomical structure, as *faits accomplis,* have passed the points of susceptibility to prediction and probabilistic outcomes to which management of disease is still amenable. They already exist as such; they are fixed.

The need to want to carve a distinction between wellness and disease is not as crucial as the need to recognize those covert forms of disease that merge imperceptibly into the overt, advanced, and more life-threatening forms such as happens with cancers and renal failure from silent renal damage caused by hypertension. Hence there is a need for defining disease in its molecular form before it manifests grossly in functional failure of organ systems. To achieve this end new categories of disease have to be created strictly on the merits of technological parameters, e.g., molecular, pre-clinical diabetes has to be precisely distinguished from gestational diabetes, overt type 1 or type 2 diabetes, and transient hyperglycemia induced by steroid use or physical stress. Although only technological probes can differentiate and define pre-clinically covert diabetes, all factors relevant to the patient's state in his milieu have to be addressed in the diagnosis and treatment of these forms of the disease, if the patient as person is to overcome the threat to his continued existence posed by them.

The hierarchy from the molecular through the cellular, tissue, organ, system, and organismic level may lend itself to arbitrary and convenient dichotomization into either/or, but it is a continuum that constitutes the whole individual as one person with emergent functional properties of mind, spirit or soul who transcends the microtome and the test tube. As Sadegh-Zadeh observed, where one ends and the other begins may be so fuzzy that the same condition can be regarded as disease or not, depending on the temporal and cultural affiliations of the persons concerned, e.g., homosexuality and masturbation, previously thought to be forms of perversion, are now accepted as variations on the statistical norm of behaviour. Perhaps the situation presents the same problem as that of when baldness sets in: after the loss of 0.1%, 1%, 10%, or 30% of someone's scalp hair. These fuzzy boundary lines among these states militate against the universal over the individualized approach to diagnosis and treatment of disease and illness adopted by biomedical physicians.

Now, to say that health is the absence of disease, sickness, and illness, is to put the cart before the horse, because it is inconceivable that disease antedated health in the ontogeny and the phylogeny of the human organism, otherwise phylogenetic development would have terminated before it took off; but to say that disease is the absence of health makes more sense from the ontological perspective, bearing in mind that health is as polysemous a concept as are its opposites: disease, sickness, and illness. On the flip side, the problem with the definition of health as the absence of diseases is that it omits the patient's total well-being and equates health with everything else in the world besides disease: if x (health) is not-p (not-disease), and y (lunch) is not-p (not-disease), then y (lunch) could be x (health), because both are defined as not-p. Ordinarily we do not make undue efforts to define health precisely, because it is the accepted norm, and thus we do not expect it to generate any concerns; so there is nothing to do about it and nothing for a healthy person to do besides ensuring that he maintains that state. We don't usually fret about why a normal situation is normal, but about why and how an abnormal situation deviates from normal, and what we can do to restore it to normality. In the case of disease there is need for precise definition so that those parameters which are deranged can be corrected and appropriate treatment instituted to restore normal function in the patient, which is always the ideal target.

Lester King defines the healthy state as "the state of well-being conforming to the ideals of the prevailing culture, or to the statistical norm."[10] This definition and those given by Boorse and Scadding, seem to converge on "*normal functional ability*: the readiness of each internal part to perform all its normal functions on typical occasions with at least typical efficiency"[11] for the person in comparison with his reference class; i.e., functioning above a certain minimum level in carrying out physiological goals, in conformity with his species form, as far as our present knowledge of form and function extends. But it differs from those of Kenneth Richman (below) and Nordenfelt who stress the second-order ability to realize vital goals. It is noteworthy also that some diseases and deformities don't affect ability to function, and that extremes of function may be unhealthy; e.g., hypo- or hyper-insulinism. The introduction of a goal or teleology into the definition of health and disease serves to avoid the circularity of defining one in terms of the other, i.e., health is the absence of disease. Hence the individual's good health is determined largely by his appropriate individual (versus group) vital goals in his life and the possibility of attainment of those goals. But many people are still healthy even when they fail to achieve some of their goals in life.

Accordingly, Richman defines health as a state in which an individual is "able to reach or strive for a consistent set of goals actually aimed at by [him]."[12] Good health therefore amounts to the positive balance between the good and the bad in a person's life, since no life can be all good or all bad. If what is good or positive in a person's life exceeds what is bad or negative, a positive balance remains, and that amounts to well-being, which can also include good health. The opposite of well-being, which comprises sickness and physical or mental ineptitude, is the abnormal, because a human organism in that state cannot function optimally, even though some so-called normal humans also fail to function optimally in what is presumed to be the normal state for all persons.

## Naturalism, Normativism, State description

The overlay of a value criterion in the definition of disease and health highlights the question whether diseases are natural kinds that exist independently in the world, without need for their reification, but with a decided need for eradication, or else that they represent objectively specifiable qualitative and quantitative dysfunctions of the organism that they affect, also raising the urgent need for corrective intervention. Most people would argue that classifying a bodily sate as a disease involves both describing and evaluating successive states of an organism as deviations from the normal form and function of its organ systems that result in dysfunction of the entire organism versus only describing a natural, eradicable state attached to that organism; and this circumstance alone imparts a normative character to disease. For instance, people with Progeria syndrome do not only look old, which is acceptable in chronological, natural old age; they are young but prematurely old and disadvantaged as a result of disease in a way that we would not attach the same label to naturally old people. We seek to correct the genetic derangement that causes Progeria while

144

we do not feel pushed to do the same for natural old age.

As we saw in the preceding definitions of disease and health, Naturalism (Descriptivism) maintains that disease is a purely factual and scientific concept concerned with statistical deviations of bodily function from the norm and is therefore devoid of the value connotations that accompany the use of the concept words "well-being" and "welfare", even as they use words like "function" and "norm" that imply a process of evaluation. In support of this stand, Boorse states that the mere recognition of deviation from species biological design (disease) is a purely scientific process that is non-evaluative, because it does not call for value-laden corrective action, e.g., a stenotic heart valve impedes cardiac function mechanically, causing heart disease; and the eradication of the disease is also mechanical, viz., substitution of an artificial valve. However, normality as conceived by naturalists is not a necessary condition for health, because unhealthy conditions like tooth decay, which is common, may be statistically normal while healthy conditions, like the absence of teeth may be statistically abnormal. Normal function of organs refers to the statistically typical contribution that they make to the fitness of the organism as favoured by natural selection; their value role in its healthy or diseased state, e.g., the already noted function of the heart to pump blood efficiently to all parts of the body, is supposedly irrelevant. This function has its origin in the phylogenetic and ontogenetic development of the heart from a single tube to a multi-chambered structure (see diagrams on page 356) in pace with the increasingly complex anatomy and physiology of the higher forms of life in their struggle for survival, and this is not a value-laden process. The question raised by this viewpoint is whether biology is an authentic index of fitness as the goal of any organisms.

Another question is whether this line of reasoning can claim to be value-free in the face of a struggle for survival in an environment that is changing rapidly, not naturally, but artificially in keeping with our ever changing value systems and our ability to transform the disadvantages of nature into advantages by gene therapy and other means of preserving and adding value to lives that would otherwise be eliminated by dysfunctional organs, e.g., transplantation of normal organs and implantation of prostheses. Proponents of what are styled propensity functions believe that some organs that previously served the survival function for which they were designed quite adequately have lagged behind the rapid evolutionary changes induced by our modern environments and can now no longer serve those functions, e.g., the sickle cell gene is a protective boon against malaria in regions where both conditions are prevalent, but a curse where malaria has not followed the migration patterns of persons with the gene, so that they now no longer need its protection and suffer only from its deleterious effects. (See chapter 21, Darwinian Medicine).

Normativists or holists hold the view that illness as the ultimate unpleasant manifestation of disease is not value-neutral, but always value laden for the patient, and thereby imposes a connotation of value on all our talk about disease itself. They maintain that disease cannot be defined solely in terms of biological abnormality to the exclusion of what ought or ought not to be by the standards of a particular society and culture at a particular time in a particular environment. Our perspectives on health and disease reflect the relative values of what our societies desire and what they don't as contributors to their continuing survival. Those conditions that are labelled as diseases because we disvalue them may later be declassified as such, because our values have changed, e.g., change in our values regarding slavery and those who try to escape from it has made drapetomania a non-disease. So disease does not only reflect disvalues, and it is not the case that all undesirable states are stages in the disease process, e.g., alcoholism may be odious, but it is not a disease in the biomedical sense, even though, according to Boorse's definition of disease, it compromises some physiological functions. This concept is compatible with the concept that not all diseases cause illness and not all illnesses are caused by disease. On the other hand, some conditions that satisfy this definition, like the prowess of the super athlete that we mentioned before, are not considered to be diseases. So disease entails much more than a disvalued state.

A third group styled Hybrid theorists, like the constructivists that we discussed before, espouse a combination of both theories, maintaining that disease is a biologically caused state of organ dysfunction and harm or disvalue to the person. According to them, health is good for the individual, because in health his organs are functioning normally to ensure his survival, while disease is bad for the same individual, because it is bad for the functioning of his organs and hence his survival. As we saw in the discussion on pages 58-59, constructivists claim that naturalists judge any bodily process that deviates from its natural form to be a disease that is causing harm to its bearer, even if it is still maintaining good objective function. For them, harm results from derangement in biological processes; mere dysfunction is not harmful. They claim that what we call disease is what we disvalue before we look for an associated or causal biological process that we then label as abnormal. So how we classify diseases is determined by our moral and social values, more than by what occurs in nature; i.e., diseases are not ontological or natural types that can be identified as existing apart from the persons who are diseased. But this viewpoint is negated by such conditions as misanthropy, which is socially condemnable but cannot be labelled a disease that needs medical treatment.

A fourth theory that enables us to help the patient achieve his valued goals of good health and function advocates embracing state descriptions and normative claims. As Marc Ereshefsky has stated,

> Instead of trying to find the correct definitions of 'health' and 'disease' we should explicitly talk about the considerations that are central in medical discussions, namely state descriptions (descriptions of physiological or psychological states) and normative claims (claims about what states we value or disvalue). . . . After providing a state description and deciding whether the state in question is desirable or not, there is a sociological question concerning which aspect of society treats (successfully or not) such states. [13]

The languages of physiology and psychology in referring to claims about levels of hematocrit and Blood Urea Nitrogen (BUN), or states of excitement and anxiety, do not involve naturalist, normative, or functional claims; they refer only to states of affairs as they exist. Whether we value or disvalue these states depends on the subsequent additional normative claims that we make about these descriptions, e.g., a hematocrit of 20% is just that, 20%, but it is bad, because it reflects a deficiency of oxygen carrying erythrocytes, which results in strained myocardial function and can result in acute heart failure. On the other hand, the language of normative claims expresses the specific value claims of these physiological and psychological states *ab initio*, e.g., deafness is a physiological state, but differences of opinion already exist as to its value or disvalue right from its onset. It need not be labelled first as a mere state of affairs or as a disease in naturalist terms, and then evaluated by its demerits and disvalue to persons to assert normative claims about it; those with intact hearing assign a negative connotation to its fabric. The same could be said about obesity, which is a state of being and not a disease in naturalist or normative terms, but can be evaluated negatively, because of its many adverse side-effects on the hearts and livers of obese persons.

The pragmatic question is not whether deafness and obesity are diseases on the basis of the biological values that we hold, but whether their effects are desirable or not, whether they can be compensated for or reversed, and to what degree that can be done. This approach is believed to eliminate the value bias that is inherent in naturalist and normativist approaches to medical problems, which assume that health and disease are natural states with a biological basis, i.e., nature and biological evolution determine which states are normal and which are abnormal, and therefore which states are diseases and which are not. If diseases were natural kinds, we would not be able to delineate and name new disease on the basis of arbitrary and convenient categorization of symptoms, signs, and laboratory criteria, because nature would have laid down the strict parameters to be followed in delineating those pre-ordained diseases. On the other hand, there are very few diseases that are man made, but many that are waiting to be discovered as our techniques and observations become more refined to be able to pick them up readily.

From the mental health point of view, naturalists maintain that mental disorders are deviations from the normal, natural design in human mental functioning that is independent of the values or norms of the society and its times. We have already seen that normativists hold the view that illnesses are a reflection of cultural norms; hence the jettisoning of mental diseases of the past, like drapetomania and homosexuality. In fact, some psychiatrists, like Szasz, have said that mental diseases are a myth, since mind is not an organ of the body and disease to him means bodily disease affecting only the body and marked by physical criteria. His view of mental disease is strictly constructivist, while his view of physical disease is objectivist, as in the demonstrated role of genetics in the etiology of schizophrenia. Like Boorse who holds an objectivist concept of disease, he also believes that diseases are somatic entities and that mental diseases, being psychological and lacking neuropathological and neuro-chemical attributes, are also not diseases of the brain but problems in living due to deviation in behaviour from established "psychosocial, ethical, or legal norms. . . . Mental illness exists or is 'real' in exactly the same sense in which witches existed or were 'real'."[14] That may be true for some aberrant mental states, but not for organic brain syndromes with mental symptoms. They are as real as witches are airy.

# Genetic Disease

The concept of genetic disease adds another perspective of conflict between the concepts of naturalist or ontological and value or normative characters of disease. If the physical structure that causes the genetic disorder (the gene) can be isolated, it can also be replaced or amended physically to avert its ill effects, in as much as microbes can be eliminated before or at the onset of infectious diseases, implying that the disorder or disease is an existent entity and not an expression of a value concept. But to think so is to conflate the cause of the disease with the disease process itself. The challenge is also to view genetic disorders as normative or having intrinsic value, because of their capability to predispose the individual to organic and mental dysfunctions, whereas they are the result of statistically random distributions of particulate genes without any teleological intent. The appearance of disease in this abnormal baseline context of gene distribution, depending as it does on a host of environmental factors that reflect the fit of the genetically abnormal organism to its environment, will alone confer a normative context on this statistically

148

naturalist disorder. Disorder in nucleotide sequences by itself does not constitute disease or illness, nor is it a value concern.

However, the genetic constitution sets the person up for one or other disease, depending on the degree of unfavourable changes in the standard functioning of the cellular elements of the person as caused by the genetic disorder. For example, the presence of a Philadelphia chromosome spells predisposition to a particular type of leukaemia (chronic myelogenous), but it does not mean that the person already has leukaemia. Although it sets the stage for the disease, it is not the disease. Nevertheless, apparently healthy persons known to harbour unfavourable genetic variations, like the Huntington gene and the gene for colon cancer, are stigmatized by that very fact of being in the preclinical or pre-symptomatic stages of some of those devastating disorders while those couples who carry similar deleterious recessive genes live in constant fear of bearing homozygous progeny with full blown clinical disease and a future life of hardship; e.g., Sickle Cell Disease. The adverse socio-economic implications of these genetic deviations, such as potential uninsurability and chance of steady employability, and the climate of uncertainty among non-affected and unwittingly affected relatives, are self-evident. On the other hand, lack of knowledge about a person's genetic status, which is preferred by many people in these pedigrees, does not expose them to the discrimination likely to be suffered by those whose status is common knowledge.

Together with questions of pre-empting the births of individuals who may be severely affected by genetic disorders and who may perpetuate the disabilities in their offspring and the community at large, these issues raise thorny ethical problems for the doctor. Should she divulge this information or not; what if she is required by law to do so and violate the patient's autonomy? What if the interests of family members are compromised by the non-disclosure of a deleterious gene that is sure to be passed on to them; does not utilitarianism subject personal to collective interests? Where does the doctor's duty lie if we follow deontology? perhaps she should follow the four principles discussed in chapter 17: Autonomy, Beneficence, Non-maleficence, and Justice.

Consequentialism, justice, and non-maleficence demand that family members be tested for devastating genetic diseases, e.g., Sickle Cell Disease and untreatable ones, e.g., Huntington Disease. This is different from population screening for untreatable disease, which expends resources pointlessly, because the yield of positives is meagre for the amount of effort expended. The obligation to inform family members should be weighed against the patient's autonomy in the face of potential discrimination in the social sphere.

According to the genetic disease concept, apparently normal

persons who harbour certain genes known to cause dysfunctions in later life but are now deemed not to be sick, are, in fact, diseased even long before they manifest the symptoms of those diseases. This situation leaves very few persons in the general population without disease as more and more genetic disorders are found to underlie many disease processes in the reductionist view, a view that is stigmatizing of persons and their families, whilst it also encourages a practice that vitiates their privacy in the interests of illuminating the medical problems of their kith and kin by exposing their disadvantageous genetic connections. Evidence for this opinion abounds in the genetic accounts of breast, colon, and prostate cancers, inborn errors of metabolism, diabetes mellitus, and virtually every organ system of the human body where efforts are made to forestall the effects of defective genes by eliminating their anticipated effects prospectively in affected symptom-free persons, or by preventing their perpetuation somehow, without going so far as to eliminate their carriers. The sequencing of families will be a revolutionary tool for medicine and human genetics in the future.

In attempting to minimize the number of persons who can be declared diseased by virtue of harbouring clinically dormant abnormal genes, Eric Juengst suggests that genes should be considered diagnostic of genetic disease only when their use is restricted to confirming the clinical diagnosis of a disease process that is known to have a genetic basis; e.g., "mutation analysis to diagnose Fragile X syndrome in developmentally delayed children."[15] Other genes will fall into the prognostic category if they can be "used to forecast the emergence of a clinical health problem with a high degree of certainty",[16] e.g., Huntington disease for which, in the absence of curative treatment, the healthy adult subject can only plan his future, including his reproductive choices, but which would only impose a variety of emotional burdens on a younger person for whom those options do not yet exist and may never exist for various reasons: premature death, or, in the case of his prospective offspring, celibacy, homosexuality, etc. Some other genes that identify potential clinical problems, if no intervention is undertaken, he calls predictive, e.g., the gene for Phenylketonuria (PKU) whose adverse effects can be avoided with the early institution of a phenylalanine-free diet; the gene for Glucose-6-Phosphate Dehydrogenase Deficiency whose hemolytic effects are expressed only after exposure of some affected subjects to triggers like fava beans, sulpha drugs, aspirin, etc., proving that it is not possible to lay down parameters for a definite species design as Boorse maintains. He calls those that confer susceptibility to diseases that actualize only in particular environments or with particular lifestyles prophylactic, e.g., emphysema in persons who carry the gene of alpha-1-antitrypsin deficiency, due to constant exposure to tobacco smoke filled environments. He also appears to express reservation in defining probabilistic gene testing and genetic profiling used in assessing population health risks as "a loose empirical association between a particular mutation and an increased incidence of a given health problem."[17] Needless to say, such ventures are open to the abuse of unauthorized and, therefore, unethical testing for uncharitable purposes, which are to be condemned.

Notably, gene therapy is designed to introduce genetic material into cells to compensate for abnormal genes or to make a beneficial protein. If a mutated gene causes a necessary protein to be faulty or missing, gene therapy may be able to introduce a normal copy of the gene to restore the function of the protein. Certain viruses are often used as vectors because they can deliver the new gene by infecting the cell. Viruses are modified so that they can't cause disease and other serious health risks when used in people, e.g., inflammation, toxicity, and cancer. These are laudable efforts to reduce the burden of genetic disease on the population that bears a heavier burden from other conditions like infectious diseases, trauma, and the unnecessary collateral damage inflicted on them by senseless warmongers.

## The biomedical concept of disease

Disease, to reiterate, is an adverse process that is devoid of specifiable onset and distinctly characterizing lines of demarcation, and which involves pathophysiological changes at molecular, organ, and system levels of the person. Disease is not static, but it evolves with time, reaching a climax after which it resolves, or else progressively gets the better of the patient until it causes his demise. Progressive changes take place in three phases: first without identifiable signs and symptoms in the preclinical induction period when causative factors are operative before the initiation of the disease, and then in the latent phase when the disease is incubating, which precedes the third phase of detection of the disease by its clinical manifestations. This dynamism of the disease process can be appreciated in the cycle of infectious disease where a pathogenic micro-organism first inhabits a host or reservoir from which it exits to spread to other susceptible hosts, repeating the cycle many times until it is broken by one or more of the following: a hostile host environment, failure of its means of exiting the original host for its subsequent transmission to other hosts, and failure to find a new host, latch on to him, and flourish in him, because his defence mechanisms overwhelm the micro-organism's penetration abilities. The dynamism is consistent with the pronouncement of Heraclitus that nothing endures but change. Disease presents as a state of constant change, so that a patient's condition is never the same at time $t_1$ as it is later at time $t_2$. This situation of flux necessitates the frequent re-assessment of patients to document clinical changes resulting from their underlying diseases and to adjust their treatment to these changing conditions; otherwise, the doctor misses the boat.

Illness, as we will have already learnt, is a personal, internal, and undesirable lived feeling of being unwell and dysfunctional that should not be confused with its biomedical causes, as in the examples of asymptomatic hypertension and in situ cancer where the affected patient who does not feel ill now begins to feel so, because the doctor has told him that he has a disease that can proceed to certain undesirable consequences. The general public comes to know that indisposition accompanied by certain symptoms constitutes a particular disease, because the science and technology of medicine has adumbrated the clinical spectrum of disease comprising those symptoms and signs and assigned a name to it. If the meaning of disease were to be left to differing cultural definitions, medical science would not be able to speak authoritatively about the role of anatomy, pathology, and other disciplines in the delineation of the concept of disease and in the international classification of diseases and their universal understanding and treatment. So the doctor is still the final authority and the expert on whether or not the illness felt by a particular patient is a disease with a biochemical basis and a predictable course, including its abortion with medical treatment, or only a feeling of indisposition with other causative factors that call for a different mode of treatment from that of biomedical disease.

The discovery of biomedical disease depends on the exposure of gross anatomical and histological changes, which cause correlative changes in the function of the affected organ and in the functioning of the organism as a whole. All these changes produce recognizable patterns, which can be unravelled with laboratory investigations, thereby making possible a rational approach to the treatment of the disease in persons who may not display the exact same elements of disease, but close enough to constitute a class in which prevention and treatment are similar. However, the introduction of the word "function" into this concept conveys a normative dimension that some Philosophers wish to deny, as we have already observed; and yet when this concept of disease is extended beyond the ambit of objective physical findings to include the subjective area of the psychological involvement of the patient, it leads to the recognition of the concept of illness. As has already been indicated, the qualitative state of illness defies quantification in the laboratory, because it is limited to personal feelings and carries a phenomenological flavour that is unique to each patient. It is a clinical state that cannot be shared among the members of the socio-cultural group.

This unique phenomenology makes it possible for someone

with or without a disease to complain of feeling ill without presenting us with any of the objective signs and measurable parameters of pure biomedical. It affirms the need for more than a mechanistic approach to the treatment of all patients, because their sickness involves more than disease as deranged anatomy and physiology with mechanical and deterministic parameters. It entails recognition by the doctor of the inaccessibility by others of illness or disease-as-lived by a person with his unique total constitution, reactions, and outlook. The suggestion is, therefore, that disease is not a thing with the kind of ontological status that permits it to inhabit the patient's body, thus allowing us to speak of its particularized transmission or carriage by him. If it has an ontological status, that status is the special reified one that we will consider in the section on ontology.

In his diseased state, the patient is unable to function normally, because of deviant changes in his physical, social, and psychological behaviour induced by disease. The effects of disease, therefore, are to frustrate our efforts to achieve the goals that we have set ourselves in life and to disvalue those pursuits by which we measure our moral worth; but it may also have effects on the integrity of the society in which as patients we function or malfunction. Ultimately, disease threatens our integrity and survival as individuals and as a species. Hence our natural reaction to it is to eliminate it and to prevent it by all means possible, via public health measures like proper sanitation, health education, universal vaccination; environmental control measures like elimination of acid rain, industrial pollution, global warming, and their ill effects on our well-being. The process entailed in this endeavour depends on our knowledge of the modes of operation of disease-causing entities like bacteria, fungi, and viruses, the effects or manifestations of disease as caused by the protective and morbid reactions of the victim's physiological systems, and the cure of disease as carried out by the actions of the doctor or by the natural processes initiated by the patient's own disease combating, immunological, and homeostatic mechanisms.

For those people who still become sick in spite of preventive measures, efforts have to be made to cure them of their diseases, and that entails knowing the causally related agents, how to neutralize their effects, and how to eliminate them by enhancing the natural defences of the body. Some protective reactions, like the phagocytic function of leukocytes, with the help of complement and the immune system, contribute to the survival of the patient while other reactions, like the autoimmune response, only serve to derange the patient's physiological mechanisms further and to hasten his demise by attacking his defence mechanisms. Treating a case of pneumococcal pneumonia requires the antibiotic to which the bacterium is sensitive; so all such patients are treated the same. But, as we have repeatedly observed, there are personal and social factors that should be taken into account if the whole person is to recover from the effects of the disease in his infected lungs, and these call for specialized approaches suited to the circumstances of each patient, over and above the classic and standard treatments. There is also the non-classic presentation of right lower lobe pneumonia with right upper quadrant abdominal pain mistaken for cholecystitis, which escapes the physician who loses sight of the whole patient in his diagnostic efforts. The same kind of error creeps in when parochial attitudes are adopted toward the demographics of disease. For instance, modern travel has made possible the universalization of diseases that were previously regional; malaria and zika are no longer confined to patients in circumscribed regions but are brought to Toronto by foreign tourists and returning locals.

As a sequel to our delineation of the concept of disease, legitimate questions arise: how far do someone's physical state, functional status, and laboratory data have to deviate from normal to be considered as constitutive of disease? When does the patient claim to have returned to normal health using the same parameters? What does the doctor mean when she declares someone cured of his disease as opposed to healed of his illness? As we saw in the discussion of normality, what is considered normal is not static, but may vary with the yardstick chosen to measure normality and with age, time, and place. In ancient Greece, for example, Epilepsy was considered a sacred disease, and those afflicted with it were thought to be holy and not sick, until Hippocrates dispelled the view. Mental illness was once considered to be due to possession of persons by demons, which is not sickness, and Diabetes Mellitus was thought at different times to be due to consumption of excess salt or alcohol. In our own age homosexuality, previously considered to be a disease, is now seen as a different way of life on par with heterosexuality.

In the end, the best operative concept of "disease" will depend largely on pragmatic considerations, which boil down to rational, targeted relief for those afflicted with it, its prevention, and its hoped for elimination. However, we should be aware of what Ray Moynihan, Iona Heath, and David Henry call selling sickness. They maintain that corporate construction of disease has replaced its social construction,, as we have been discussing, by labelling innocuous conditions like baldness, or non-lethal conditions like Irritable Bowel Syndrome and social problems as diseases, because "there's a lot of money to be made from telling healthy people they're sick"[18] and making lots of money selling unnecessary drugs to them at the their expense and at the expense and "viability of publicly funded universal health insurance systems."[19]

# 9

## Identifying and managing Disease

## Disease terminology

### Symptoms, signs, natural history

According to Wittgenstein, the meaning of a word is its use in language by people: patients and doctors. This favours idiosyncratic and parochial uses of the words 'disease' and 'illness'; but we need a standard usage so that we can all mean the same thing when we talk about these conditions. The technical meanings of these words, which are derived from the need for accuracy and specificity in talking about disease and illness, answer to this criterion better than their common usages.

To begin with, the concept of disease is firmly based on those of clinical symptoms and signs as its starting points. Without them there can be no talk of disease, or syndrome, for that matter. A symptom is defined as a new subjective feeling or experience of feeling unwell that often triggers a person's desire to consult a doctor, such as a severe headache, blurred vision, ringing in the ears, or acute abdominal pain. Some people believe that symptoms can be appreciated by persons other than those whose feelings and experiences they are, but that is hard to swallow. Symptoms prompt the doctor to elicit signs causally related to them; e.g., bloody cerebrospinal fluid associated with subarachnoid hemorrhage, choked discs (fundi) and visual field defects associated with brain tumour or subdural hemorrhage, Murphy's sign of acute cholecystitis, or Rovsing's, obturator, psoas, and McBurney's signs associated with acute appendicitis, and vomiting from many causes. A sign is thus an objective feature that underlies the symptom and may lead to the recognition of one or more types of abnormal conditions or diseases that are causally related to particular constellations of symptoms. On the other hand, uncharacteristic signs like the right shoulder pain of acute cholecystitis and the dysuria or diarrhea of acute appendicitis may challenge and trip the unwary diagnostician.

Disease as a process has a natural history that consists of its course uninterrupted by treatment. In the case of infectious disease, the course of this process depends on the ability of the infecting agent to survive or succumb in the environment provided by its host, whether it is conducive or hostile. The host's defence mechanisms, including its immune system, also trigger adaptive mechanisms in the invading agent that have the paradoxical effect of aiding its survival (see chapter 21). In non-infectious conditions the course of the disorder evolves spontaneously, e.g., anaplasia of cells, carcinoma in situ, or regional cancer progressing to metastatic cancer; or else the disorder stagnates at the in situ stage or regresses under the influence of the patient's immune system or in response to treatment. The patient's favourable prognosis depends a lot on the point in the natural history of the disease at which it is detected with accurate tests and interruption of its subsequent progress with treatment.

## Screening tests

Screening tests thus become necessary for the early identification of disease generally, and for the isolation of diseases that have rapidly serious consequences, as indicated by their natural histories. The purpose is to avert their unfavourable effects by identifying the victims of these diseases and by beginning to treat them in the preclinical or pre-symptomatic state, thus preventing the pathological changes that result in clinically adverse manifestations, e.g., neonatal screening for Galactosemia and PKU permits early and simple institution of dietary restrictions of galactose and phenylalanine before they begin to cause brain damage and severe mental retardation. An adaptation of these tests to the identification of asymptomatic persons in the community is styled case finding. In both situations, however, both positive and negative tests for abnormal conditions must be interpreted with caution, so that affected subjects are not missed where a condition is rare and unaffected subjects are included where it is prevalent.

So, in order to refine the selectivity of the screening and diagnostic tests, sensitivity and specificity of tests, as reciprocal states reflecting the accuracy with which a test predicts the reliability of its positive and negative values, have been devised. Nevertheless, it remains true that there will always be a grey zone of overlap around the line of demarcation in which any discriminatory test will be unable to separate affected (diseased) from unaffected (well) persons; hence there will be true positive and negative test results, and false positive and negative test results.

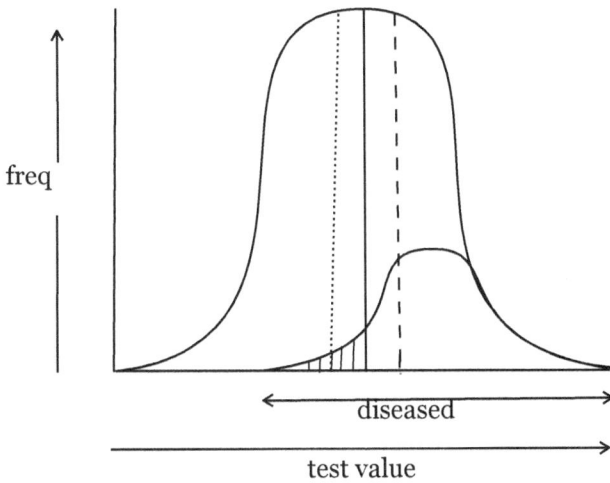

freq=population frequency of normal and diseased individuals

vertical broken line in graph=cut off value for diseased and normal individuals as defined

zone below big bell, outside little bell=true negatives; zone above little bell bounded by big bell and broken line=false positives for normals; zone under little bell excluding hatched area=true positives for disease; zone in hatched area=false negatives for diseased, i.e., false normals for diseased.

In the above diagram used for matching test values for normal and diseased persons against their distribution frequencies, moving the vertical line in the centre of the above graph further into the disease area (broken line) improves the sensitivity of the test; it picks up more false positive cases among the well subjects. Moving it into the opposite area (dotted line) improves the specificity by picking up fewer false negative cases among diseased ones. In the context of disease, specificity is an index of unaffected persons who are thus identified by a negative test result, i.e., true negatives in the absence of disease (true negatives+false positives); sensitivity is an index of affected persons who are thus identified by a positive test result, i.e., true positives in the presence of disease (true positives+false negatives); affected ones who test positive have true positives; unaffected ones who test positive have false positives while affected ones who test negative have false negatives, and unaffected ones who test negative have true negatives. The utility of and need for highly sensitive tests becomes evident in conditions that require urgent treatment, as in the cases of PKU and Galactosemia, because when they are negative they rule out the disease since they have relaxed criteria of inclusion. Their highly inclusive character makes them less specific, though. Highly specific tests, on the other hand, are needed

157

for conditions that expose patients to arduous diagnostic and treatment procedures, because their stringent criteria for inclusion rule in the disease in question and justify these arduous procedures. As screening tests in symptomatic patients with positive tests, sensitivity and specificity tests importantly demarcate affected from unaffected patients, but specificity tests are the crucial ones in population screening.

The preponderance of normal persons in the population, which predisposes to more false positive than true positive tests for a prevailing condition, e.g., HIV, necessitates a trade off between sensitivity and specificity or false positive and false negative at every point of demarcation. Receiver/response-operating characteristic (ROC) curves for each test are designed to circumvent these problems, and to reduce the influence of "chaos" on the measurement of both parameters.

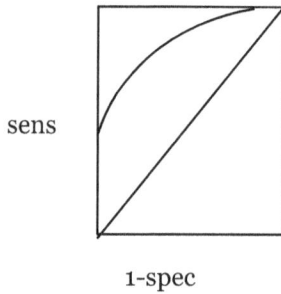

sens

1-spec

sens=sensitivity/true positive rate    spec=specificity/false positive rate

In the above ROC curve, the area under the curve represents the degree of accuracy of the test; the diagonal represents a very poor test with an area of 0.5, (50% specificity and 50% sensitivity), eg., WBC for bacterial infection; the curved line represents an excellent test with an area of 0.9. (almost 100% specificity and sensitivity), e.g., procalcitonin test for same infection. Small false positive and negative rates indicate a good test; low false positives and high false negatives indicate a bad test.

Another refining statistical tool consists of predictive values, which assess the effectiveness of the testing procedure in categorizing patients by their clinical presentation: those with positive predictive tests are truly affected, and those with negative predictive tests truly unaffected. Positive predictive value implies probability of presence of disease when the test is positive; negative predictive value implies probability of no disease when the test is negative. However, low positive predictive values have been found to be problematic in cases of mammography for women under age 50 years, HIV screening, and prostate cancer screening with Prostate Specific Antigen (PSA), until the advent of measurement of free and protein bound PSA—excess free PSA was found to favour benign prostatic hypertrophy, while excess PSA bound to anti-chymotrypsin and alpha macroglobulin favours prostate cancer). Positive predictive values were found to be low when the prevalence of disease was low, while negative predictive values rose reciprocally when fewer people had disease. The practical significance of all these tests depends on their utility for screening or for diagnosis of clinically evident conditions where their positive predictive values are high.

# Classification of Diseases

Diseases that carry disvalue for persons in whom they are manifested, e.g., lysosomal storage diseases, deserve to be systematically classified for recognition, definition, study, treatment, and prevention more on a universal level than a parochial one. Some other conditions that bear the dubious label of 'disease', e.g., Bornholm disease, do not, however, call for intervention. Since variability obtains within the heterogeneous collection of all diseases, an all-inclusive classification becomes a serious difficulty. It may be too narrow and thereby exclude some diseases, or too wide and thereby include even non-diseases. Edmond Murphy lists what he considers the essential elements of a disease classification system as: natural (vs stereotypical), exhaustive, non-disjointed (limited by shared boundaries), simplicity, constructability (demonstrable), free of disciplinary conflict of objectives and subject to multidimensional analysis,[1]. Other methods of classification vary with the specialty interests of the medical personnel concerned, e.g., internists, surgeons, radiologists, pathologists, oncologists, physiologists, geneticists. Some others are based on age (Paediatrics, Geriatrics), clinical signs and symptoms (limping, headache, tremors), organ systems (Renal, Hematological, Orthopaedic), etiology (bacterial, nutritional, genetic, neoplastic), laboratory presentation (anemia: microcytic, macrocytic, hemolytic; proteinuria, hematuria, hyperammonemia, hyperbilirubinemia), or combinations of some of these categories. The ease with which most diseases seem to fall into groups like immunodeficiencies, collagen disorders, and allergies suggests that diseases are also natural kinds waiting to be discovered and classified according only to their grouped natural occurrences.

Ontologists of disease, as naturalists seeking only the bare facts about the internal constitution of existent groups of entities named diseases, regard diseases in that light. But the real scenario is that the tissues concerned react similarly to the same stimulus by virtue of their basic structure. Clinicians, on the other hand, are concerned with nosology (designation of disease) as a pragmatic, but arbitrary, invention grounded in normative concerns surrounding a patient's struggle for survival in an unfavourable physiological or social environment. The point of diagnosing a medulloblastoma is primarily to remove it and save the patient's life, not to provide the pathologist with a specimen to study; hence the preference for a classification of neoplasms that takes this fact into account over one that is concerned only with their causes or structure. Admittedly, knowing the cause will contribute to efforts at preventing the problem and also sometimes assist in planning treatment, but for the patient who already has the problem, the only useful measure is elimination of the neoplasm, not the theories of its causation.

In all cases, the tenets set out in the section on definitions should be applied for the sake of maintaining clarity and distinctness of diseases thus classified, although overlaps cannot be completely avoided, owing to the commonality of some of the defining features of the various diseases, e.g., disorders of chromosome 22 produce systemic morphological and functional states with different clinical presentations, viz., De George, Cat Eye, velocardiofacial, and Opitz syndromes. At the same time, diseases with several modes of presentation are classified as one disease on the basis of their etiology, e.g., rheumatic fever presenting as arthritis, carditis, neuropathy (chorea), dermatitis (erythema marginatum), while some diseases with similar clinical presentations are classified separately on the basis of their different etiologies, e.g., neuropathy of diabetes or Riley-Day syndrome, and arthritis of rheumatic fever, or gout, or trauma. In the end, all classification of disease is pragmatic, enabling us to deal successfully with the threat that it poses to our survival. Contrarily, psychiatric disorders are more amenable to conceptual classification, because they lack the biomedical handles by which most diseases are classified, relying more on observations of variable behavioural patterns and their responses to treatment.

# Diagnosis of disease

## Hypothesis

Before the doctor can cure or ameliorate the patient's disease, he has to diagnose it. Using the hypothetical-deductive method that we will discuss in Popper's method (chapter 13), she gathers information from the patient's personal story and demeanour, epidemiological facts, and her own preliminary observations to help her recognize a familiar pattern whereby she can formulate a hypothesis for the disease label that she will attach to the patient's condition after she has rationally considered, by deductive, inductive, and abductive methods, all the contextual evidence necessary for

160

reaching the correct diagnostic conclusion. Without putting the evidence in the context that will enable her to advance a hypothesis that will allow her to arrive at only one diagnosis, the doctor will dissipate her effort in chasing will-o'-the-wisps by entertaining several possible diagnoses on equal footing. As she logically follows her hunch, or as she goes fishing for rational supportive evidence, her selection of historical, clinical, laboratory and other tests will be systematically guided by her postulated hypothesis, and hence by her provisional diagnosis. She will avoid haphazard search for evidence.

In this process she will rely more on the method of forming explanatory hypotheses (abductive) that conform to known patterns of disease from the relevant data provided by the patient's historical facts and by her physical examination of him, i.e., inference to best explanation. This method of unraveling and labeling the patient's history, symptoms, and signs in search of an explanatory hypothesis differs from the deductive method where she is called upon to test and confirm the diagnostic label as applicable to a particular instance of an existing hypothesis about conditions like it, or the inductive method whereby she engages in perpetual accumulation of similar cases to increase the probability that an already entertained hypothesis applies to the case at hand.

Although deductive reasoning plays a role in the making of a diagnosis from prior evidence, the multiplicity of variables—genetic, physical, psychological, and socio-economic—that go into making it permit mostly an inductive conclusion based on the strength of accumulated circumstantial evidence, which is always less than 100%. So the diagnostic process cannot always meet the rigorous demand of passing the paralyzing test of falsifiability before a diagnosis can even get off the ground or be finally accepted. The more the amount of confirmatory evidence that can be adduced in support of a particular diagnosis, the higher the probability that the diagnosis will stand as the best possible guide to further action to improve the patient's condition, despite the few contradictory pieces of evidence that may often threaten to discredit it, and the more likely it is that the patient will receive needed treatment, which is the ultimate goal of all clinical diagnoses. Some diagnostic labels are non-controversial and only need fine tuning, e.g., Trisomy 21, 18, or 13, in the sense that the conditions picked out by them are immediately recognizable, although finer details of conditions like translocations must be obtained before genetic counselling can be initiated.

## Bayes's theorem and probability

The essential question in diagnosis and treatment relates to the statistical probability of the presence of a disease $d$, given symptom $s$, and taking into account the prevalence rate of that disease in the community and the likelihood of the occurrence of the symptom in people who do not have the disease. A high prevalence rate increases the probability while a high incidence of people who are not similarly affected lowers that probability, and "the strength of a physician's initial disposition to regard his patient as having a certain illness is directly proportional to the incidence of that disease in the population he serves."[2] As well, the statistical reliability of the laboratory tests used in confirming the presence of that disease and excluding other similar diseases will influence the probability of the presence of the disease. As we saw in the discussion of false positive and negative test results and predictive values of positive tests above, the problem of the probability of the correctness of the test compounds the problem of its reliability in truly reflecting the presence of disease, since the test is likely to show disease where it does not exist and fail to show it where it exists. All these considerations are tempered with the fact that symptoms and signs occur more in suggestive clusters than singly, which only helps to confound our statistical calculations.

Thomas Bayes's theorem tries to correct for these anomalies by facilitating computation of the actual probability of the presence of disease, given its probability as measured by the test. It states that

$$P(d/s) = \frac{P(s/d).P(d)}{P(s)}$$

$P(d/s)$ = prior or posterior probability that the patient has disease $d$, given that he has symptom $s$, or a positive test result. It depends on the incidence of the relevant disease in the population under consideration. The higher the incidence (prior probability), the higher the predictive value of a positive test, and the more likely it is that a patient presenting with symptoms and positive tests matching that disease will have the disease.

$P(s/d)$ = likelihood of the patient exhibiting symptom or positive test result $s$ given that he has disease $d$.

$P(d)$ = prior probability that the patient has disease $d$.

$P(s)$ = probability that the patient will exhibit symptom $s$ or positive test result whether or not he has disease $d$.

It is noteworthy that $P(d/s)$ is not the same as $P(s/d)$.

Exponents of the method argue thus: If 1% of children in a population of 10,000 with fever >40°C are septic, and can be identified by White BloodCell count >15,000 with an accuracy rate of 80% (true positive test), but an inaccuracy rate of 10% (false positive test), then a child with a positive test or WBC >15,000 has a sepsis probability of

$$P(d/s) = \frac{80 \times 1}{80 \times 1 + 10 \times 99} = 7.5\%,$$

as against the prior probability of sepsis of 1%. 7.5% is the posterior probability of sepsis, which is more accurate than the random 1% of any one of the children being septic when all the circumstances are taken into account. 80% is the conditional probability of sepsis given that the test is positive in the presence of sepsis, and 10% is the conditional probability of sepsis given that the test is negative in the presence of sepsis.

Unfortunately, this method of calculating probabilities and avoiding false beliefs in the face of new evidence has not yet become standard practice; it remains as an ideal epistemic pursuit used in Evidence-Based Medicine. Nevertheless, employment of the method helps to avoid the use of unnecessary tests carried out when the evidence is clearly against the presence of a particular disease that presents with some of the signs and symptoms that it shares with other clinical conditions.

## Semiotics

In the final analysis, medical diagnosis is based on semiotics (from sema, meaning sign), the philosophical theory of the arbitrary production of meaning from the interpretation of signs and symbols. Signs are the basis of verbal and non-verbal communication in the understanding of the world of people and things. According to one semiotic theory their nature is as varied as the words, images, objects, etc., that we invest with meanings, which they do not possess intrinsically, as long as we do so in context and without personal or other bias (see Humpty Dumpty, page 26). For instance a piece of paper of certain quality with certain specific inscriptions becomes money with a specified value by virtue of the meaning that we give it; otherwise it does not represent anything and lacks meaning and value. Signification is the relationship that exists between signifier, or the form which the sign takes, and signified, or the concept that it represents; e.g., the four letters e, e, r, and t, which have no semantic meaning in the English language constitute the linguistic sign for the object "tree" when arranged in the order t-r-e-e, which is a material instantiation of the abstract concept "tree".

No interpretation can, however, be regarded as conclusive in the face of accumulating confirmatory and contradictory evidence relating to it, and difficulties can arise in the interpretation of signs because of the ability of one sign to convey many meanings that are subject to misinterpretation deliberately or through ignorance. A rapid heart rate can mean, hemorrhage, fever, exercise, fright, or excitement, depending on the context in which it occurs. The doctor's knowledge and experience in knitting together divergent data from history, physical examination, and the laboratory will also guide her in the interpretation of this and other signs. Seeing the typical facies of an older child with Down syndrome immediately suggests delayed mental development or the possibility of congenital heart disease if he is cyanotic, while the same facies in a neonate with persistent bile-stained vomiting will suggest duodenal atresia or stenosis. The sight of a child with a normal facies who suddenly becomes distressed, cyanotic, and assumes a squatting position suggests Tetralogy of Fallot; one who suffers a sudden wheezing spell may have aspirated a foreign body, if he has been mouthing objects, or may be suffering an asthma attack, if he has been exposed to cat dander or grass pollen, in the absence of other possible causes of wheezing. The doctor learns to read her patient as she reads her textbook of signs and symptoms. All patients are always her best textbook; they provide data for textbooks.

In all the above cases, there is always a margin of error; but these signs still form the bases of presumptive diagnoses and act as guides to the doctor in her quest for more information to assist her in arriving at the final diagnosis that will point to the most rational treatment for her patient in his life situation—whether she will give remedial, radical, or palliative treatment to suit the circumstances of her patient's existence. No doctor will jump to the conclusion that hematemesis is a sign of peptic ulcer without looking at the symptom in context, because it could be due to swallowed blood from epistaxis or buccal laceration; no doctor will diagnose renal failure from the elevated BUN of a dehydrated patient without first rehydrating him and then repeating the test (and matching the BUN with other parameters like glomerular filtration rate and serum creatinine); and no doctor will be so naïve as to think that a normal test result means that all is well with her patient in the face of clinical evidence to the contrary and test results that change with time.

Selection of tests should, therefore, be guided by clinical data, and batteries of irrelevant tests should be discouraged. The doctor should always remember that a single laboratory test reflects the state of her patient at time $t_1$, which may be different at times $t_2$ $t_3$, and $t_4$. Each subsequent test is a cumulative expression of an increasing degree of probability in support of the entertained diagnosis, or a pointer away from that diagnosis. Tests are not information proxies; they are picked out by pointers from the narrative and the physical examination; they do not determine these procedures.

# Prevention of Disease

The concept of disease with its many etiological factors entails a multiplicity of reactions over time of the many facets of human physiology and psyche to the onslaught of a wide variety of stresses from the environment: toxins, physical factors, and microorganisms. The problem is to prevent these stress factors from taking their toll

164

on persons generally. In the case of micro-organisms as living units, there is an immediate clash between many aspects of their own physiology and that of the human host, which is another living unit with its own physiology. This contest issues in the clinical symptoms and signs that alert patient and doctor to the existence of disease and the need to eliminate it before it causes harm. But some of the subtle physical and stress factors that disrupt human physiology, do not permit that opportunity, e.g., undiagnosed hypertension where the dysfunctional state remains silent as an indicator of illness until it constitutes a threat to the person's well-being and survival. Many a patient with hypertension has become aware of this danger only when he presents with symptoms of uremia. At this late stage, he can only find relief in renal dialysis and a possible cure in renal transplantation. Prevention is no longer an option for him, but his plight serves as a useful measure for educating the rest of the people, and as a stimulus for instituting screening procedures to avoid the catastrophe that results from the disease. Other diseases like cancers also have this stealthy mode of attacking the human organism, often becoming manifest too late in the day for any efforts to save the patient's life. These conditions indicate the need for vigilance and periodic screening for disease in all members of the population.

On the other hand, the overt limitations imposed by diseases like angina and pneumonia on a patient's regular activities, and their effect on the integrity of his personhood by promptly hampering the realization of his goals, *ipso facto* proclaim their threatening presence and prompt the institution of immediate remedial measures and preventive action in the community at large against them where possible. The lessons learnt from the presentations of both categories of disease and others that behave like them have inspired efforts to educate the general public about them and to institute public health measures for their prevention. The target of this education is not so much the ones who are already afflicted, but those who are still free of symptoms and signs but may yet develop them if they do not adopt early preventive measures. Such are the measures undertaken by health care givers of the paediatric population against the remote effects of atherosclerotic disease, for which there is autopsy evidence of its childhood onset in the fatty streaking seen in some aortas at post-mortem for unrelated causes of death. Such also are the efforts of clinical geneticists to map the genealogy of breast and ovarian cancers through generations of certain pedigrees in their attempt to anticipate and treat these disorders successfully.

## Treatment of Disease

Prevention of disease is the ultimate goal of medical endeavour, but where it is not possible or timely, the next best course of action after diagnosis is treatment. Such treatment must necessarily be differential, because all diseases manifest differently in different people, and even responses to drugs vary within and across ethnic boundaries as a result of the statistical selection and distribution of genes related to the metabolism of these drugs in keeping with the different environments in which they have evolved through the ages. But these facts do not grant a license for discriminatory treatment on the basis of economic, social, gender, age, racial or other artificial and unfair biases. The question of the apportionment of treatment modalities in accordance with how much cost the state can bear is unfairly biased against what appear to be dispensable sections of the population based on their inability to afford costly private medical treatment. In all situations, however, if complete cure cannot be effected, then therapeutic modalities and rehabilitative measures aimed at the reasonable recovery of the patient with a capacity for independent existence should be the goal; no one should be written off, as so often happens in some societies. Examples abound of patients who are several years post-write-off after they were given the proper care and treatment and given the chance to be healthy enough to live for many more years.

Pharmacological methods of treatment do not all need to be active; there is also room for treatments that appear to be pharmacological, but are inactive or inert, and yet are administered with the express humane purpose of bringing about a change for the better in the patient's condition. These are placebos that have a positive psychophysiological effect on the patient's feeling of well-being and his symptoms, simply because he believes that they will benefit him, although they are known to have no specific pharmacological activity on the condition being treated. Hence their use in new drug trials as blind controls to test the efficacy of the new drugs to which the randomly selected recipients and testers are blinded to authenticate the responses of the recipients to both placebo and active drug administered to them. This concept, which underlines our ignorance of functional disorders and how they respond to purported corrective action with placebos, can be extended to the management of psychological illnesses like conversion and anxiety reactions and to non-pharmacological treatments of physical infirmities with equal effect.

Sometimes patients believe that the effects of active or non-placebo treatment are harmful to them, and they develop symptoms to match their beliefs, a condition called nocebo effect. This may be the kind of situation that truly justifies the use of a placebo, unlike its use in some less than ethical clinical trials where known effective treatments exist, but patients are subjected to placebo trials to test the efficacy of new drugs—a futile exercise that does nothing to justify the superior effectiveness of the new drug over older drugs, because it was not compared with them, but with an inert preparation, in contravention of suggested guidelines for the use of placebo such as those listed by Benjamin Freedman, which include the following:

(1)   lack of standard therapy;

(2)   standard therapy is placebo or is no better than placebo;

(3)   new evidence has called the therapeutic index of standard therapy into question;

(4)   cost and short supply have made optimal therapy unavailable.

Discussing clinical research, he goes on to state that "a state of genuine uncertainty on the part of the clinical investigator regarding the comparative therapeutic merits of each arm in a trial"[3] allows her to have a "treatment preference", but it is not the ethical *sine qua non* whereby the investigator can fulfill her moral obligation to offer all patients the treatment methodology that will provide them with better therapeutic effects. He suggests instead that a different concept of clinical equipoise be followed that justifies a preferred treatment only when the vast majority of experts in the medical establishment have reservations of uncertainty about available treatments. In that case the investigator and the clinician can offer their preferred treatment as freely as they would to their individual patients, because it is the best available treatment for the clinical condition under consideration as established by the findings of Randomized Controlled Trials and in keeping with the demands of the Hippocratic Oath. Nothing should tie them down to one or the other method of delivering treatment to their patients.

In both clinical practice and blinded treatment trials, there is no justification for subjecting patients to treatment with placebos or experimental drugs in place of known effective treatments that are accepted and approved by medical experts. As stipulated by the uncertainty principle,

> Physicians who are convinced that one treatment is better than another for a particular patient cannot ethically choose at random which treatment to give, they must do what they think best for the patient. . . . On the other hand, if the physician is uncertain about which treatment is best for a patient, offering the patient randomisation to equally preferred treatments is acceptable and does not violate his or her duty.[4]

# 10

# Disability

## Perspectival delineations and definitions

How society defines disability will determine its attitude toward people with disability. Essentialists define disability as the temporary or permanent, physical or cognitive state of functional limitation and dysfunction in the performance of a person's activities of daily living. Progressivists see it in the context of social injustice inflicted on the less generally abled of its fold by society. The focus is on what this person is or has become against the background of a statistically and biologically normal and unyielding world that fails to recognize the burden imposed on him by a situation that demands adjustments, sometimes radical, from him if he is to survive in the role that his disability has created for him. The selfish attitude of society toward the individual with limited ability is that he belongs to a socially and morally inferior class that can be brazenly discriminated against and thwarted in the achievement of its goals like some other minorities. The misguided and offensive assumption is that the contribution of people with disabilities to the economy is negligible, and so they do not deserve equal rights and privileges with the rest of the population.

As it is wont to do, society places the burden for his limited ability on him as the one who bears the disabling factors; all it does is pity him for the existential predicament in which he finds himself. His physical or mental state is solely responsible for his restricted ability to run, play soccer, or ride a bicycle with an amputated leg; to enjoy music, use the telephone, or modulate his voice, because he has lost some of his hearing or he never had any; to see his newborn baby, navigate his way around strange places, or gauge distances, because of defective vision (cataracts) or congenital blindness; or to learn or engage in conceptual thinking because of mental ineptitude. The blame needs to be shifted away from the limitations of the persons who are settled with an extra burden in their daily lives that is not of their making. The social and other remediable impediments constructed by able-bodied society for its able self and carelessly placed in the path of someone who does not meet the ideological standards imposed by it are the ones at fault.

Agreement on the definition of disability is plagued by problems

168

resulting from subjective and objective attitudes, concepts of normality and abnormality, biomedical (mechanistic) and bio-psychosocial views, and difficulties of deciding who has disability—only some persons with obvious impairments, every obese person who cannot fit in an airplane seat, or society as a whole. Society decides at its whim what it will label as a disability and on whom it will pin that label. Perhaps it is time to replace the label 'disability' with 'different ability', especially since no one person has 100% ability all round, and suppression of any one of the five senses sharpens the acuity of some of the other four, making ability a relative term with graded applicability from 1 to 10.

Jonathan Wolff has proposed a working definition that is similar to the one outlined above that locates the disability, not in society, but in the subject of impairment: "To be disabled is to be in a position where *one's internal resources are impaired* and do not provide one with adequate genuine opportunities for secure functioning, given the social and material structure in which one lives, together with the external resources at one's disposal."[1] (my emphases—to stress Wolff's perspective on the role of the impairment of the individual's subjective resources, which limits his ability to function in his social milieu, more so than on the limitations imposed on him by that social milieu because he functions differently from its other members in at least one respect and at most many respects). Hence, disability is seen not as the socially constructed result of interaction between a person's physical or mental shortcoming and his social environment, but as a limitation imposed by his biomedical condition on his access to what society has to offer to everyone alike, without any concessions.

The doctor should, however, be concerned with both conditions if she intends to assume a holistic approach to her patient, because she can, in some situations, not only agitate for change in the social and material structure in which her patient lives but also assist him to gain easier access to, and control of, his environment on par with everyone else. As well, she can also help to reduce the restrictive load imposed on him by the disabilities enumerated above, even if she cannot help to eliminate them, e.g., by professionally facilitating access to an artificial limb, hearing aid or cochlear implant, and corrective eye glasses or cataract surgery to meet society's demands on him. Nevertheless, the person with disability should not be left with the impression that he will be fit to participate fully in society only after he gets rid of his different ability, when it is the moral obligation of society to make his life easier for him in his present condition. Society should realize that many impairments that lead to disability can't be eliminated, much as the people who have them may also often wish that they did not have them because of the disrespect with which they are sometimes treated. Society should therefore spare no effort to recognize the impaired person's different ability in a positive way and to create opportunities and an atmosphere that will assure him that his humanity is equal to, and respected like, that of every other person who appears to be unimpaired. The fault for limiting his scope of operation lies not with him; it lies with our otherwise paternalistic society for not creating those opportunities for him but instilling a sense of nihility into him.

The problem of correctly conceptualizing disability is reflected in the many definitions of it from different sources as objective (and subjective) impairment. They vary from 1) "the functional limitation within the individual caused by physical, mental, or sensory impairment"[2] and 2) a non-ability to perform a basic action in a particular context for the full realization of Nordenfelt's "one or more of one's vital goals" cited on page 139, to 3) "disability results from the interaction between persons with impairments and attitudinal and environmental barriers that hinders their full and effective participation in society on an equal basis with others"[3], and 4)

> A person's functioning and disability is conceived as a dynamic interaction between health conditions (diseases, disorders, injuries, traumas, etc.) and contextual factors. As indicate above, Contextual Factors include both personal and environmental factors. . . . Environmental factors interact with all the components of functioning and disability. The basic construct of the Environmental Factors component is the facilitating or hindering impact of features of the physical, social and attitudinal world. . . . Environmental factors include the physical world and its features, the human-made physical world, other people in different relationships and roles, attitudes and values, social systems and services, and policies, rules and laws.[4]

Like Wolff's definition, the first two definitions objectify disability and restrict the responsibility for his disadvantaged state to the person who is rendered unable to find his niche in society by virtue of his physical limitations, when it should also impugn society for imposing limiting hardships on the person. The latter definitions rightly stress the role of society in the making of a disabled person with its prejudices and the barriers that it creates against someone who has a disability, a society that too often inculcates in him an attitude of negative self-regard and the inclination, in some cases, to want to use his disability as an excuse for failure. He may also deny it to escape isolation or to prove that he is equal to the able-bodied persons with whom he spends his time, because he lacks the support of equally disabled persons to fall back on and buttress him in his self-assertion as just another person with different abilities.

Entailed by these definitions is the need for the delineation of impairment, which is variously stated in the medical model as 1) "a perceived or actual feature in the person's body or functioning that may result in limitation or loss of activity or restricted participation of the person in society with a consequential difference of physiological and/or psychological experience of life"[5], and 2) "a biomedical condition that is presumed to be subtypical of the human race, without any assumptions about the disadvantages that might accrue to individuals who possess impairments."[6] This last definition is reminiscent of Boorse's definition of disease mentioned on page 137 as a reduction of one or more functional abilities below [species-] typical efficiency, which implies an abnormality in the form and hence the function of the individual by stressing

> "the impairment that the person lives with rather than their abilities.
> The model does not take into account issues such as: the role that
> barrier-free environmental access can make to the independence and
> human dignity of persons with disabilities; the human, social, political
> and economic rights of persons with disabilities; the rights of persons
> with disabilities to full inclusion and integration into mainstream
> society; and the abilities of persons with disabilities.[7]

The emphasis should rightly shift from impairment solely as a person's fixed attribute to include limitations imposed on him by society. The introduction of the value word 'abnormality' in the above reference in turn calls for its definition on the understanding that the concept of abnormality is parasitic on that of normality, as we have already discussed in chapter 5(a). Besides, what is normal may be relative to the cultural standards of different perceivers in diverse communities, like the differing attitudes toward illiteracy or dyslexia as disabilities in highly literate societies but not in illiterate ones.

Nevertheless handicaps, defined as "the loss or limitation of opportunities to take part in the normal life of the community on an equal level with others due physical or social barriers"[8], can be created and augmented by some societies for persons who are either not physically or intellectually impaired or disabled, or are minimally so without the impediments that their societies have thoughtlessly created for them. In that sense, most of us are variously handicapped by our societies whose abiding influence we cannot escape, although we are not simultaneously disabled by them, unless they have forced us to live in poverty, subjugation, and despair.

Societal barriers actively impose and augment disabilities on persons with handicaps and impairments more than they merely disable them by placing amenities and conveniences out of their reach, so that even if they could partially overcome their physical limitations to participate in the regular activities of their able-bodied peers they are disabled by the barriers to the point of frustration and hopelessness. They blame society for their predicament, while society exonerates itself by blaming them for their inability to avail themselves of existing amenities and conveniences. Ultimately, it appears that there is no way of separating the two perspectives, even though they have different phenomenologies; they appear to be two sides of the same coin that are often viewed differently by different persons with similar disabilities but different interests, activities and degrees of disability. So, society sees the person as disabled while the same individual sees society as disabling.

## Social implications of Disability

The biological basis of the person's disability resides in the alteration and acquired impairment affecting his body and its parts, or in the impairment with which he came into the world, both of which limit his functions and single him out for easy recognition, branding, discrimination, and oppression. Such a person may look and act abnormally, and be labelled as less adequate by the many others who consider themselves to be "normal" by their own standards. Society forces persons with disabilities into the dilemma of choosing between accepting their being-in-the-world as they are and trying to change their physical form to approximate the normal for the sake of societal acceptability. Because of the conflict between the self-regard of the person with disability and the condescending attitude of society toward him, the term 'disabled', which is used to stigmatize these persons who are disadvantaged by disabilities, should be applied to what society does to them (disabling them), not to denote their condition. As it has been said, a person with a disability only feels she is disabled when confronted with discrimination. So, philosophical considerations of disability should include its meaning to the person with disability *per se,* its implications for the authenticity of his conception of his personhood, his general health and its maintenance; its meaning to the items that together constitute the quality of his life; and its meaning to his full integration into the society of which he is a devalued member as a result of his disability, a disability that is presumed to make him contribute less to the economy of his society than he reaps out of it in some instances, but mainly when he is not given a fair chance.

The problem for society becomes more complex when the disability is cognitive and the disabled person does not have any understanding of his existential predicament or the ability to even attempt to alter it for the better. Then it is entirely up to his society to evaluate and direct his life as it sees fit, which sometimes happens without regard for his felt needs, desires, and limited aspirations that he can't always communicate intelligently and adequately to those who are looking after his interests and welfare, assuming that both parties know exactly or even vaguely what these are. These are the disabilities that are not as obvious to discern as those considered in

172

the previous paragraphs, because they originate from mental changes that are always difficult to quantify but are discovered only by behavioural observations or by special tests, e.g., mental deficiency, as defined by society's standard Intelligence Quotient tests, coupled with impaired cognitive performance or conduct disorder.

Other covert disabilities include 1) colour blindness in someone who gets by until he is challenged to make crucial distinctions like having to interpret different traffic light colours, thus posing a handicap on his ability to drive a car in the city; 2) genetic disorders that have a delayed clinical onset where the person moves from latent to overt disability, and where questions arise as to the ethics of avoiding the extreme difficulties of living associated with some of these predictable adverse conditions by aborting affected fetuses—a case of persons with able bodies and minds making life and death decisions for potential persons with anticipated disabilities, as some people will argue. This kind of predicament is often presented to prospective mothers who are found to be pregnant with some identified fetal congenital anomalies—trisomies, malformations of the neuraxis, etc., which portend immense emotional and financial burdens for the parents and for the society that will care for the individual who is disabled by any of these conditions. So, "instead of seeing the child with impairment as being in need of help and in terms of a social problem, the 'handicapped' infant is understood in terms of an individual problem"[9] that should be eliminated.

By their nature, all forms of disability reflect a crucial misfit between the person and the world in which he participates, a state in which his body and mind become objective liabilities in that world, not the subjective assets that we all take for granted until they fail to respond to the challenges of daily living. All of them call for radical changes in how society can genuinely attempt to help the affected person to fulfill his new and often permanent role if his disability cannot be reversed or his previous life-style recouped. Society is not called upon, as is often its practice, to consign all disabled persons to a lower rank in life on the grounds that they are less than perfect, thereby aggravating their emotional burdens, because there is no perfect human being; every one is impaired in some way. Instead, it should appreciate the fact that in most situations the person with disability never had any freedom to choose his lot, even as he is unable to contribute to the economic welfare of the society that is now obligated to look after his interests. He should be distinguished from the person who disabled himself by living recklessly and thereby rendered himself deserving of censure. In the end, the person with disability, although he has a problem, is not the problem; the problem is society's attitude toward him and his disability, which is not readily acceptable. The unwillingness and professed inability of some societies to provide easy access to public buildings and vehicles, adequate housing, educational, and training facilities, and tools to counterbalance lack of dexterity in gross and fine motor skills are some of the common shortcomings that society imposes on those with disabilities. There are many other less known hardships imposed on these citizens by society, such as long waits for wheel chairs and for access to appropriate health and other relevant care, and lack of regard for and relief of their state of poverty.

The foregoing are oppressive situations from which the great majority of persons with disabilities cannot emancipate themselves without the help of the society that tends to shun, devalue, and marginalize them, sometimes unintentionally, because they take their own able-bodied status for granted as the one to which the person with disability should aspire. Little wonder that in their heterogeneity the less abled are under-represented in professional and managerial positions where the standard is a generally homogeneous ableism in all its ramifications and with all its social advantages. For instance, it is relatively easy to mass produce any kinds of conveniences and amenities for the able public than it is to do the same additionally for each one of the few extra persons who have different kinds of lesser abilities where any one type of modification will not be suitable for all. So society ignores them ostensibly for shortage of funds.

Effective social change remains largely the domain of socio-political forces with a minor role for rehabilitative medical endeavour, since the change has to provide arenas of education and training that will prepare persons with disabilities to seize opportunities where they can successfully exercise their restricted abilities to their full potential. Progressive social change provides the kind of productive atmosphere into which one who is thrust by disability will not be placed at a disadvantage by lack of facilities but will be able to compensate for his deficiencies by using the already existing means created to bring into his grasp the ability to manipulate his environment for his enjoyment and survival, even if he may not have the economic means and social standing to make it on his own. The doctor must empathetically advocate for these changes, in spite of society's self-regard and the recalcitrance of the bureaucracy to countenance such change. As per Foucault, "The first task of the doctor is therefore political: the struggle . . . must begin with a war against bad government."[10] If she fails to do that she will be abandoning her patient to the ravages of the vulturous markets that devour everything in their path regardless of the questionable ethics of this kind of behaviour or concern for the unfortunate victim of their inhumane disregard of his suffering. She answers to a much higher morality than the pursuers of financial gain at all costs, and she should not allow herself to be intimidated and dissuaded from her duty by spurious remonstrations for her to stay

174

out of politics and concern herself with treating patients, as if she can treat them in a political and social vacuum when political and social influences are the ones that shape unsympathetic public policies.

# Cosequences for the disabled

Some able-bodied persons and some people with disabilities believe that disabilities make for a life that is less than maximally good, and which would have been better without the disability; but many persons who live with disabilities think that if society could accord them the respect of always remembering to provide for them also when they are providing amenities and conveniences for the non-disabled, the apparent disvalue of their disabilities would largely disappear. Many other people would soon realize that the holistic quality of the life made possible by these provisions to some disabled persons can be far better than that of some "able" persons who claim entitlement to these provisions but fail to use them sensibly. For instance, someone with paralytic poliomyelitis or Down syndrome can achieve much more from getting the help that he needs, and consequently be happier with his life than someone else who is able-bodied but perpetually under the influence of drugs, whether these are liquor or hard drugs, regardless of whether the drug addict trudges through many more years of life than those that the disabled person lives through. He adds on more years of misery and oblivion to his life than the short years of happiness enjoyed by the person with disability within the circumference of those things that matter most to him. The standard here is not the goals of everyone else, but those of the person with disability in relation to his being-in-the-world as exemplified by the keen senses of hearing and touch possessed by blind people, which exceed those of sighted persons and make for a more sensitive receptivity to external sensory stimuli. So it is arguable that the life of the person with disability could have been better if he did not have the disability, even if his present state exposes him to far fewer opportunities to encounter many more gratifying situations in life. In the end, even the lives of those without disability could have been better in some notable ways.

So disability is not always an impediment to the good life, regardless of how we choose to characterize the good life. But it can diminish the worth of the person with disability in the eyes of those who mistakenly regard physically able persons as the only ones who can adequately meet the production needs of their society. Moreover, the pity, patronization, malicious paternalism, and disdain with which some of the more able-bodied members of society tend to regard the disabled ones is more fitting of pity and contempt for them than those for whom these attitudes are reserved, because our moral obligation to them is not lessened but only made different. We learn such lessons from some of our patients who are afflicted with conditions like Down syndrome and Autistic Spectrum Disorder who also happen to be the most affectionate in our practices, like the two who were in my practice. They render dubious the questioning of the wisdom of permitting the survival of fetuses or potential children with different types of recognizable prenatal indications of disability in later life; instead, they teach respect for the dignity of human life in all its forms where there is even the minimal residuum of cognition.

## Philosophical entailments

The fundamental philosophical question about disability concerns its phenomenology, what it feels like for the person with disability to have a disability that sets him apart from the "normal" crowd, resulting in the attitudes that we have already discussed. To the person born with a disability the answer is simple, because that is his way of being—until he is made painfully aware of it by able society—in the same way that the able person does not have to wonder about what it is like to be his able self until he can no longer do the things that he was accustomed to doing. The person who was born blind does not miss sight, because he never experienced it, as much as many persons who have appeared on television screens with disabilities that would overwhelm able-bodied persons execute actions with their feet only with a dexterity that many of the latter would have difficulty executing with their hands, or they achieve feats like climbing mountain peaks solely on their hands when their able-bodied counterparts fail to achieve those feats with their feet and hands. Phenomenology of disability has real meaning only for the person who was previously able and is now limited by the lack of his erstwhile abilities. He is able to compare the two states and describe his contrasting feelings to be able to formulate two phenomenologies, one of which will necessarily be retrospective. The able-bodied person who temporarily assumes the disabled role to get a taste of what the other is experiencing, sometimes permanently, can only mimic this experience with limited authenticity to arouse public conscience to its disreputable treatment of persons with disabilities. That is why, while paying heed to his representations of what it is like to be disabled, society should pay more heed to what people with disabilities have to say about largely self-centred public policy that is of minimal benefit to them—they owe them that respect. We have already discussed other aspects of this phenomenology in our consideration of the enhanced sensory and other compensatory abilities possessed by persons with disabilities that are less equally developed in able-bodied persons. Finally, there cannot be a

176

phenomenology of disability when people are disabled in so many different ways; so what we may logically venture to enquire about are the many phenomenologies of different degrees and states of disability in different circumstances.

Disability is related to disease and illness, because some states of disability may result from disease and illness. But disease and illness are not necessary conditions for disability, although they may be sufficient conditions, because disability can still occur in their absence, even though they can precipitate it. Disability does not entail complete lack of abilities or dependency; many other abilities may still be present and operative, so that the disabled person can live an independent, healthy, and autonomous life. Having a disability is his way of being-in-the-world, his existential mode with impairments, limitations in his capacity for action, and restrictions on his freedom to participate in the daily pursuits of living. Disability does not place less moral and economic value on his life or stigmatize him as less of a person than the non-disabled individual whose way of being-in-the-world also entails a lack of some other abilities, including some that the disabled person may possess, as already indicated.

On the other hand, impairments, as reflected in incapacitation or personal impediments to the successful performance of tasks, are necessary but not sufficient conditions for disability, because there can be impairment without disability in the sense here defined, but there cannot be disability without prior or concomitant impairment. If a person is disabled, he is impaired, because his disability imposes limitations on him; it impairs his ability to perform certain tasks. Disability may be said to guarantee the prior or concomitant presence of impairment, although it can be precipitated by sufficient causes other than impairment in someone who is not physically impaired; e.g., social neglect. Where impairments are present without disability, adjustments and corrections have been added to transform the impairment into a near normal state that enables the person to lead a happy life and pursue his vital goals; e.g., myopia or astigmatism corrected with lenses, conduction deafness ameliorated with hearing aids, limb deficiencies corrected with prostheses. Physical appearance may be improved without providing for the many other factors that contribute to enhancing the quality of the person's life; and, although one may wonder if anything positive has been achieved by that effort, it seems that the person affected derives emotional consolation from this maneuver by not looking obviously different from other people and thereby being a target for bigotry. Correction of visible disabilities may, howbeit, derail the expected flow of social support to the person who is now conveniently regarded by our insensitive bureaucracies as normalized and no longer eligible for his social privileges despite his residual disabilities, and this alone will downgrade the quality of his life by neutralizing his newly acquired goods and advantages.

From the above discussion, it is clear that disability has both individual and social causes and consequences. Emphasis should not be placed on the disabled person as the source of inadequate function without correcting the disorganized role of his social environment in this inadequacy, including radical changes in attitudes of the able population toward the disabled—treating them fairly and with the understanding and respect that moral equals deserve. The practice should not be to assume that the existing social environment is the standard of normality in everything, and that the person with disability is the abnormal variable in this equation, thus placing the onus on him to conform to the environment when the environment also requires change to accommodate his needs. Failure or reluctance to change the environment while expecting the disabled person to subject himself to structural and other changes to befit him for a comfortable place in that environment can only be interpreted as selfishness and arrogance on the part of able-bodied persons. Society can improve this disadvantaged state for him by providing interventional training and support facilities expressly to alleviate the impediments that further limit his ability to access common facilities, even to some doctors' offices—wheel chairs can't roll up stairways. However, disability that accompanies lack of cognitive capacity entails concomitant lack of general ability and demands more extensive modifications in the environment of the disabled, but it also places bigger demands on their training and physical ability to meet that environment. The doctor should assume a crucial advocacy role in ensuring the realization of both remedial measures for her patient.

In the end, no disabled persons should be doomed to eke out a subhuman life, simply because they are looked upon as elements of society that demand ongoing care and special consideration, which they cannot reciprocate because of their lasting physical (not moral) limitations. Disability need not impose an indefinite and unwelcome worldview of passivity on them as they struggle laboriously through every moment of their lives to achieve the minimum that the able-bodied and able-minded take for granted. Besides confronting limitations of personal ability that constantly discourage their intentions and frustrate their efforts, they also suffer from limitations on their ventures into their contracted world caused by failures of those technical extensions of themselves without which they cannot engage it (motorized wheel chairs, hearing aids, blind man's stick), and whose dysfunction adds to their own physiological dysfunction. The world can be a better place for them too if we care enough to make it so by treating them justly and with respect. The loneliness of being-disabled-in-the-world-alone can become the joy of simply being-in-the-world-with-others-in-a-less-able-way.

# 11

## Chronology of Disease

### Philosophy of Time

In addition to the familiar trio of disease: host, pathogen, and environment, time as the fourth dimension in human biology and disease permeates our being and actions from conception to death. Disease, sickness, and illness occur in time, and physicians employ the concept of time in their routine and specialized procedures, such as tracking a patient's vital and other signs. They monitor heart beats per minute to track dangerous arrhythmias such as supraventricular tachycardia, heart block, or ventricular fibrillation, and respirations such as Cheyne-Stokes respirations, in addition to less deadly tachycardias, tachypneas, bradycardias and bradypneas. They track wide fluctuations in temperature q24h to follow the course of septic infections with their quotidian temperature swings that are exaggerations of normal circadian temperature variations; they follow the 48 hour or 72 hour cyclic fever spikes of Plasmodium vivax or P. malariae, and the q7-20 day fever spikes of Hodgkin's lymphoma. They study the electroencephalogram (EEG) for wave frequencies to spot the 3-Hz spike-and-slow-wave patterns of absence seizures and the hypsarrhythmia patterns seen in infantile spasms. All these observations in time help them to narrow the focus on a patient's medical diagnosis and to follow the course of his disease and illness with the passage of time. But what is the intrinsic nature of this temporal parameter by which they measure their patients' well-being or ill health, and which they use to help steer their patients' courses to desired ends and away from undesirable ones, and what do they understand by the passage of time? How does time pass?

The difficulty of defining time is as old as Philosophy, but its recalcitrance may simply be due to asking the wrong questions about time. Perhaps we should not be asking what the nature of time is and ending up with a reification of time as an existent, which appears, passes, and disappears; we should be asking how the concept of time is applied to daily events, or how we use the word "time" ordinarily and technically. A fitting answer to the question "what is time?" is: "time is δ". "Time" as grammatical subject of this sentence is then mistakenly presumed to represent something, an existent that is described by the predicate "is δ", which can be partitioned into past, present, and future, and which has a direction or arrow of flow or passage from past to future. In fact, δ only represents a concept and not an entity in this context, since it is not the case that every referring expression names something, an existent.

Some philosophers have argued that the attribution of past, present, and future tenses to the same event endows it with logically incompatible properties—which is what happens when one asserts that the birth of a baby that was future is now present but will be past after a while. The same event is described as having three temporal modes at the same time, and that baffles the imagination. But others maintain that the argument misses the point, because the same event can have past, present, and future (anticipated) perspectives predicated of it simultaneously in the milieu of time without contradiction, since it can occur earlier or later than, or simultaneously with, but independently of the vantage perspective of the observer, which is what gives it its irreconcilable aspects. The everyday viewpoint of time as tensed, i.e., as having a past, present, and future is the unfortunate result of regarding it as an ontological entity that is segmented into past, present, and future. But time is not an entity of that sort and it cannot have pastness, presentness, and futurity, nor does it have an arrow that points to its direction of flow, except in a metaphorical sense. In this regard David Mellor proposes an alternative theory of time, viz:

> There is in reality no such thing as being past, present or future. . . . The question is, what makes a statement like 'e is past' true when it is true, namely, at a later date than e? . . . what makes 'e is past' true at any time t is the fact that e is earlier than t. Similarly, what makes 'e is present' true at any time t is e's being located at t, and what makes 'e is future' true at any time t is e's being later than t.[1]

The time-worn expression that no one can perceive the future as it is not yet real, nor can any one perceive the past, because it is no more is meaningless in the absence of an entity named "time", which possesses a past, present, and future. Objects and events can have past, present, and future predicated of them but not time; to do that is to confuse the object with the medium and to attribute predicates to the medium that erroneously objectify it. But even if such an entity did exist, any talk about the past or the future must necessarily be done in the present tense, because our perception of past events consists only of memories occurring at the moment of their recall. Future events cannot be called in, because unlike past events, they do not yet offer any memories, only anticipations, which occur in what time ontologists call "the present", and not in "the future". So we can never traverse the metaphorical past or future; we remain mired in the present, and for us the only reality is the one from whose

180

perspective we can envisage what has been and what is. What has been has vanished and what will be is still non-existent, and we can never assume a stand outside of time to view them objectively; we always speak of time from within its ambit, dividing it up into past, present, and future only for ease of categorizing events temporally.

These conceptual temporal limitations cut against the entire purpose of the endeavour of clinical medicine and the significance of the history of its beginnings and progress up to its present, besides spelling doom for its future and the future of the salvation of mankind from the ravages of disease. We need an expanded time spectrum in which to pursue medical enterprise in the service of humanity, so that we can carry past achievements into the structure of present events and still project them into future events as the bases of further advances. This concept is reflected in the different series of medical periodicals entitled "Advances in Paediatrics, or Surgery, etc".

In later years, St. Augustine epitomized the problem thus:

> What then is time? If no one asks me, I know well enough what it is, provided that nobody asks me; but if I am asked what it is and try to explain, I am baffled. All the same I can confidently say that I know that if nothing passed, there would be no past time; if nothing were going to happen, there would be no future time; and if nothing *were*, there would be no present time. . . . how can we say that even the present *is*, when the reason why it *is* is that it is *not to be*? . . .we cannot rightly say that time *is*, except by reason of its impending state of *not-being*[2]

He knew how to use the word "time" and to conceptualize about it, but he could not define time. He noted that describing an event or an interval of time as short or long could not possibly refer to what is past and is no more, because what is non-existent cannot have the property of being short or long. Furthermore, it could not refer to the present, because the present has no duration, since it is only the boundary between past and future. Besides, duration cannot be assigned to an incomplete event, only to one that has been completed, by which time it is past and can only be assessed from memory. He therefore concluded that time is an unreal mental phenomenon that lacks objective reality, because it cannot be measured; only the experience of the event occurring in time can be preserved in memory; so what is measured is memory. John Mctaggart came to the same conclusion:

> Past, present, and future are incompatible determinations. Every event must be one or the other, but no event can be more than one. This is essential to the meaning of the terms. . . . But every event has them all. . . . all the three . . . are predicable of each event which is obviously inconsistent with their being incompatible.[3]

181

With regard to the present having technically no duration, we know that usage permits us to characterize the present as being an hour, a day, a year, or a decade long, depending on the nature of the event whose duration is being measured. The said periods are present for as long as the last event in them has not terminated, at which point the period and its constituent events will be over and past; the entire period does not have to be present in toto. If we are watching a soccer game and someone asks us when the game took place, we would reply that it was still taking place then, meaning that it was undergoing a prolonged presence, since instantaneous occurrence would be impossible and absurd for a soccer game. A flash of lightning or an explosion might be instantaneous by their nature, but not a soccer game. Usage should, therefore, not be allowed to confer an ontological status on time, because such a lapse will make time existentially polymorphous and conceptually unmanageable.

In clinical medicine all the temporally measurable events mentioned above (past and present) are regarded as present during the time span that they are under observation for the purpose of arriving at a diagnosis of the patient's medical problem. If that were not the case, we would not have access to historical facts and events that bear directly on the problems that we are engaged in trying to resolve at any time, simply because they repose in future or past oblivion where they are naturally inaccessible. Every physician knows the importance of delving into a patient's past history of illness, injury, travel, social relationships, and his plans for the days, months, and years after his illness, because these facts help to place in perspective the probable genesis and the anticipated prospects of his present existential predicament and how to go about resolving them at the same time that his disease is being cured.

Long before St. Augustine, Aristotle had observed that we never perceive time as such, but only a succession of events or changes in time; from which he concluded that time is the measure of change. It cannot be change itself, because, while change can be slow or fast, time itself cannot be slow or fast; only its passage may feel so subjectively, depending on how much pleasure or suffering a particular state of health is causing for someone. Time grinds slowly, metaphorically, for someone who is constantly in pain or distress if there is no relief in sight, while for the healthy person it may pass quickly as he flits from one pleasant event to the next. After St. Augustine, Kant postulated that time is a product of the projection of the perceptions of the mind upon the objective world of things, again emphasizing the point that we do not perceive time as such, but we experience objects and events in time. Events come into and pass out of existence in time; but we misrepresent this phenomenon as the passage of segments of time, in my opinion. His contention draws

182

support from a commonly observed illusion about which train is in motion when one is sitting in a stationary train and the train in the adjacent rails starts moving in the opposite direction, creating the impression that one's train is the one in motion.

# Medical perspective on time

A science-based conception of time recognizes that perception is never strictly perception of a present event in a present perspective. It is mostly perception of a past event in a present perspective, because there is always a lag period between stimulation of the sensory receptor and transmission of the impulse to the decoding centres in the brain, however brief that lag period may be. The situation is highlighted in the perception of stars that are many light years away and may be perceived as present long after they have ceased to exist, constituting a kind of past perceiving, or perceiving past events as present. One who sees the star several light years after it has emitted its light rays may be seeing as present a past star that has already passed out of existence. So what looks like present reality is, in fact, no reality, only the illusion caused by a past reality. The paradox of time is also compounded by the dawn of the New Year as it is documented from east to west, being acknowledged at a point $Y$ just east of the International Date Line some 24 hours after it was acknowledged at point $X$ just west of the same line. So a sick baby born at point $X$ at 0005h on January 1, 2007 who is transferred to a special care unit at point $Y$, two hours away, and subsequently dies there one hour later will have died on December 31, 2006 some 20 hours before he was born on January 1. The upshot is that at $Y$ at that time by the clock the baby's birth was a future event while his death was a present event—which is absurd.

A further twist in the complexity of our concept of time is seen in the indisputable fact that unless we perceive two objects $A$ and $B$ simultaneously, our perception of $B$ after $A$ consists only of a comparison of our present perceptual experience of $B$ with the memory of the perception of $A$, which is not now before us but only persists in the memory. This latter situation can be extended to the case of comparing a pain $B$, which is now being experienced by a patient, with pain $A$, which the patient experienced yesterday. The two pains may be physiologically the result of the same intensity of response to instrumentally the same intensity of stimulation of his $A$ or $C$ fibres, and they may have identical physical characteristics; but they have different phenomenologies. Thus it will be as impossible to describe them as the same pain qualitatively as it is to describe them as the same pain quantitatively, even if it were possible to experience both pains at the same time from the same site.

Whatever the nature of time, the fact remains that we operate in tensed time and we track the sequence of events in the patient's course through his illness in terms of past, present, and future, because that is the easy or pragmatic way of dating events. If we did not operate in this fashion, the confusing and sometimes contradictory conclusions that are forced upon us by our concept of time as an existent would paralyze our thinking and severely impair the doctor's ability to minister to her patients. This perspective on the role of time in the medical history of her patients constitutes one of the two basic modalities of the doctor's experience of time—subjective for the patient and objective for the physician—which present to each one of them as differences in duration. For the patient, time moves slowly as he grinds through the tedium of his illness, always anticipating a change for the better with each passing day, but sometimes suffering setbacks; for the physician the clock ticks on time for some aspects of her patient care, but it also seems to tick ahead of itself when she cannot keep pace with the demands of the same patient care and appears to be losing control of situations.

So both situations reflect the influence of subjective, internal, or perceptual time, as opposed to objective, clock, or conceptual time on how each person operates. The implications of these perspectives for patient management become even more concerning in those cases where patients suffer from illnesses that distort their concept of time as discussed below. The pitfall associated with our reliance on subjective time is that our subjective estimate of the passage of the same duration is about five seconds longer than that recorded by clock time. So we cannot ordinarily rely on this estimate to define the lapse of time between any two states or the duration by which we measure and compare the lapse of time between two or more sets of similarly spaced events in time, viz., past, present, and future. On that account, subjective time cannot be relied upon to be the standard for defining and measuring time, because it defies consensus. This consideration gives clout to the objective theory of time as accounting for the temporal order of public events about which we all agree. A converse difficulty arises when we define change as the sequence of successive events in time, all of which are regarded as successive states of the thing that changes over time. We create a vicious circle, defining change in terms of time and time in terms of change.

So from the point of view of the doctor and her patient, the compartmentalization of events-in-time into past, present, and future probably makes much sense as she follows and compares the daily course of her patient's signs and symptoms, such as the pain of yesterday, which is now no more and thus portends improvement, or the appearance of deteriorating coagulation parameters over time, which spells the onset of Disseminated Intravascular Coagulation and a worse prognosis for her patient, calling for vigorous remedial action now and not later. An acute and contrary example arose when during my heart attack the attending physician kept asking me where I rated my chest pain on a scale of ten with each passing minute. I was being asked to compare the memory of several past experiences with a present sensation of pain in the crushing period of the most globally incapacitating kind of pain on the assumption and in the hope that my memory would serve as a reliable guide to my response to treatment. Time was of the essence if my coronary arteries were going to be unblocked to save my myocardium from severe ischemia and necrosis resulting from my coronary thrombosis; but the race against time also depended, among other things, on the reliability of my memory during those moments of crisis, which cannot be relied upon in those dire circumstances anyway.

I wondered afterwards how frustrating this method of assessment will be in the case of someone with dementia who has lost the concept of time; or one with schizophrenia who has merged past, present, and future perspectives into a single "timeless" period; or one like the lobotomized patient with a defect in recent and remote memory recall but only a perception of the present to rely on; or the bipolar one for whom time is moving (figuratively) at varying speeds, disrupting his long and short term memory—for the manic subject experience of time speeds up, while the depressed subject experiences a slowing in time; or one in whom an overestimation of time is induced by hallucinogens and related drugs; or even one whose sense of time is dulled by morphine. These different cognitive states with their unique interpretations of duration suggest that doctor's need a better yardstick to assess subjective impressions of duration if their use of the present concept of time is to serve the clinical purposes to which it is being put.

The patient has been described as one who conquers time (or is conquered by time) as he waits for his illness to take one of three courses: resolve, linger, or end his life rapidly. He takes his medicines at the rate of one or two tablets once or more times daily while he waits patiently for the desired results. In keeping with his designation as a patient, he has to be patient and take his medication over a stretch of time to avoid a lethal overdose that ends up defeating the purpose of his treatment. He cannot take all fifty to five hundred of his tablets at once to get an instant cure; they have to be taken patiently over periods of days, weeks, months, or years. Time is always of the essence in patient care, but biding his time is more essential for the patient's survival in this case.

On the other hand, it also makes sense for the doctor to survey in one swoop a panorama of her patient's illness in her phenomenological approach to the totality of the patient's problems to avoid the inevitable temporal reductionism inherent in the first method of approach that predisposes her to react to situations that call for being proactive. Hence the importance of her departure from the tensed perspective of time that relies on the arrow or the march of time to unfold events to which she reacts as they occur. She embraces the panoramic view, which does not rely on the present being carried forward in the flow of time, but rather seeks out the course that events are likely to follow in the anticipated sequence of before, simultaneous with, and after the present. She is able to reflect on the meanings of her patient's symptoms, signs, and laboratory data in the context of a continuum, not as occurring in unreal segments of time, and is thus able to plan his treatment prospectively with anticipation.

Reference in the opening paragraph of this section to heart and respiratory rates and to EEG frequencies necessarily raises the question of the mystery of biological clocks and circadian rhythms, which happen to be closely tied in with these phenomena. How is it that these and other functions have set frequencies by which we can determine the malfunction of the systems assessed by their means? These phenomena must have their roots in the order and systematizations that we see in our world and solar system, which we cannot pursue in the present context. Suffice it to note, however, that biological clocks follow closely the 24 hour cycle of the earth's rotation (sleep-wake cycles), the 28 day cycle of lunar rotation (menstrual periods), the 365.25 day revolution of the earth around the sun, and seasonal variations under the synchronization of light and other factors (phases of depression).

# Circadian rhythms

More precisely, scientists have observed and studied circadian rhythms, operationally defined by Isaac Edery as biorhythms that run continuously over 24 hour cycles "in the absence of external time cues"[4], and have the capacity to respond with constancy to cyclic variations in temperature and to illumination or its absence.

These biorhythms are controlled via a pathway that starts with the response of the retina to the stimulatory effect of bright light and the inhibitory effect of dim or no light. Impulses are then relayed via the retino-hypothalamic tract to the suprachiasmatic nucleus (SCN) in the hypothalamus, and then to the superior cervical ganglion, and finally to the pineal gland from where melatonin is secreted through a process beginning with activation of the $\alpha_1$ and $\beta_1$ adrenergic receptors in the gland. The end result of these processes is a cyclic influence on functions in the brain, heart, lungs, liver, thermostat, sleep and wake cycles, and circadian rhythms generally. In addition, the daily rhythm of melatonin secretion is also controlled by a free-

186

running pacemaker located in the SCN as detailed by Amnon Brezinzki in "Melatonin in Humans", *New England Journal of Medicine* 336 (1997): 186-95.)

Ian Hickie, Sharon Naismith, Rébecca Robillard, Elizabeth Scott, and Daniel Hermens have stated that the auto-regulatory network operated by the SCN is controlled by regulatory genes that lose their regulatory function and communication with the pineal gland and its release of melatonin after destruction of the SCN without disrupting the circadian system itself. Through its regulatory effect on the pituitary gland, SCN affects the release of ACTH and thereby the secretion of corticotropic hormones with their various sites of action. It is also subject to the feed-back effect of Melatonin, which "inhibits the circadian signal for increasing wakefulness."[5] Importantly, also, SCN affects autonomic neurons and the neuropeptide hypocretin/orexin whose role in the sleep-wake cycle is to counter the sleep debt towards the end of the day, and whose deficiency is connected to the narcolepsy/cataplexy syndrome.

> the circadian system coordinates all other hormonal, metabolic, immune, thermoregulatory, autonomic nervous and other physiological processes to optimize the relationships between behaviour and body functions. At the cellular level, almost all individual cells and, hence, organ systems have their own intrinsic clocks. As these cellular (for example, fibroblasts, fat cells, muscles) and organ-based (for example, liver, pancreas, gut) clocks run to intrinsically different period lengths, the differing physiological systems need to be aligned in coherent patterns.[6]

Among the varieties of physiological effects of the circadian rhythm and its induction by cyclical levels of melatonin, seen alone or together in any single individual, are the following:

(1) Patients with derangement of dopamine secretion from the basal ganglia, as seen in ADHD, Parkinsonism, Huntington's Disease, and right parietal lobe stroke specifically have confused and defective time perception.

(2) Heart attacks occur mostly in the morning, while asthma attacks occur mostly at night when secretions of epinephrine and cortisol are at their lowest.

(3) The incidence of depression rises in the winter months when periods of reduced daylight and melatonin secretion are longer. Furthermore, Hickie et al. state that post-mortem histological anomalies have been found in the SCN of patients with depression, while "changes in the rate of rise of evening melatonin and the amplitude of the melatonin response have been associated with various mood disorders."[7]

(4) Constant lighting in newborn nurseries depresses secretion of melatonin resulting in disturbed sleep and slower weight gain in neonates. Variation in the intensity of lighting in these nurseries to mimic day and night offsets these setbacks.

(5) Night workers, under the effect of increased levels of melatonin, are more prone to serious errors than daytime workers, as seen with on-duty physicians (medication and other errors), motor vehicle operators (accidents caused by sleeping while driving), air controllers (near collisions of airplanes), and operators of oil tankers (Exxon Valdez oil spillage), nuclear plants (Chernobyl explosion), and chemical factories (Bhopal explosion).

(6) Office workers who are sleep deprived or have a tendency to postprandial somnolence are kept awake by Activia lamps, which emit light of 17,000 Kelvin whose effect on their hypothalami is to simulate daylight, thereby decreasing their secretion of melatonin.

(7) Cyclic physiological and pathological phenomena are seen with biological clocks that regulate body temperature, the menstrual cycle, migraine, cyclic neutropenia, PFAPA syndrome, isolated nocturnal hypertension, and some forms of epilepsy.

(8) Recently a single case of an adult "patient" with recent onset of allegedly spontaneous, cyclic, insurmountable closure of her eyelids for 72 hours and spontaneous opening for another 72 hours has been reported. Whether this is another case of natural circadian rhythm remains to be seen.

(9) Another recent speculation surrounds the case of the timing and frequency of infant colic and migraine, both of which are closely related to disruption of the sleep-wake cycle and show a diurnal pattern. The suggestion here is that altered sleep-wake pattern may be the trigger for both.

The ingrained regularity of these biological clocks and circadian rhythms has further been observed to be beyond permanent disruption by simple interruption or by the use of drugs. The clocks always revert to their previous rhythms regardless of the length or intensity of interruption.

The concept of time thus has an important role in shaping our management of disease and illness by unraveling the relations of patient and doctor to each other and to the clinical problem that is facing both of them from their respective perspectives. It also affects our attitude to death as we dreadfully anticipate this terminal event in our lives that we suddenly realize are heading toward oblivion.

# 12

## The Final Phase

### Concept of Death

In Webster's Dictionary death is defined as "a permanent cessation of all vital functions: the end of life." Robert Veatch states his matching concept of death as follows: "death may be formally defined as the irreversible loss of that which is considered to be essentially significant to the nature of man" [1]; which means the irreversible loss of consciousness and cognition, and also implies the irreversible loss of integrating capacity for experience, physical, psychic, and the social bodily functions mentioned below, since these functions cannot be integrated without the rational direction of consciousness. It also means PVS patients are dead, because "somatic integrating capacities" without a whole brain and consciousness are not possible in nature, although they can be mimicked artificially. These integrative functions must be distinguished from the residual physiological activity of the brain and other tissues that persists for a while after the demise of the person as detailed by Amir Halevy and Baruch Brody, that Veatch considers irrelevant. He says, "Because these isolated nests of neurons no longer contribute to the functioning of the organism as a whole, their continued functioning is now irrelevant to the dead organism." [2] James Bernat defines death as "the permanent cessation of the critical functions of the organism as a whole" [3] i.e., the irreversible cessation of all clinical functions of the entire brain—those functions necessary for the maintenance of life, health, and unity of the entire organism. 1) The cerebral cortex as the locus of irreplaceable higher functions, 2) the brain stem as regulator of heart beat and circulation, breathing, and excretion that can be replaced with artificial hearts, ventilators, dialysis, and 3) the integrating neuraxis are pivotal in this respect. In this way a niche is created for persons with dementia and those who have suffered permanent loss of consciousness but are not dead as Veatch's definition would have it, e.g., PVS patients.

Dying is the reversible event that marks the change undergone by a person from being alive to being dead, and hence the event that also marks the beginning of the permanent or irreversible cessation of the cellular organization and homeostasis that supports responsiveness of the organism as a whole to stimuli, metabolism, growth, adaptation, and reproduction. That means persons and organisms are not distinct parts of the same entity that terminate at different times in different ways and for whom the concept of death is different. "Since the organism constitutes the person, it is not the case that a property inheres in one object and then there is a second property possessed by the other"[4]; so death of the organism=death of the person; i.e., the concept of death is univocal; when the person dies, his Strawsonian M and P predicates terminate together. It is also the case that "a creature has died just when its vital processes are irreversibly discontinued, that is, when its vital processes can no longer be revived"[5], and

> Persons cease to exist when the organ of thought—or more precisely, the parts that make self-consciousness possible—are destroyed. So the onset of an irreversible coma or permanent vegetative state would mean that the person has ceased to exist. And when persons go out of existence, they cease to be alive. They no longer exist, a fortiori, they no longer instantiate life processes. Thus the criterion of death for organisms does not need to be altered to apply to persons that are derivatively alive.[6]

The consciousness endowed person exists in the unconscious organism in the same inseparable way that a representative statue exists in a mere lump of rock; his consciousness imparts personhood to the organism as the form of a statue confers import to a piece of rock. In Aristotelian terms, the material and formal modes of being of matter are represented by the organism and its related/derivative person. Also implied by these concepts of death is the absence of intermediate state between alive and dead in keeping with the law of excluded middle whereby the organism that is the person must be either alive or dead, although a process of dying during which the person is still alive is undeniable. But that is not a recognized state of human existence, in spite of the saying that in the midst of life we are in death. To call the event of death the antithesis of the process of life is, therefore, not strictly correct, although it is true that death can take place only where there is life as a necessary pre-condition by which it is defined, whilst life does not need the pre-condition of death for its definition.

Definitions of death are necessarily couched in negative terms, because of this parasitism of the concept of death on the concept of life. Besides, the characteristics of death as a state, and not an event, blend in almost imperceptibly with its observable criteria, because these criteria are the connotation of the denotation "being dead" and whoever satisfies them has experienced the event of death, whether as a sudden event or as the terminal event in a short or long process of dying. People can recover from the process of dying as in the many cases of heart attacks, which, if left untreated are fatal, but there is no way that once dead the person can recover. However, if the process of

190

dying is treated appropriately, recovery is possible and the person can continue to live. These criteria serve as the sufficient conditions for the state of being dead, not for the event of death, with preceding life serving as its necessary condition, because only living organisms die. The absence of these criteria of identification, most people believe, nullifies the application of the concept "death" as it is commonly used to signify the state of being that is representative of "the dead". These considerations appear to provide room for different societies and cultures to entertain their specific concepts and practices of what constitutes the death of a person, because not everyone is convinced that death is a real natural event (rather than process), or that death means the clinically irreversible loss of integrative functioning of the organism as a whole as seen in brain-dead individuals, and for which diagnostic criteria can be proffered. The importance of these criteria will become evident when we consider the problem of euthanasia.

The difficulty of distinguishing between death as a state and death as a biological event in the integrated functioning of the human organism marking the permanent loss of embodied consciousness that occurs at an empirically unspecifiable time $t_1$, or a period $t_1$ to $t_2$ during which body cells successively lose their viability as a result of the dissolution of the integrated organism or the organism as a whole must be acknowledged. Dying is regarded as the reversible process occurring over this variable time interval (apoplectic or prolonged) during which the person is permanently losing consciousness and finally satisfying the criteria for being dead and in a state where he has already lost consciousness permanently and all his cellular activity has ceased and decay is setting in. This viewpoint is in contrast to regarding the cessation of respiration and heart beat as the events that mark the occurrence of death, based on the inability of brain and other tissues to maintain viability in the absence of a steady flow of oxygenated blood to them as a result of cardio-respiratory failure.

As stated above, the overall conceptual definition of death has focused mainly on irreversible coma or the permanent loss of consciousness, the faculty that integrates the functioning of the organism as a whole. This is the discriminating faculty that belongs to the higher brain centres (cortex and reticular activating system), as distinct from the reflex, integrated function of the lower centre or the brain stem. The definition is based on the idea of personhood as essentially embodied consciousness with the brain as the seat of that consciousness and the director of integrated functioning of the entire organism; but it overlooks the possibility of substituting another mechanism for executing the same function and thereby disqualifying the failed functional state of the brain as the logically sufficient criterion for this definition of death. It also ignores the continuing life of the body as object after the death of the subject (person) caused by the permanent loss of consciousness that he suffers, raising questions about what to do with the still alive body and what value to assign to it now. Besides, there is no test for permanent loss of consciousness, although there may be tests for loss of consciousness *per se,* and only time can assure us that the person has lost consciousness permanently, especially in the face of current reports of people regaining consciousness after many years in vegetative states.

Long ago, Epicurus stated the essence of death in words that really amount to a tautology: "Death is nothing to us, since when we are, death is not come, and when death is come, we are not."[7] Epicurus is here referring to that moment which separates the living from the dead organism as being (we are) and not-being (we are not). There is no suggestion here of a third state of existence or halfway house that is styled "purgatory" in some circles, and, as we have already observed, death as the permanent cessation of all brain functions, cortical and brain stem, is consistent with its characterization as a transition point marking the end of brain function and life, and the beginning of oblivion or a state of being with hard factual criteria of identification, mainly neurological, but also cardio-respiratory. When this transition takes place has traditionally been difficult to tell and has been made more so by our methods of keeping brain-dead individuals "alive" with respirators, thus creating an intermediate state that defies the law of excluded middle that a person is either alive or dead. We now have a category if semi-dead persons, the so-called "alpha period".

This technology has created the quandary of what to call the organism that is thus kept alive—dead person, living body with dead brain, or corpse in transition with or without a chance of reversal of that state, and when to institute all the statutory or customary procedures associated with persons who have died, without being paralyzed by differing opinions of different people's perspectives on when the person has died. The old method of diagnosing death by cessation of heart beat and respiration, which made a functional brain stem as the centre for the control of these functions the necessary and sufficient condition for the occurrence of death, has had to be amended to include loss of function of the whole brain, cortex and brain stem, as that condition to obviate the dilemma of cortical awareness in a person declared dead by brain stem criteria. But there still remains the question whether one organ in the body (brain) and its integrating function can be equated with the whole organism and its integrated function to the extent of bestowing on it a decisive role in the life or death of the organism. Equally, the question can be asked if a ventilated human body without a functioning brain and whose circulation is being maintained artificially can be called a living person. Also, if an anencephalic who never had a brain but has normal breathing and circulation and a PVS patient whose brain has lost its function while his breathing and circulation remain normal are living persons who should not be counted among the dead simply because they lack the organ of consciousness and integration. Assuming also that assimilation and excretion are proceeding normally in these bodies, is the first one a mere artifact, or the second one a kind of humanoid organism that represents the person that might have been, and the third one a mere living vestige of the person that was?

## Timing of death

The modern definition of death omits any reference to the presence of a beating heart or spontaneous respirations, and thereby immediately creates an ethical dilemma about the disposal of the body or corpse of the erstwhile alive person who is now dead, but is still breathing and has a heart-beat. If consciousness has been irrevocably lost and the cardio-respiratory functions are still naturally active, then we are dealing with a corpse according to this definition. But the brain does not exclusively integrate every bodily function, even though it integrates the functioning of the entire organism, e.g., metabolic, kidney, lung, and liver functions can continue in the absence of a functional brain. Can this corpse, therefore, be buried with a heart that is still beating and maintaining an active circulation and with respirations that are still spontaneous? Can his organs be harvested for transplantation with these two functions still ongoing? Most people will recoil at the thought of subjecting this kind of "dead" person to these procedures. Even those who believe in the permanent loss of consciousness and goals in life as indicative of the death of the person will harbour some misgivings about carrying out these procedures while those vital functions are still in progress, attesting their regard for cardio-respiratory integrity in the life of a person.

Therefore the question of death and its timing, and the specific criteria employed in particular situations, becomes all important in decisions concerning when to harvest organs from recently dead persons whose cardio-respiratory functions are still active enough, either spontaneously or through artificial support to maintain the cellular structure and function of those organs in intact form for the purpose of transplantation. This perspective is only possible if we define death as the permanent loss of consciousness, even as the heart and lungs are still physiologically viable to sustain the viability of other tissues. Defining death in terms of cessation of all the vital functions of the biological organism defeats the aims of organ transplantation, because by then the tissue cells will be non-viable and useless for transplantation. Contrarily, the rights of personhood cannot be withdrawn from one who still retains some vital functions relating to his personhood in the Strawsonian context, and his organs can therefore not be harvested without thereby killing him, whether he is in a permanent vegetative stage or he is an anencephalic. Such human beings will first have to be declared non-persons who lack moral standing before their organs can be harvested, regardless of their grossly impaired ability at or absence of mental integration in the presence or absence of somatic integrative abilities.

Much also rides on concordance of perspectives or some kind of forged agreement between physicians and families of deceased persons about when death has occurred in their loved ones, if organ donation for transplantation is to happen. The family will have to bring themselves to a point where they can appreciate the transplant doctor's definition of death as brain death, "lacking in *both* bodily and whole brain integrative function"[8], even if the patient's heart is still beating and he is still breathing or his breathing can be artificially sustained—so-called living cadaver. Sometimes their outlook can accommodate only permanent cessation of all of the patient's vital functions as indicative of death, and brain death for them implies a stage in the process of dying, not the terminal event, so that to them arbitrarily harvesting organs in this phase amounts to immorally killing the patient, like burying the patient alive. The terminal event in this process is reached only when spontaneous respiration and heart beat cease, but by then the body organs have lost their viability and suitability for donation. Important offshoots of the discussion of the concept of death and its meaning for physicians, erstwhile or extant patients, their families, and the general public relate to the aversion of some people to the idea of their loved one's body desecrated and mutilated by the harvesting of organs or by being subjected to dissection in an Anatomy laboratory, even if all parties agree to the definition of death as embracing the criteria of either permanent loss of consciousness in one case, or cessation of all vital functions in the other case. Some others may feel that the recipient of the decedent's organs helps to perpetuate his life, making him somewhat immortal, or else that the organs have been wasted on an undeserving recipient like the alcoholic whose own liver failed from the effects of alcohol, but who continues to indulge in excessive consumption of alcohol after he has received a transplant.

The situation is rendered more complicated by the ability of medical scientists to sustain these cardio-respiratory (and nutritive) functions artificially for indefinite periods; i.e., keeping a "corpse" alive for as long as desired. So it becomes legitimate to ask if the definition of death is a mere convenience, because a sharp line can't always be drawn between when someone is alive and when his dying process has finally been accomplished, as the use of the appropriate standard tests can demonstrate some subtle persisting functions in the corpse whose gross physiological functions can be shown to have

194

ceased. Furthermore, there does not appear ever to be a definition of death that represents a logical necessity in that it does not admit of contradiction, nor is there a more convincing criterion than the onset of decay of the decedent's body. The problem appears not to be with the definition of death, but with the timing of the application of its criteria. Those patients whose death is accepted are declared dead when their cardio-respiratory functions cease, but those whose death is not accepted are kept "alive" by artificial means until their deaths can be accepted. Therefore some (or most) of the time organs are harvested from "live" donors or should we call them "live corpses?"

## Phenomenology of death

The burning question about death is that of its nature or its phenomenology, what it feels like. Unfortunately no one living can answer that question, and no one who has died has come back to describe that phenomenology to us, because, by its nature, death is not the kind of experience through which any one can live, nor can its phenomenology be gleaned from watching others die. In fact, it is not an experience, but a cessation of all experiences that can be thought of only in terms of a perpetually impending possibility for oneself and for others, because its actuality (finality) can never be retrospectively recounted. Those who compare death to near-death or to deep sleep are still not describing death *per se*, because these are states from which recovery is possible. In my own case, I have had occasion to plummet into temporary unconsciousness without even being aware that I was losing consciousness. I became aware of it only after I regained consciousness and finding a crowd of medical personnel around me, with intravenous drips running, oxygen flowing from tubes into my nostrils, and someone getting ready to intubate me. If I had not been resuscitated, I would have been a dead person or human organism, and I would not have had even a remote chance to relate my experiences of the time that I was unconscious. Even now, I am still unable to relate any of those experiences, because there are no experiences to relate; only temporary oblivion, which is this side of the permanent oblivion that follows the event of death. There is no phenomenology of unconsciousness, and although there may be a phenomenology of dying, which the dying have never survived to tell us about, there is also no phenomenology of death; period. Not even the death of other people will reveal it to us, since we have no way of finding out what goes on in their minds to read their experiences.

## Body, person, and death

In the meantime, it is important to decide if death is a purely biological phenomenon that overtakes all categories of organism alike: humans, dogs, fish, protozoa, and even plants, or whether there is a special legal, socio-cultural, and personal aspect to the death of the human organism as an agent and a subject of embodied consciousness, which, however, excludes anencephalics and persons in persistent vegetative states. Anencephalics, some people believe, exist in "a state of vegetative human life" with a brain stem but no cognitive function, and since they have never lived as persons, they can never die a brain death like persons, but only a death of human organisms. Besides, their potential for brain development to bring them to the level of self-conscious persons is zero, because failure of their neural tubes to close (at 30 days post-conception) deprived them of the potential for further brain development and brain life, but endowed them with an overwhelming potential for early demise. (The term "vegetative" suggests that the person has become a vegetable, implying, against our better judgment, that vegetable matter is always dead by the current definition of death as applied to humans, since it does not possess consciousness). These categories of humans may still have integrated brain stem functions, even though they lack potential for consciousness; but they are not considered to be persons by those definitions of the concept "person" that focus only on consciousness as being the *sine qua non* of personhood.

If death can be predicated of organisms only, including animals that are questionably presumed not to have the higher brain function of consciousness, or self-consciousness (like neonates), then human organisms that have lost cerebral control and consciousness permanently from birth (anencephalic) or during their lives (PVS), are alive only as bodies with cardio-respiratory (natural or artificially supported) and low level reflex functions until these functions cease. They will be declared dead only after these functions have ceased, not before; but one who retains consciousness and higher level functions while depending on artificial support for his lower level functions is dead by these criteria. Nevertheless, we know that this is not always true, because persons with a parallel disability who suffer from a dissociation of their active upper motor neuron functions from their inactive lower motor neuron functions, Locked-in syndrome, are still regarded as real live persons, by virtue of the intactness of their conscious activity, although we would not regard them similarly if they had intact lower motor neuron function without upper motor neuron integrity. Locked-in Syndrome is defined as "a neurological disorder characterized by complete paralysis of voluntary muscles in all parts of the body except for those that control eye movement. . . . The patient is alert and fully conscious but cannot move. Only vertical movements of the eyes and blinking are possible."[9]

As we will see in a later discussion of the right to life, psychic experience consists in attaching meaning to the raw impulses that

196

persons receive from their environment, such as the qualitative conscious experience of pain and the concomitant dysfunction that accompanies the quantitative firing of $A$ and $C$ nerve fibres. In the life of the person, it is the qualitative experiences that count in his struggle for survival, even if he has no clue about the quantitative physical phenomena underlying those experiences. He does not react to his deranged physiology but to the effects of that physiology on his consciousness. His life derives quality from the conscious experiences of what happens to his body and how it affects his goals in life. Without that responding consciousness, he might as well be dead.

As a further consequence, the absence or destruction of certain brain cells should not define the person's death, just in case they can be replaced. Death of the whole organism (brain included) as the source of all biological functions, including consciousness and the readiness to have experiences, should define death of the person, while death of its by products is discounted, otherwise we have to postulate two deaths in some cases, viz., death of the person (consciousness) and death of the organism. On this view, it is the criterion of the body's ultimate loss of its ability and propensity to integrate its functions that defines death and saves the PVS patient who lacks integrative cerebral activity from being declared dead and being buried alive, unless we engineer his death by depriving him of nourishment, or by failing to treat the infections or other potentially lethal conditions that he develops, or else by euthanizing him.

From another perspective, the person who has suffered cerebral damage or loss of his cerebral cortex dies or ceases to exist or be alive subjectively, because his source of self-consciousness is destroyed by irreversible coma or PVS; but he leaves behind an objectively living organism. The person was erstwhile subjectively alive, because the conscious organism that constituted him was alive, until it ceased to be an essential part of him. Loss of his cerebral cortex and hence his consciousness terminated his existence as subject with interests in his own welfare, although it did not terminate his objective existence, because his bodily functions are still going on much as they did before his loss of consciousness; i.e., the destruction of his brain cells does not necessitate, mean, or serve as a criterion of his death as a biological organism, as James Bernat has argued in his rejection of the concept of two deaths, viz., death of the person and death of the organism: "the concept of death is applicable only to an organism because death fundamentally is a biological phenomenon."[10]

This univocal concept of death—that death is the same for all living animals—ensures that the person dies when the organism that undergirds him dies by any criterion, on the premise that whole brain death is not a necessary and sufficient criterion for the death of the person and loss of function of the "organism as a whole", versus the "whole organism". The reason is that it is logically possible to replace the brain with a mechanical device of equal integrative proficiency to fulfill those same critical functions carried out by it to keep the organism as a whole integrated and the non-person alive, even though his self-consciousness has not been restored. Hence it is that PVS patients are labelled as living non-persons in some circles.

On the preceding view, brain death is only a temporary test of the death of the person, but in the absence of a functional device to replace the brain, it will have to do as also a definitive criterion of death. Death of the whole organism implies cessation of all the physically recognizable life processes in that organism; but death of the organism as a whole implies the existence of transcendent, unifying parts of the organism that dies versus its concrete parts, which may remain alive after the death of the former without necessarily implying that the organism is alive, e.g., residual cellular activity that is either spontaneous or ventilator maintained. On this latter view a functional body that lacks a functional brain to maintain integrated activity of the two systems is a dead person for most practical purposes. In direct contrast, the Strawsonian person needs both his primordial M and P predicates to retain his integrity; therefore loss of either category equates with death of the person. This is a tricky situation, because non-functional somatic elements entails a non-functioning brain and hence no consciousness, which is in fact death of the whole organism with its M and P predicates.

The distinction between persons and human organisms extends also to the legal identity of a decedent. If it were possible to transplant brains, who would be legally held liable for a crime committed by individual A after he has switched brains with individual B? If bodily characteristics are the essential guide in determining identity, then A will still be held responsible, even if he genuinely disclaims all knowledge of the event and the totality of circumstances surrounding it, because his brain truly has no engrams of the event. On the other hand if B admits to the crime and can relate its attendant events in fine detail, his bodily characteristics will disqualify him from being the culprit, because they will not match the descriptions of eye witnesses to the event, regardless of how vehemently he claims to be the same person, A, who committed the crime, even though he does not fit the description of eye witnesses and his DNA and finger prints don't match those found at the site of the crime. On the other hand, the law does not recognize brain death as the death of the individual for the purpose of enacting his last will and testament. His death is taken as the terminal one when all vital functions have ceased and not the short or long process during which they are terminating and may sometimes be maintained by artificial means. Only after the loss of consciousness, heart beat, and respirations is he declared dead.

198

So, when someone dies, it is the embodied consciousness with all its memories that has demised, and not just the body that we subsequently identify as the corpse of the person who was and is no more. Sometimes we loosely say "Jack was buried last week", meaning that his corpse and not the person was buried last week. But we never say Jack was exhumed yesterday; we almost invariably say his remains were exhumed yesterday. In this manner of speaking we are tacitly implying that there is a difference between the person as embodied consciousness and his body, which can be deprived of that consciousness and remain as a mere corpse or "his remains". We will not try to find out where his consciousness went or whence it came; those may not be appropriate questions in this enquiry. Asking the wrong kinds of questions always generates a pursuit of will-o-the-wisps and generation of false theories, whereas the right kinds of questions, which are often not asked, generate a quest after genuine solutions to the problems at hand.

That said, the fundamental fact still remains that our bodies are essential as foci from which our consciousnesses are launched, and without them consciousness would be a nebulous concept without a locus of identification. How else could we identify it, except as the conscious experience of person *A* or *B*, or of body *A* or *B*? So far, there is no known means of restoring consciousness once it is lost, or of substituting another faculty like it. We can substitute for respiratory functions that are failing by using respirators, and extracorporeal membrane oxygenation and lung transplants in the worst scenarios of respiratory failure. In the case of hearts that have been temporarily disabled for surgery we employ cardiac bypass, and for those that have failed completely to maintain adequate circulatory functions, ventricular assist devices, SynCardia Total Artificial Heart, and heart transplants. The permanent loss of irreplaceable consciousness and the ability to interact with the environment spells the death of the person in a way that replaceable pulmonary and cardiac functions do not, and the permanent loss of conscious functional integration of the whole organism spells its death in a way that the loss of some replaceable organ or regenerable aggregate of cells does not. The philosophical question still remains: at what point do we say that consciousness has been lost permanently?

# Criteria of death

Technological advances have blurred the line of demarcation between life and death by giving us the power to postpone the event of death defined as the permanent cessation of all vital functions, thus leaving the person in a prolonged phase of dying, if he has no chance of recovery, and leaving us trapped in providing potentially endless and futile treatment. Where previously we did not have the know-how to alter cessation of respiration and circulation, which were closely followed by permanent loss of consciousness, we can now artificially defer the necessity of at least the first of the sufficient criteria by which the death of a person is recognized, as detailed in the Report of the Medical Consultants on the Diagnosis of Death:

> a. an individual with irreversible cessation of circulatory and respiratory functions is dead.
> 1. cessation is recognized by an appropriate clinical examination
> 2. irreversibility is recognized by persistent cessation of functions during an appropriate period of observation and/or trial of therapy.
> b. an individual with irreversible cessation of all functions of the entire brain, including the brainstem, is dead.
> 1. cessation is recognized when evaluation discloses findings of a and b: cerebral functions are absent, and . .
> c. brainstem functions are absent.
> 2. irreversibility is recognized when evaluation discloses findings of a and b and c:
> a. the cause of coma is established and is sufficient to account for the loss of brain functions, and. . .
> b. the possibility of recovery of any brain functions is excluded, and . . .
> c. the cessation of all brain functions persists for an appropriate period of observation and/or trial of therapy.[11]

Observations include absent heart beat or flat ECG, apnoea with arterial $pCO_2$ of >60mm Hg after hyperoxia test, deep coma or isoelectric EEG, absent reflexes (except spinal cord reflexes). Special precautions have to be observed in applying these criteria to cases of drug and metabolic intoxication, hypothermia, shock, and in children. One arm of the sufficient conditions for the death of a person, as defined, can be defeated by the artificial maintenance of circulatory and respiratory functions in the "dead" person, but the other arm involving consciousness cannot be artificially controlled. Therefore, since we are left with a body without consciousness, the sole persistence of this latter condition (circulation and respiration) still fails to satisfy our concept of a person as a biological organism with a material body on one hand, and brain-based integrated consciousness, sentience, cognitive function and subcortical (brain stem) reflex functions on the other.

According to the Harvard Criteria for Brain Death, substituting a respirator to assume the function normally performed by spontaneous breathing in a body with a permanently non-functioning brain does not make any difference to the fact that the person is dead. Nevertheless the logical possibility remains that a brain transplant could revitalize the dead person; but that raises ethical questions of personal identity (previously discussed) and it throws a wrench into the euthanasia question: is the brain-dead body an appropriate subject for euthanasia or is the term being misapplied in this case of an already dead person? The answer provided by these criteria is that irreversible coma and total non-functioning of the whole brain means that the person has died and does not need resuscitation; instead any of his organs, including his still beating heart, can be harvested for transplantation if artificial respiration is maintained to preserve those organs. The criteria for being dead, in the absence of CNS depressants, a flat electroencephalogram over at least ten minutes of recording and no hypothermia or drug intoxication are listed as:

1. *Unreceptivity and unresponsitivity*—patient shows total unawareness to external stimuli and unresponsiveness to painful stimuli;

2. *No movements or breathing*—all spontaneous muscular movement, spontaneous respiration and response to stimuli are absent;

3. *No reflexes*—fixed, dilated pupils; lack of eye movement even when hit or turned, or ice water is placed in the ear; lack of response to noxious stimuli; unelicitable tendon reflexes.[12]

Canadian criteria for brain death also state that "a person is dead when an irreversible cessation of all that person's brain functions has occurred"[13]. What seems to follow from these definitions is that the death of a person is essentially different from the death of an animal by virtue of the human consciousness that is lacking in animals. Therefore a person may die while the biological organism that constitutes him continues to live by means of artificial support, or else he may lose his personhood without dying, as in the case of demented, amnesiac, and severely mentally retarded persons whose residual consciousness lacks goals to which it is directed. These categories of persons are considered to be alive in spite of their lack of the complete cognitive and sentient functions that constitute persons, because they can satisfy the conditions that are negated by the above criteria. Recently researchers have demonstrated activity in the brains of persons who were presumed to be in vegetative states, and who have gone on to regain consciousness after being in coma for many years. These cases stress the importance of caution in applying brain death criteria in the absence of anatomical disintegration of the brain, which may prove to be the only certain indicator of irreversible cessation of cerebral function. Physiological dysfunction alone does not appear to be any longer enough to claim irreversible coma. Furthermore, this last criterion does not include anencephalics who, nevertheless, exhibit sub-cortical functions in the absence of a cerebral cortex and integrative cognitive function.

In their article on brain death, Amir Halevy and Baruch Brody suggest a different approach to the standard criteria of death, viz., absence of heart beat and respiration, and absence of brain stem and voluntary movements. Their suggestion is based on reports of residual brain functioning in persons who have allegedly satisfied the conditions of whole-brain death. These "dead" persons have been found to have persistent neurohormonal regulation (growth hormone and anti-diuretic hormone secretion), cortical functioning (non-isoelectric electroencephalograms), and brain stem functioning (evoked responses), which Bernat regards as non-critical, non-integrative cellular activity for the survival of the person, although it implies that the whole brain is not dead. The authors therefore suggest what amounts to a pragmatic definition of death based on criteria for absence of active functioning of the agent, not his body cells, thus permitting the withholding or withdrawing of medical care after consideration of "stewardship of social resources", and harvesting organs for transplantation only after "cessation of conscious functioning with apnea". In support of this position, they further arguably maintain that "if medical care, including artificial hydration and nutrition, is unilaterally withheld or withdrawn, the vegetative patient will satisfy the classic criteria of irreversible cessation of respiration and circulation within 7 to 14 days, whereas the patient who is in addition apneic will satisfy the criterion within an hour"[14]. It is clear, however, that if we go by residual cellular activity in some brain-dead persons, then these persons are still alive, and these criteria require modification. Although it may be equally clear that mental functioning is necessary for organismic integration, such an organism cannot be arbitrarily considered to be dead on that account. There may yet be other centres that are essential, although not critical, for sustaining cultural life, and that obfuscates the determination of the death of the moral (mental) person, his integrated body, and the biologically integrated organism that he is.

As we saw when we discussed personhood, the set of criteria mentioned in these definitions of death are not completely consistent with theories of the person as a primary complex of what are styled $M$ (material) and $P$ (psychic) predicates, so that loss of any one category entails disintegration and loss of personhood, and if loss of personhood is equivalent to dying, then the person dies by losing one or the other of his personal predicates. Therefore any one of the human organisms described above (anencephalics, those who are in persistent vegetative states, or those who have lost every integrative function but their consciousness) are dead, and by parity of reasoning locked-in patients should also be counted among the dead, because we cannot separate persons from their bodies. According to this concept of a person, the human biological organism exists as an integrated whole; therefore death of one component should spell the death of the whole organism, whether it is loss of the brain as the centre of integration or dysfunction of the parts that are integrated by the brain; but that is not how the whole brain death criterion spells out the death of the person. Other concepts of personhood discussed before would also persuade us to think otherwise, since they also relate personhood to a combination of functions—of body and mind.

## The nature of death

Philosophically, the question can be asked whether death can be accepted as a purely natural event, a biological phenomenon that occurs in time as it has been portrayed in the preceding discussion, or a socio-culturally disruptive event whose definition depends on the religious, cultural, and traditional beliefs of the bereaved members of a particular society, since differing western and oriental cultures have different end points of life that can be precise or nebulous. In Japanese culture, consciousness and the brain are not the sole seat of personhood, but both mind (brain) and body befit the individual for participation in the community where he realizes his personhood. As Paul Tillich has aptly expressed it, "the courage to be as a part is the courage to affirm one's own being by participation. . . . Only in the continuous encounter with other persons does the person become and remain a person. The place of this encounter is the community."[15] This is also the Strawsonian view of a person that we have already encountered, and the point of view outlined by Rihito Kimura in his article on Death, Dying, and Advance Directives in Japan: "Historically, death was a natural event, and the criteria for death—cessation of heart beat and respiration—was (sic) unquestioned. This is no longer the case."[16] The death of a person might have been a personal and private matter in the past, but it is now also a familial, communal, and social one by virtue of his participation in the dynamics of community life, as John Lizza also points out: "Advances in medical technology make us particularly aware that human or personal death is not a strictly biological matter. . . . human beings and persons can now die in entirely new ways."[17] Furthermore, certain rituals must be completed before the person can be declared to be medically, legally, and socially dead.

Our traditional socio-cultural understanding of human life is well accepted as admitting the natural fact of death as a specific event marking the end of life—the final episode in the drama of life. John MacQuarrie quotes Martin Heidegger's description of death as "the probability of the impossibility of any existence at all . . . the last possibility of all. [And he goes on to add that death is] the possibility that makes impossible any further possibilities whatever, . . . not an end in the sense of a goal or a fulfillment, [but] a limit to existence."[18]

This definition says essentially the same as his characterization of death as "the possibility of the absolute impossibility of Dasein" into which it has been thrown without its choice. In this situation Dasein is gripped by an authentic being-toward-death by attunement to death as an existential possibility of its annihilation, with resulting Angst or fear of what is not objectively present; i.e., death. (Dasein is a term used by Martin Heidegger to characterize conscious human being which understands the meaning of its own existence and being and can also understand the existence of beings other than itself because of its particular factuality as subject in a world of objects and its ontological priority over them—as the superior being). No one wants to actualize this last possibility, nor does any one spend all his time waiting for it to strike, but it happens to be an inevitable part of the human condition that we try to avoid and flee from, and which we deny, because we fear it, in spite of the fact that our body cells die every day without eliciting the same sentiment of fear from us. We do not feel threatened by the death of our body cells, because our cognitive functions are not affected, and we still continue to live without being aware of that loss; but the oblivion of death hangs like the sword of Damocles over every moment of our lives, and that fact alone casts an inescapably ominous shadow over them, because in death there are no further possibilities—we have reached the end of the line, "the last possibility of all".

Still, we continue to live and plan for the future, because our essence transcends our existence and it evolves with time, reaching its climax only at the end of life, never before. We pack quality into our short life-spans to compensate for their lack of longevity. Perhaps we feel re-assured by the fact that we still continue to exist as new cells are generated after our old cells die and are cast off, although death of the person or the whole human organism precludes the possibility of any regeneration and return to life as we know it. Someone who merely loses consciousness but regains it after a while, or one who continues to replace his dying tissue cells, is not similarly concerned, because his loss is temporary. In death consciousness is lost permanently; all options are extinguished with finality and only oblivion prevails. In response to this kind of fear styled "Angst" or dread of something indefinite that is not present to the senses but is anticipated with ill-understood anxiety, which is lacking when we reflect on the deaths of other persons, Heidegger remarked: "If I take death into my life, acknowledge it, and face it squarely, I will free myself from the anxiety of death and the pettiness of life—and only then will I be free to become myself"[19], even if I am still a "being-towards-death" or a being who is authentically attuned toward death as the existential possibility of no longer being in the world. If we anticipate death, we exist authentically; if we fearfully expect it, we exist inauthentically and contrary to Epicurus's advice not to fear biological death.

## Temporal dimension of death

We should, however, be aware that this is the same oblivion that prevailed before the birth of the person who is now terrified of oblivion after his death. It did not affect him then, because he was not there to be affected by it, in the same way that he will not be there to be affected by his post-mortem oblivion whose anticipation is now affecting him adversely. Even if he had been there he would most likely not have been the present existing product of a fertilized ovum of twenty or so years later. Perhaps his fear arises from the fact that he has had, on balance, pleasant experiences in life that he does not wish to see curtailed for an unknown future, even as he would not entertain the thought of a boring longevity during which his body will become a burden on his life. He hopes to enjoy enough future life to fill a lifespan that he could not prolong by being born very much earlier than he was, because he thinks he now has the scientific wherewithal to prolong his existence, even though he does not have any assurance that his future life will be as pleasant and free of an excess amount of pain as his present life may be. Joys and pains that are not the subjects of present or past experiences are not real, and they are not states to fret about at any time before they actually occur; but we do it all the time, because of our future-oriented disposition. Besides, some people prefer life to death, even if their lives are filled with pain and suffering, which can be paradoxically relieved by death, and this is where euthanasia comes into consideration, as discussed in chapter 20.

The foregoing attitudes also relate to the concept of time that we have already discussed. On one concept of time, the person cannot change past events, like his birth, because they are no longer a part of present reality; time has flowed over them and left them to oblivion; but he can attempt to change the courses of present and future events, like his time of death, through the use of life-sustaining technology, because his present is real and his future still has a potential for reality. For the patient who subscribes to the other concept of time, however, all times are equally real; the only thing that matters is his lifespan. Both of them try to pack into life as much good as they can, for the sake of those who will succeed them after their deaths, since they have no idea how long their lifespans will be, and they may not be able to do all that they want to do before their deaths, in as much as they could not have started doing it before their births. Of course, some do it for the vanity of the personal glory that they hope will succeed them even after they have made a mess of their present lives. While he lives, every person lives in hope; but the ultimate threat of death hangs over him during every moment of his life, and he has no way of divesting himself of this existential threat that has dogged him from the day he was born. Birth and death serve as the temporal limits of his life, and as the necessary beginning and end in what has been called the recycling of life through death. He can beat death figuratively only with his legacy; never with the the doctor's drugs.

# Death phobia

It is noteworthy that ancient philosophers like Epicurus tried to dispel this fear of death, which he also thought was merely based on human anxiety about the oblivion of an uncertain afterlife—a fear and existential anguish during which some persons neglect to live, forgetting that once they are dead, they will have no opportunity to indulge their hopes and wishes or to recoup lost time spent in entertaining needless and unrewarding fears of non-existence. In his letter to Menoeceus, he states:

> Accustom yourself to believing that death is nothing to us, . . . death is the privation of all sentience. . . . Foolish, therefore, is the man who says that he fears death, not because it will pain when it comes, but because it pains in the prospect. . . . Death, therefore, the most awful of evils, is nothing to us, seeing that, when we are, death is not come, and, when death is come, we are not. It is nothing, then, either to the living or to the dead, for with the living it is not and the dead exist no longer.[20]

Epicurus wants us to understand that as long as we are alive, death is not bad for us, because it has not affected us, otherwise we would be dead; and when we are dead it is still not bad for us, because we are not conscious of it, since only the existent (living) can be ontological subjects of harm, benefit, or loss (of life). The non-existent can only be syntactical subjects of sentences about harm, benefit, or loss; never ontological subjects of direct harm, benefit, or loss. So we have no reason to fear death and to fret over it, because it is not bad for us alive or dead. This logical tautology does not concern itself with the fear of the pain that precedes dying, as in some cancers, or of premature death, but only with the ontological fact of non-being or non-existence. Oblivion is scary when we think of it while we are alive; but it has no meaning when we are dead, because at that point life, time, and space have disappeared and all possibilities have ended. Harm from death can accrue only to existent beings; the non-existent can't be harmed by death; so any concept of posthumous harm relates only to the sentiments of the living *vis à vis* the decedent, but it does not retroactively affect the now dead person,

206

because there is no way that his corpse can react to emotional harm. What remains, however, are the effects of the event and our absence from the niche that we occupied in the society to which we used to belong. Whether we played a major or minor role in the life of that society, our passing leaves a unique void that only we could fill while we were alive, in the same way as only we can die our deaths.

In his account of *An Existential Understanding of Death* James Park expresses a view that is similar to that expressed by Menoeceus. He describes the fear of death as

> a composite experience encompassing:
> (1) the abstract, objective, external, empirical fact of biological death;
> (2) our personal, subjective, emotional fear of ceasing-to-be,
> which arises from our awareness of our own finitude; and
> (3) our ownmost ontological anxiety,
> our Existential Predicament disguised as the fear of ceasing-to-be.[21]

We try hard to distance death from us and relegate it to other forms of life first before it touches us, because we realize that when it comes we will not be spectators any longer, since this event that comes to us once in our lifetimes is indubitably the final event of our lives, as Menoeceus and Heidegger have already informed us, and as Diane Zorn has observed: "In the everyday mode of being, Dasein interprets the phenomenon of death as an event constantly occurring in the world. It is a 'case' that happens to others. The general comment is 'One of these days one will die too, in the end; but right now it has nothing to do with us.' Dying remains anonymous and it has no connection with the 'I'."[22]

From the perspective of one theory of time as a wide expanse without the presumed demarcations of past, present, and future, we should not be afraid of death, since we will still exist at some point on the time scale even if it is not now. A bird's eye view of the time field will be able to pinpoint our position to other observers, even if we may not be able to do so, and that should give us comfort. On the other theory of time, when we demise at this moment that has real existence, we immediately become part of the past that is no more, and that gives us reason to fear the oblivion of death, making the harm that we sustain from death a bad thing for us living beings whose interests have been prematurely terminated in situations that survive our demise.

## Embryos, stem cells, human life

We come now to extending the concept of death to the case of human embryos and the use of their stem cells for therapeutic and research purposes. As we continue with the controversies surrounding these procedures, let us remember the advantageous capacity of

207

embryonic stem cells to live indefinitely in tissue culture and their ability to develop into any kind of body tissue that is needed to repair the damage that has been done to the form and function of any of our tissues, and especially to the tissues of the potential transplant patient. Unfortunately, much controversy has been created by certain sections of the community, religious and secular, about the personhood of embryos and hence respect for them as potential, not actual, persons who have a right to life, although they lack the bare essentials of full personhood, viz., sentience, rationality, and autonomy, which unconscious people admittedly also lack during that time of their unconsciousness without losing their personhood. But the claimants of personhood for the fetus forget that the personhood of the unconscious individual was not in question before, nor will it be after he regains consciousness and displays again those faculties that were suspended during that critical time. The personhood of embryos will, however, always be in question.

The claim for personhood of the human embryo begins in less audacious terms at its very inception as a claim for its moral status and right to life as a human being like other human beings, presumably by virtue of its chromosomal constitution that makes it distinct from other animals and determines that it will develop into one of us. The argument that the embryo is entitled to full moral status because it is alive and therefore satisfies the necessary and sufficient conditions for claiming moral status does not hold water, otherwise all other living organisms from which we withhold moral status would deserve the same regard, and that is not the case; we even eat some of them. From here, the claim jumps to the rights and interests of the embryo as a person who has a right to life, in spite of the fact that being human is not a necessary and sufficient condition for being a person, and also despite the fact that the embryo lacks the neural apparatus with which to cherish hopes, desires, and intentions that can form the core of its interests in its life. If any deference is due to embryos, it is the same deference that is lacking for those living animals and plants that end up gracing our dinner tables, gratifying our appetites, and filling our bellies so merrily. If we can override their right to life with such ease for our nourishment and flourishing (eudemonia), why can we not do the same for the sake of preventing human suffering and sometimes alleviating and curing conditions like Parkinson's disease, Cancer, Spinal cord injury, Diabetes, Batten's disease, etc., and save lives like that of the infant born without a trachea for whom one was built out of stem cells and successfully implanted? Instead, we compromise our ethical obligations with red herrings and contrived piety about entities whose moral standing is unquestionably unequal to that of the real persons who must perish while we quibble. Deontologists who insist that it is our duty to "respect" the personhood of embryos and not use them as means for our benefit will reject this utilitarian statement that advocates for the welfare of the entire human race at the cost a few embryos that do not possess the full moral status of all the people, if any. But they also have to face the bare fact of what is morally pragmatic for living persons versus potential persons .

True, the harvesting of embryonic stem cells from the inner cell mass entails disruption of the trophoblast and hence destruction of the embryo, which some people call murder; but until the embryo is about 14 days old at which time it is implanted in the uterus and before it develops the primitive streak, it does not have a nervous system nor is it an integrated system that is already destined to become one or more fetuses that can suffer harm any more than inanimate objects can be said to suffer harm. This is rendered more so by studies which show that fertilization is not an instantaneous occurrence but a process that extends over a period of time, and it is only after this stage in its development that the embryo undergoes "substantial cell differentiation and organization" towards forming a human organism that can suffer harm. Besides, many embryos don't make it past 14 days; they suffer organismic death by losing what Donald Landry of Columbia University calls their "integrated cellular division, growth, and differentiation", and that is the point at which their stem cells can be harvested without stirring controversy and without running the risk of being accused of disrespect for a human being in its formative stages (This, of course, is a backdoor insinuation of the presumed moment of fertilization and ensoulment as the dawn of personhood). If, as we will see later, we can harvest organs from brain-dead persons who have been kept alive artificially, why not stem cells from dead embryos?

Frederick Grinnell notes that arguments about the personhood of embryos should be abandoned in favour of reaching "a consensus about what defines embryo death, [for the sake of avoiding the exploitation of] one human life for the benefit of another. . . . under appropriate ethical and regulatory guidance, to use dead embryos and their cells for research to advance scientific knowledge and medical treatment"[23] The unproductive conflict that is being perpetuated by parties with vested interests should be replaced by ethically regulated efforts to formulate rational policies that will respect human life by using artificially fertilized (in vitro) embryos and their cells or spontaneously aborted fetuses for critically needed research to advance scientific knowledge and medical treatment of the citizenry, including those sections of the religious community that are opposed to the utilization of embryos for this utilitarian venture. They have yet to refuse any of the life-saving procedures that are the result of the limited stem cell research that they will "permit", and they have yet to indicate how they will preserve these elemental forms of human life and personhood forever, or how they intend to dispose of them when they are no longer in use.

Relatedly, Landry and his research colleagues found many non-viable embryos that were "hypocellular and lacked compaction on embryonic day 5 (ED5). All of the hypocellular embryos failed to progress to compacted morula or normal blastocyst when observed further."[24] This important research observation clearly indicates that the controversy over the use of live cells for creating stem cells from embryos older than ED5 that have suffered developmental arrest, and which have died due to sustaining "irreversible loss of integrated organic function" is without foundation. It is merely an exercise in a barren ideological fancy that should not preclude the harvesting of surviving cells from these types of embryos, under the guidance of the appropriately adapted ethical framework, based on the existing practice of harvesting essential organs from dying and dead persons for transplantation. Abandoning them to inevitable decay as is being done with embryos is selfish and of no benefit to anyone.

The concept of death still admits of differences among its legal, socio-cultural, philosophical, and purely scientific delineations, as the traditional definition is being moulded in adaptation to evolving methods of postponing death in its refined form as other than the cessation of all vital functions. This perspective must be accommodated in the definition of death if medical science is to achieve its objectives of organ transplantation and stem cell research and therapy at the same time that the public conception of death is being gradually turned around from the traditional one to one that is consistent with the aims of modern medicine. The concept of death as defined medically is certainly more critical than its legal or socio-cultural implications, if we take into consideration the overriding importance of time in making the strictly primary medical decision of pronouncing death for the sake of procuring transplant organs and the secondary decision of instituting procedures like enforcing wills and performing burials and cremations. The philosopher can help in refining this concept and ensuring that it does not fall victim to counterexamples that will nullify it.

The philosopher and the physician can also help to divert a good deal of the attention that is being directed to many theoretical hairsplitting issues while millions of people suffer neglect and loss of life due to starvation, controllable and eradicable diseases, because those who are comfortable and untouched by these adversities fiddle with imaginary religious issues like the sanctity of stem cells that have no foundation in fact.

# Philosophical Systems
# in Medicine

# 13

# Logic and Scientific Method

## Logic

In the search for knowledge of the world, for which we employ many methods of gathering, sifting, and validating information, we are also engaged in the search for truth. Logic is one of our best tools for assessing the validity or invalidity of reasoning that we have employed to draw conclusions from the information that we have gathered; but the truth of our conclusions depends solely on the truth of the premises that lead us to those true conclusions and the validity of these methods of reasoning. Where these conditions obtain, we have a sound argument. False premises cannot produce a true conclusion, even if the argument used to arrive at that conclusion is valid, as in the example: All men are cats; John is a man, therefore John is a cat. The major premise (All men are cats) is not true; the minor premise (John is a man) is true; the conclusion (John is a cat) is not true, because, even though the structure of the argument is valid, it and the major premise do not truly reflect what obtains in the real world. Irrelevant premises also fail to support their conclusion; e.g., the sunshine is bright; bright is an adjective, therefore sunshine is an adjective. Here a category mistake has been perpetrated through irrelevancy caused by the fallacy of equivocation on the word 'bright', i.e., arbitrarily changing the meaning in use of the word 'bright', whilst preserving the valid form of the argument. Similarly, true premises cannot yield a true conclusion if the argument is not valid, e.g., if he ingested poison, he will get diarrhea; he did not ingest poison, therefore he will not get diarrhea. But he could get diarrhea for any one of several other reasons.

These are the methods that we should employ at all times in dealing with medical problems and our attempts to resolve them; truth should be our goal.

## Logical axioms

Logical concepts are based on three fundamental axioms, like the axioms on which Euclidean Geometry is based. These axioms provide our handle on reality as experienced, versus stating what is logically possible, which is anything that is not irrational or self-contradictory, even if it does not exist:

(1) The law of identity states that $A$ is $A$; i.e., $A$ is identical with itself, and not with $B$, which is different. Cancer is cancer; it is not glomerulonephritis.

(2) The law of non-contradiction states that $A$ cannot be not-$A$ at the same time; i.e., it cannot be true to say of the same thing that it is both $A$ and not-$A$. Cancer cannot be cancer and not-cancer at the same time.

(3) The law of excluded middle states that something is either $A$ nor not-$A$; i.e., it cannot be neither; therefore it is false to say of something that it is neither $A$ nor not-$A$—neither cancer nor not-cancer. It has to be one or the other, and the other includes the rest of the world that is not $A$.

Another basic tenet in logic is implication. Implication makes it possible to validly infer the meaning of one statement from knowing the meaning of another statement, or the probability or certainty of one set of events from another: p implies q, if q must also be true when p is true; i.e., q cannot be false when p is true. It can also hold between particular propositions and between particular and general propositions, thus validating an inductive inference from particular to particular and from particular to general propositions; e.g., if it is true that John sustained a traumatic amputation of his finger, it is also true that he has one less finger on his hand and the same holds for all persons who sustain the same kind of injury. Also, if we define fever in terms of the number registered on the thermometer under certain conditions, then if it is true to say that $x$ has a fever, it is also true to say that his temperature is elevated. But if his temperature is not elevated, then it is not true that he has a fever; i.e., if not-q, then not-p. To say he has a fever and his temperature is not elevated is to utter contradictory statements, one of which must be false, or else both may be false; i.e., if p, then not-q; and if q, then not-p represent incompatible propositions. On the other hand, it cannot be false to say either he has a fever, or his temperature is elevated. One of these statements must be true, and both could be true; i.e., if not-p, then q, or if not-q then p. In medicine, in keeping with statistical principles, we infer inductively from observations on small samples to similar effects on large populations, relying on p values to exclude pure chance in our results; but we also infer deductively from general principles to specific cases, as in the taxonomy of diseases.

## Logical formulae

Logicians employ formulae to represent their arguments; e.g.,

p⟨?⟩q;
p,
therefore q. (true).

That means: if the truth of p entails the truth of q, and p is true, then q is also true (of necessity); e.g., it is true that if Tommy has a ruptured appendix he will have peritonitis; it is true that Tommy has a ruptured appendix, therefore it is true that he has peritonitis.

p→q;
not-q,
therefore not-p. (conditionally true)

In this example, if p is exhaustive, or refers to every known p, then the conclusion is true; but if p has singular reference, then the conclusion is not necessarily true of every case of p. If I state that a ruptured appendix implies the presence of peritonitis, but Tommy does not have peritonitis in spite of having a ruptured appendix, then I should conclude that not every case of ruptured appendix causes peritonitis. My initial statement of implication states a conditional truth which depends on the truth of every possible case of ruptured appendix causing peritonitis. But if cases of ruptured appendix are considered individually, some will have peritonitis, and others will not, depending on a host of individual circumstances. This is an important formula in Science, because if a hypothesis which is meant to be all-inclusive implies the truth of even one situation that proves to be false, that fact alone is enough to falsify the hypothesis; e.g., if every ruptured appendix that I have seen over twenty years has been followed by peritonitis, and if I then formulate the hypothesis "all ruptured appendices are followed by peritonitis", but in the twenty-first year I see one without peritonitis, my hypothesis is nullified.

p→q;
not-p,
therefore not-q. (false).

If p entails q, and p is not true, q need not be false. q can be true for other reasons. Tommy can have peritonitis for reasons other than a ruptured appendix; so I cannot conclude that he does not have peritonitis because he does not have a ruptured appendix. It has to be reframed to include the exceptions(s), actual and possible.

p→q;
q,
therefore p. (false).

If p entails q, and q is true, q can be true for reasons other than its entailment by p. The argument is therefore false. I can't conclude that Tommy has a ruptured appendix, simply because he happens to have peritonitis, which could result from a penetrating abdominal injury.

In the traditional formal logic, similar arguments apply. The argument, known as the syllogism, was structured as follows:

All S are P, e.g., all viruses are micro-organisms (a universal statement);

therefore

some S are P is true; i.e., some viruses are micro-organisms (a particularizing statement),

some P are S is true; i.e., some micro-organisms are viruses,

all P are S is not true, unless S is co-extensive with P; i.e., not all micro-organisms are viruses unless viruses are the only micro-organisms in the world.

some P are not S is true; i.e., some micro-organisms are not viruses.

no P are S is not true; i.e., it is not true to say that no micro-organisms are viruses, because some micro-organisms are viruses and all viruses are micro-organisms.

These principles ensure that fallacious reasoning does not creep into our arguments, permitting us to misrepresent the truth, as in the examples given in the first paragraph of this section. They also facilitate the preservation of logical implication as the foundation of our reasoning and classification of information, e.g., avoiding the category mistake of classifying disease with trauma. Besides, we are able to eliminate vagueness and ambiguity from our reasoning if we follow these principles and match our assertions with what obtains in the experienced world. Only the logically possible can be ontologically so, although not every logically possible situation does obtain in the world. Round squares are not found in the world, because they are logically impossible as conceivable or describable objects. Their properties cannot coexist in the same object, although the phrase "round square" can serve as the subject of a sentence and has done so in this sentence, and although individual objects that are either round or square do occur. Unicorns are logically possible as conceivable and describable objects, but the concepts that embrace them are empty of real objects that can instantiate those concepts, because unicorns are not found in the world of real things. The word 'unicorn' serves well as the grammatical subject of the sentence in which it occurs in a story relative context, but it does not name an existing (ontological) entity in our real world; although it may do so in a possible world.

216

# Deduction, Induction, Abduction, Analogy

Logic also employs three principal methods of reasoning:

(1) Deduction: drawing conclusions with certainty from two or more premises using the valid steps outlined above. This method extracts an occult conclusion contained in the premises, and does not really add much to our store of new knowledge about the world, but it allows us to attain certainty about our conclusions, barring any mistaken evaluation of the premises; e.g.,

all human tissues are composed of cells,
the skin is a human tissue;
therefore
the skin is composed of cells.

If we know a) and b), then we already know c) with certainty. In the case of invalid arguments, we know that their conclusions do not follow from their premises, even if these are true, because the method of reaching the conclusion is faulty.

(2) Induction: drawing valid conclusions with varying degrees of probability of truth from a collation of similar cumulative data and predicting what lies beyond our knowledge and experience; e.g.,

$S_1$ is P, $S_2$ is P, $S_3$ is P, $S_4$ is P...,
all known S are P;
therefore it is highly probable that
all S are P.

Inductive arguments provide new and uncertain information at every stage; therefore the evidence thus accumulated can never achieve certainty, as further new evidence may contradict existing conclusions. The truth of the premises does not ensure the certainty of their conclusion, because these arguments are assessed by degrees of strength for the support of their conclusions; hence the arguments cannot lay claim to soundness or validity, but only to degrees of probability of their conclusions. Their conclusions outreach the evidence; so any decisions made on their strength are arbitrarily determined by the pragmatism of their attendant circumstances and our discretion. Nevertheless, they are often used as such in the formulation of laws of science and nature, and of statements of the form: All S are P. Even in the case of 'All S are P', the world does not come with its laws of nature inscribed or with S already being P; we observe the regular, sequential association of events, coupled with some basic scientific principles, from which we derive the laws and universal statements that guide our method of deductive reasoning.

We leap from particularizing statements of the form: some S are P to universal statements of the form: all S are P, allowing room for one contrary example, or counterexample, to falsify the assertion that all S are P, while many more examples of the same datum will never prove with certainty the truth of the same assertion, because it is always possible that the next S will not be P. On the other hand, one S that is not P will never invalidate the proposition that some S are P. If I see 15 children with red rashes that prove to be viral, it does not follow that the next child's red rash will be viral; it could be bacterial or rickettsial; so not all red rashes are viral, but some red rashes are.

(3) Abduction: In contrast to 1 and 2, this is a method of inference that yields an explanatory hypothesis that best explains the observed facts without providing conclusive or probable explanation. This method (Inference to Best Explanation) is the subject of much controversy; so we will not discuss it.

(4) Analogy: determining the truth of the conclusion to an argument from the overwhelming correspondence of data between two or among many items of comparison. The number of samples involved and the number and variety of apparently similar features in them should exceed their differences in a relevant manner for the success of the analogy. These conditions make possible the hypothetical argument and conclusion that if item $A$ is similar to item $B$ in 85 relevant features and behaviour, and item $A$ has 15 more features that have not yet been defined in $B$, then $B$ most likely will or should also have those other 15 features. It is not imperative that $B$ have those features, only anticipated with high probability that $B$ will have them. The strength of the conclusion of an analogical argument does not depend on its validity, but on the relevance, number, and variety of cases counted, and on their relative similarities and dissimilarities; more similarities confer greater strength on the conclusion. For example, $A$ concludes that other people have minds like him, because they behave the way he does when he is angry, sad, happy, or in pain, assuming that they are not faking; research into the structure of the DNA molecule was spurred on by models which were presumed to bear an analogical similarity to it; laboratory animals are used for testing drugs on the presumption that their reactions to them are analogical to human reactions to the same drugs.

Gottfried Leibniz tried to capture the analogical principle thus, "it is not true that two substances may be exactly alike and differ only numerically"[1]. This ontological principle, *Identity of Indiscernibles,* states that if two objects are exactly the same in all their properties, so that they cannot be told apart, then they are one and the same object. (The converse principle, *Indiscernibility of Identicals*, states that identity entails indiscernibility of qualities; i.e., two objects that are identical share exactly the same qualities). However, we know some instances of identicals that are not one and the same item, such as clones and golf balls produced from the same mould. These items (analogues) have indiscernible qualitative identity, but not an indiscernible sortal or numerical identity, as Leibniz maintains, because they exist as individual spatiotemporal particulars of a particular sort; e.g., golf ball $x$ is of the same colour, shape, etc., as golf ball $y$, which means that they are identical in qualities, but they are numerically different or discrete objects, otherwise *they* would be *one* golf ball. Colin McGinn has summarized the argument thus: "if you find that for any property $F$ that $x$ has there is an identical property $G$ that $y$ has, and vice versa, then $x$ is qualitatively identical to $y$."[2] This argument implies that statements about the qualitative identity of $x$ and $y$ are really about the numerical identity of their properties. They are two "distinct kinds of entity" related by "a unitary notion of identity". Their identity consists in their sharing the same (kinds of) properties and not in two kinds of identity. So, to say that $x$ has property $F$ is to assert its identity as apart from $y$ that has the different property $G$ and occupies a different location. If $F$ and $G$ are (qualitatively) identical, that fact alone does not add numerical identity to their qualitative relation to make them the same object. So, disease A may share the same qualities as disease B, but if they occur in different subjects at the same time, they may be subject to many constitutional factors of the subjects that render their manifestations different while also requiring different modes of treatment as determined by the reactions of the subjects to available modalities of therapy; e.g., A's leukaemia will respond well to chemotherapy and radiation, while B's leukaemia will respond best to stem cell or bone marrow transplant.

# Scientific Method

As a result of the preceding considerations, when the scientific method is used in the delineation of clinical syndromes and diseases, observations are made of constellations of signs and symptoms; hypotheses are formulated about what they represent; predictions are made about the nature of future concurrences of the same signs and symptoms; these predictions are subjected to tests to prove their consistency with the hypotheses; and when they satisfy these tests, theories that explain disease patterns and predict the conditions under which similar diseases will occur are formulated, thereby also making it possible to institute preventive measures against these diseases where practical. All these steps are based on the validity of universal laws as buttresses for particular cases; but medicine, in its inclusiveness, deals with individuals first and universal populations second, making the inductive method the method of choice to hypothesize about and unravel medical problems. When predictions do not satisfy tests, the hypothesis is dropped or modified, and other explanations are sought to account for the clinical presentations.

Science depends on the probabilistic conclusions of typically uncertain inductive arguments for its progress, because the more certain deductive arguments do not add any new facts to scientific hypotheses; their conclusions follow necessarily from their premises. Similarly, many of our daily endeavours are based on the pragmatic and fundamental character of inductive inference. We sit on chairs without expecting them to disintegrate, because they have not consistently done so in the past, and we drink water from the tap with confidence, because it has not poisoned us in the past, until these bad things happen to us. The accumulation of these statistical facts would appear to strengthen the role of inductive arguments in scientific endeavour, but some philosophers of science, like Karl Popper, do not think so, as we shall see in the next section.

In scientific theories presuppositions are entailed by the theory they presuppose; therefore the truth of these presuppositions or necessary conditions is essential to the truth of the causal relations that generate the theory, as we saw when we discussed Mill's criteria. Sufficient conditions, on the other hand, entail the theory, and their refutation constitutes a refutation of the theory resting on them; e. g.,

if p, and only if p,
then q;
p,
therefore q.

(if the theory is true, and only if the theory is true, then the presupposition is true; the theory is true, therefore the pre-supposition is true).

But if
not-q,
then not-p.

i.e. if the presupposition is proved to be not-true, then the theory is not-true, and it falls, because it has been disproved.

## Karl Popper

According to Karl Popper, "induction simply does not exist, and the opposite view is a straightforward mistake."[3] Therefore in employing the logical formula " All S are P", science does not thereby imply the existence of Ss, all of which are Ps, because to arrive at the universal proposition "All S are P", we need to identify an infinite number of particular instances of existing Ss and then use induction to arrive at the universal proposition, which is not possible. For practical reasons, therefore, the assertion implies that if anything at all is a cancer (S), it is also lethal (P). Therefore, if we assert that there exists a cancer that is non-lethal, when the general statement (All cancers are lethal) implies that there cannot be a cancer anywhere that is other than lethal, then the assertion that there is a non-lethal cancer is false, because all cancers must be lethal, or else the general scientific statement that all cancers are lethal is false; i.e., the counter-example consisting of the existence of a non-lethal cancer has falsified the general statement. We will not discuss the problem posed by use of the logical equivalent of the hypothesis that all cancers are lethal and its implication that all non-lethal things, like a toothache, are not cancers as confirming the hypothesis that all cancers are lethal, because it does not. The sum total of lethal things in the world does not necessarily make cancers lethal. Similarly, if all Sickle Cell Anemia patients are anemic (by definition and factually), then if Jim has the disease, he is anemic. If he is not anemic, then he does not have the disease or he has been treated, or else it is false that all Sickle Cell Anemia patients are anemic. This is the argument represented by the formula: if p, then q; not-q, therefore not p.

Popper proposed his method of falsification as an alternative to the inductive method that he saw as inadequate for demarcating science from non-science and for sustaining scientific research, because it could be justified only through the infinite regress of further and further induction. He maintains that we cannot justify our scientific theories by demonstrating their semantic or objective truth, but we can at least demonstrate the probability of their truth, thus justifying our belief in them. Hence in his hypothetical-deductive method he proposes the positing of hypotheses (conjecture) to explain the causal relationships among observed phenomena, the scientific problems that they generate, and the predictions that they make possible. The predictions and the phenomena are then subjected to rigorous testing in an open-ended fashion that can be carried on indefinitely until they are falsified (refuted) or corroborated. If they are corroborated, they can be used for making further predictions and justifying more theories, although confirmation, which is unattainable, would provide a more secure footing than corroboration. Predictions that fail the test falsify the hypothesis; those that pass the test only tell us that the hypothesis might be true, but it has not yet been falsified and is still a target for falsification. To him, any discipline that is not amenable to this method of empirical testing and potential refutation by observation is not science, in the same way that logical positivists condemned as meaningless any statement that could not be empirically verified, without including their own principle of verification, which also failed to satisfy those restrictive conditions. According to Popper, hypotheses can only be falsified or corroborated, not justified or verified; and he defines corroboration as: withstanding serious attempts at falsification at and up to a particular time, with respect to prevailing accepted systems of basic concepts. Despite these criticisms, we know that all the practicalities of the scientific world are built on induction, and this edifice of inductive probabilities does not appear to be ready to collapse at any time, leaving humanity in its dust.

But as we will see when we consider Duhem's criticism, a consistently false hypothesis implicates other initial conditions in addition to proving the falsity of the hypothesis under consideration. Furthermore, one can falsify a conclusion that claims certainty for all time, but not one that claims only timed probability without also committing itself to timeless certainty. However, support by consistently true predictions from a hypothesis should strengthen confidence in it and its initial conditions while bringing a probable conclusion closer to certainty; and this is a more pragmatic result than consistent refutation which Popper did not favour. Besides, past falsifications do not guarantee the falsifiability of similar future hypotheses, in the same way as past sunrises do not guarantee future sunrises; so a hypothesis that has been falsified in the past can yet be corroborated by future events, and that marks scientific progress.

In keeping with this ever present possibility of falsification, the scientist is forever checking his hypotheses against reality and amending them when they fail to reflect reality as it exists. Therefore at any point in this search his explanatory pronouncements serve only as deservingly workable corroborations of his hypothesis, because confirmations only enhance the probability of truth without ensuring its certainty. In employing the method of falsification, as it refers to his hypotheses and theories, not to his statements and propositions, the scientist should be aware that it does not discredit only conclusions from hypotheses, but also the foundations of those hypotheses, if they were not derived by this method. The reason is that they could possibly be biased, if they have not passed the falsification test for their authentication. So, when two or more hypotheses are in competition, the favoured one must be more precise and explain more than the falsified hypotheses, including all their falsifying elements and the causal assumptions related to them.

Researchers in all scientific fields, including medical science, employ these methods of falsification and corroboration to assess the claims of other researchers in a public forum where phenomena predicted by their scientific hypotheses should be accessible to observation, so that they can be publicly confirmable or refutable. Those theories that cannot be falsified derive the best possible support for their truth from this failure to falsify them, as long as the methods used do not employ irrelevant observations. In this environment the intuitive biases of persons and their beliefs as

222

prompted by their faith in the pronouncements of all types of authorities, genuine and spurious, have no place; only the quest for truth prevails. If the theory resulting from this quest is falsified by failing the test of truth, it is less often discarded and more often modified to cohere with the rest of the established facts in that discipline. This is in contrast to those theories that seem to be able to explain everything without really explaining anything. They claim universal verifiability by extracting confirming evidence from every possible situation without being falsifiable by any situation. Such theories do not help to advance science and medicine.

Popper's method has been dubbed its own enemy because it, like the verificationist (positivist) theory that preceded it, is not amenable to falsification. Besides, Popper still faces the pervading problem of justifying the veracity of our perceptual statements, which are the foundations of our knowledge of the external world.

## Duhem-Quine thesis

The Duhem-Quine thesis casts doubt on the logic of falsification, but before Duhem and Willard van Orman Quine, Henri Poincare first observed that when predictions based on a theory that is constructed on multiple hypotheses do not stand up to experimental proof, it is impossible to tell which of the premises must be changed. Duhem subsequently drew attention to the failure of a single hypothesis to predict the outcome of an experiment, because no theory can ever be tested alone, in isolation; other hypotheses relating to the same data are also implicated in the testing. So if the outcome of an experiment is other than as predicted, it is still logically possible that the hypothesis in question is sound, but that one or more of the other related hypotheses is at fault, because different theories can be used to explain any set of data without relationships of entailment between that set of data and a particular theory; i.e., theory is underdetermined by the data, and can never be conclusively verified by any set of them. So the theory of interest can still be retained while any one of its rivals is jettisoned. One hypothesis can never demonstrate the falsity of its rival, because the fault can be in any one of the competing hypotheses within the web of beliefs that are under scrutiny. Also, verification of a single hypothesis can't give ultimate meaning to the proposition that states it, à la positivism, because it is only a facet in the wide web of belief postulated by Quine in his support of Duhem's point of view. In this wide web of belief, which impinges on all of our contact with reality, "the more complex the hypothesis, the more and wilder ways of going wrong."[4]

The same holistic reasoning procedure is employed when clinical problems are encountered and a conceptual framework for dealing with them is being developed, as per the presuppositions entailed by the proposed hypothesis. If micro-organisms can be the cause of one disease then it may well be fruitful to investigate whether they cause other diseases. Sometimes, however, our nicely directed research encounters the difficulty that the theory does not match the practice. This is the point at which, to borrow Thomas Kuhn's analogy, we realize that either the jigsaw puzzle we were putting together comes from two different boxes, or the box that we possess does not contain all the pieces. Just because we have all the pieces, we do not necessarily have the full picture if they are not put together in their correct relations.

Every hypothesis rests on presuppositions about expectations, which are also based on the questions prompted by the problem under consideration and its background. We do not order a hemoglobin electrophoresis to distinguish between a fracture and a dislocation; we order an x-ray and reserve the electrophoresis for a case of microcytic hypochromic anemia which is not decisively due to iron deficiency and could be due to Thalassemia.. Relevant data are assembled by asking the right questions about the problem in relation to its background subject matter (does this 6 month old child diet consist of only cow's milk, or is this 6 year old child of Mediterranean extraction?); they are then analyzed and interpreted in light of the relevant hypothesis; predictions are made, and tests are carried out. Those that confirm the hypothesis elevate it to the status of a theory, which serves to explain the problem under investigation and similar problems in etiological terms. Those that fail to confirm the hypothesis undergo further scrutiny to determine if errors in judgment were committed in formulating them, or if the basic tenets from which they originate are at fault. Sometimes the unexpected results are signs of the variability in outcome that can occur even from secure premises, because of the so-called chaos effect (page 226).

A verified hypothesis is not one that has been proved with certainty; it is still liable to falsification by even a single counter-example, or by a simpler hypothesis, which is more comprehensive and permits more truthful and accurate inference and prediction from past and present data and events; e.g., the geocentric theory of the universe did not explain existing and predict anticipated events as accurately, comprehensively, and as simply as the heliocentric theory; so it was jettisoned for the latter, which is constantly proving to be the right theory in novel situations. The same fate overtook the flat earth theory, although it still has die-hard adherents who do not support the spherical earth theory. They are probably hanging on to the idea that if x is right or true, it does not mean that all not-x is wrong or false; facts that make not-x right or true may still be discovered in time. Thus they might be hoping for the day when facts will be discovered which will prove them right. That, after all, is how knowledge grows.

## Thomas Kuhn

Kuhn rejects the notion of falsification as proposed by Popper on the grounds that normally scientific theories do not become otiose by falsification, but by the ascendancy of competing theories through a replacement of one paradigm by another. If a theory deviates fundamentally from established practice but proves to be superior to existing theories in explaining the facts, it displaces the old theory or incorporates it; and that constitutes a shift in paradigm. Paradigms are defined by Kuhn as "accepted examples of scientific practice . . . [that] provide models from which spring particular coherent traditions of scientific research."[5] That means paradigms determine the bounds of inquiry in a particular domain, and what lies beyond those bounds is not worth pursuing, even if it may have comparable or superior value to the paradigm. It also means that sciences without unifying paradigms lack direction, making for widely varied interpretations of the same observed data in the same sphere. However, paradigms like those that attributed the etiologies of cervical cancer and Burkitt's lymphoma to other factors have been shattered by the postulation and proof of a viral etiology for these conditions, proving the futility of adhering to a paradigm by denying what is outside of it for the sake of conforming with the popularly held opinions of experts within the particular scientific community.

The pursuit of paradigms tends to suppress the kind of quest for falsifying hypotheses that Popper believes is the justifying feature of science, because hypotheses guide the choice of problem and the fact gathering that is compatible with that choice of problem. Forsaking the paradigm amounts to forsaking the foundations and practices of the discipline defined by it, since, as Kuhn maintains, by studying these paradigms under mentors with whom he will thus share the same basic information, the student will seldom be at variance with them over the basics and methods of practice of their common profession. Paradigms are presumed to help direct research to relevant subjects and to keep researchers focused on them to the exclusion of irrelevant and elusive projects. Their lack of constant liability to falsification and refutation by all types of anomalies in the style of Popper's hypotheses is believed to confer upon them a stability that can resist disruption by minor anomalies, yielding only to those that produce crises, and thereby ensuring the stability and progress of science at the same time.

Some philosophers argue, however, that Kuhn's entire exercise only to a terminological one and not a metaphysical one, because the referent of the terms used in the paradigmatic description remains metaphysically the same while the referential terminology shifts in meaning, as expected, to create a paradigm shift in concert with what Kuhn styles a scientific revolution, defined as a "community's rejection of one time-honoured scientific theory in favour of another incompatible with it."[6] This happens when the accepted theory presents anomalies that cut against the grain of accepted scientific practice that is in operation. Such anomalies confound all reasonable forecasts from existing theories, resulting in shifting allegiances to traditional practices in the search for new ones.

It is tempting to think that such a shift in paradigm could very well be taking place in orthodox Medicine with the ascendancy of Complementary Medicine—Darwinian and Holistic—except that in this instance it is not terminological but ontological, since the character of this other discipline differs in essentials and not in mere referential terms from that of orthodox medicine. The proponents of Evidence-Based medicine (EBM), which we will consider in the next chapter, also claim that it is essentially a shift in paradigm from the old established practice of basing diagnoses and treatments on the tenets of basic science and pathophysiology; but that is incorrect because the new system of medical practice has not replaced the old one on the basis of complete incompatibility. If anything, it, like Precision Medicine, serves a complementary function by ensuring the realization of the principles of the Duhem-Quine thesis.

In the paradigm shift, new minds that are not bound by traditional practices precipitate a shift in the kinds of questions posed and the methods used for solving them, in spite of the general tendency for the scientific community to "suppress fundamental novelties because they are necessarily subversive of its basic commitments."[7] The anomaly that pulls off this heist is one that cannot be ignored, because it presents a crisis by 1) threatening the foundations of the existing theory that is failing to solve all the problems that the new theory that heeds the anomaly promises to solve, or 2) predicting events better than the old theory, as we saw with the heliocentric and geocentric theories. It is a falsifying anomaly. Where there is no gross incompatibility between the old and the new paradigm, the old one is retained to serve alongside the new one. Rejecting all theories that do not fit the facts (or hypotheses) leaves nothing to work with; so we should retain the theory that fits the facts better to guide further research and the management of patients. Popper said essentially the same thing.

## Non-linear dynamics

After all the necessary precautions have been taken to ensure foolproof theorizing and deduction making, error still creeps into our

226

operations, because we cannot control initial conditions so precisely that their consequences will always be predictable to the last detail; e.g., measurement values at the start of any process. This is because even the most perfectly designed instruments used for measuring these conditions are not 100% precise. The outcome is progression to a state of "chaos" resulting from extreme sensitivity of the system to these conditions and its inevitable, inherent dynamic instability, as Poincare initially proposed. Such chaotic systems are called non-linear, as opposed to deterministic linear systems where output is theoretically linearly and predictably related to input, where every event is the imperative and undeviating result of preceding events, and where the same initial conditions generally produce the same results. The behaviour of a linear system is the expected behaviour of the sum of its constituents; nothing more. In non-linear systems composition does not occur; hence the chance of sudden unpredictable changes in the response of organisms to controlling environmental conditions that would otherwise not disturb linear systems; e.g., sudden onset of supraventricular tachycardia in a heart that appears to be in normal sinus rhythm.

The widespread occurrence of chaotic behaviour seems to be the norm throughout the universe where apparently random data are seen to yield orderly results while apparently orderly data yield chaotic results, producing some of the widest fluctuations in the behaviour of natural systems. If that is the case, there is no room for linear systems in practice; every process is chaotic. The deterministic chaos seen in the normal behaviour of the human heart is a case in point. Grossly, the time between beats appears to remain constant even with normal sinus rhythm variations caused by breathing, the clash between the sympathetic (accelerator) and parasympathetic (decelerator) systems, and the normal acceleration in heart rate that occurs with activity; but an analysis of the ECG will show variability, which becomes even more accentuated as smaller sections of the tracing are examined under magnification. An underlying chaotic pattern becomes apparent. Ordinarily, therefore, comparisons between two apparently similar phenomena, like regular heartbeats, may reveal vast differences in ECG patterns. At the same time, abnormal looking patterns with large variability may suggest abnormality when compared with regular looking patterns that really reflect decreased reactivity and the decreased variability that we know can culminate in the flat line of absent reactivity indicating cardiac standstill. But these abnormal looking, chaotic patterns are more re-assuring of the absence of pathology than the smooth (flat EEG and EKG), regular, or periodic ones.

The physiological-stability-imparting irregularity seen in non-linear systems is similar to those irregularities seen in the structure of natural geometric forms called fractals as illustrated in line ab below. The typically smooth lines described in geometry textbooks turn out to be irregular when viewed under the microscope, and further irregularities can be exposed by the electron microscope, producing a series of self-similarities known as fractals. Anatomical examples of fractals are seen in the branching structure of the enteral, bronchial, vascular, and neuronal systems, which are thereby adapted to provide maximal nutritive absorption, gaseous exchange, blood supply and drainage, and neuronal integration within those systems. Imagine how inefficiently a single saccular lung would be able to hold large volumes of air, especially during exercise, without overreaching its tensile strength and bursting, or how easy it would be to deflate the entire lung with just one puncture wound with disastrous consequences to the victim, or how limited its capacity for diffusion and gaseous exchange its surface would be, and also how precariously a single delivery tube would function in case of obstruction. Now compare this with the normal lung comprising a honeycomb of multiple blisters of 300 million air sacs sharing the function of one large sac. It presents an infinitely larger surface area of gaseous exchange for the same perimeter, because a fractal line (unfolded version of the regular line), being irregular, is longer than a straight line drawn between the same end points, a and b, as shown below:

The lung can therefore hold very much larger volumes of air in these saccular pockets while still retaining its tensile strength and limited size. One of these sacs can sustain a puncture wound without incapacitating the entire lung and jeopardizing the victim's life, while the multiple rigid airways provide channels, structural support, and safety valves for continuing pulmonary function in case of injury or blockage to any one of them.

In the case of the gastrointestinal system, a single smooth tube from mouth to anus would fail to absorb nourishment in the same way as the denuded intestinal villi of celiac disease and the haustration-lacking mucosa of ulcerative colitis. A smooth tube also lacks the ability to churn and propel bowel content, and valves, like the lower esophageal sphincter and ileo-cecal valve, to prevent reflux. Parallel arguments can be advanced for lack of arborization in the vascular system, with predisposition to gangrene and edema in tissues farther away from the area of diffusion, which would be the only means of distributing the blood supply in the absence of increasing branching of the arterial system. In the case of a myocardium lacking coronary blood supply, massive myocardial

228

ischemia and infarction would result from dependence on diffusion from a saccular chamber with a single vascular trunk exiting to supply the rest of the body, while the lack of a His-Purkinje system would result in mass myocardial contraction, which is vulnerable to massive failure from interruption at any point in the impulse conducting system—a situation which does not now exist with the staggered distribution of electrical impulses through distinct nodes and bundles. Similar catastrophes would await the person who has a neurological system with a single non-segmental trunk that conveys sensation inefficiently and produces a massive reaction to localized sensory stimulation or deficit disorder, and to a motor systems that is similarly ill-equipped. As for the person with a smooth brain without convolutions, limited surface and volume area would limit all his function severely, and reduce his cognitive level to that of smooth brained persons with lissencephaly. Illustrative examples follow.

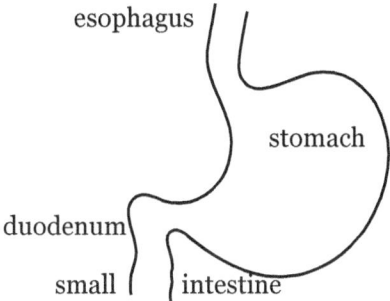

Upper gastrointestinal tract: smooth sac

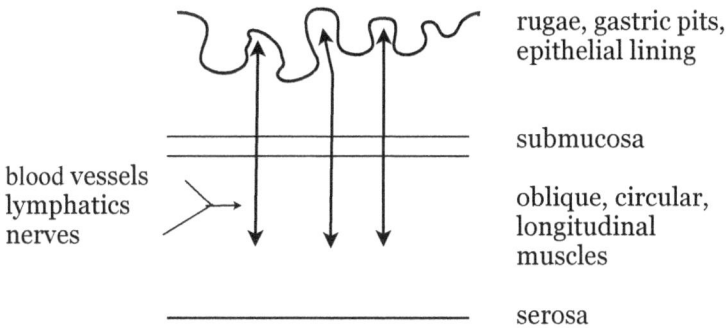

Cross section of stomach
Fractal lining versus smooth lining above

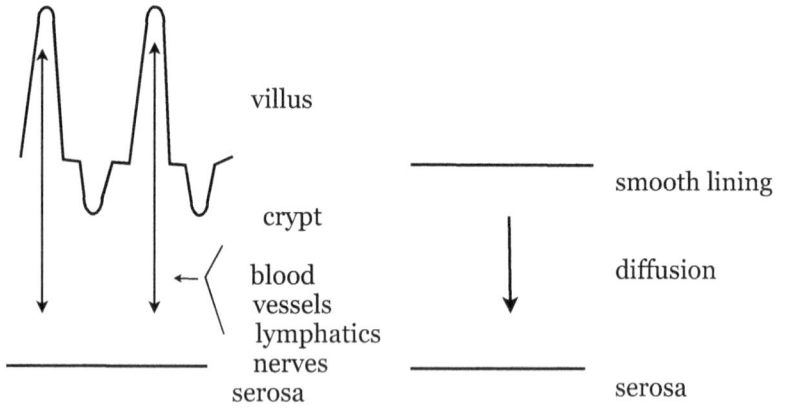

Cross sections of small intestine
Fractal lining vs smooth internal lining

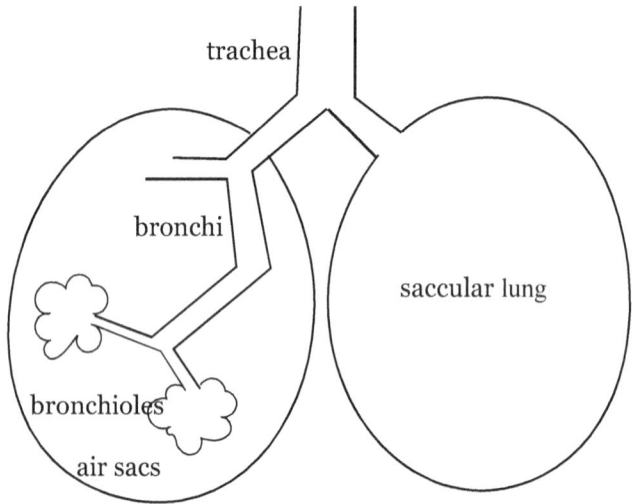

Respiratory system

Fractal arborization vs single lung sac

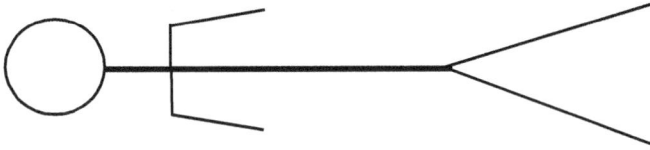

smooth brain　　single nerve trunks, lymphatics, and
blood vessels

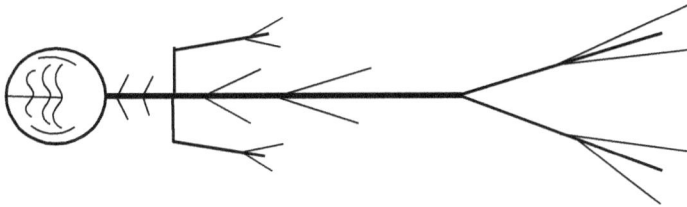

brain with convolutions　　arborizing nerves, lyymphatics, and
blood vessels

Neurological system

All these considerations of the need for balance between order and variability in the maintenance of proper physiological well-being have a direct bearing on our understanding of form and function in the human body, and they will certainly influence our ability to monitor our patients and to anticipate dysfunction and rectify it when it occurs. With regard to the desirability of substantial variability in normal physiological function, Bruce West and Ary Goldberger noted that as functional variability decreases, the frequency spectrum also narrows, thereby decreasing spectral reserve and impairing the margin of safety for reaction to inimical influences on the organ system implicated in the disease process—"The paradoxical appearance of highly ordered dynamics with pathologic states ('disorders') exemplifies the concept of complexity loss (decomplexification) in aging and disease.[8]" For example, the decrease observed in normal variability in heart rate induced by imbalance between the effects of the sympathetic (rate increasing) and parasympathetic (rate slowing) divisions of the autonomic nervous system on the sino-atrial node resulting from the effects of diabetes mellitus on patients. The same decrease in heart rate variability, loss of spectral reserve, and onset of low frequency periodicities, all of which replace the chaotic processes of a healthy heart, is seen in patients with congestive heart failure, fetal distress syndrome, and sleep apnoea, as well as in adults and

231

infants at risk for sudden death. Interestingly, progressive loss of variability and complexity has also been noted in aging, where we also know that cortical neurons lose their fractal dendritic arborization with advancing age, and in "vascular 'pruning' in primary pulmonary hypertension, and possibly, alterations in distal airway architecture with chronic obstructive lung disease" [9].

Irving Dardik[10] supports these viewpoints in his contention that well modulated increased variability and heart rate is directly related to good health, and inversely related to disease. He argues that heart beat and heart rate are continuous waves whose crests correspond to energy expenditure during systole while the troughs correspond to energy recovery during diastole. Analogous cardiac wave patterns occur with maximum exercise and resting, and with sympathetic and parasympathetic stimulation. He calls the entire complex "heart waves" whose function is to sustain heart rate variability. Decreases in heart wave range result in decreased heart rate variability; the narrower the heart wave range, the greater the decrease in heart rate variability, and the less responsive and adaptable to change the heart becomes; i.e., there is less chaos and more vulnerability. This process can continue to the point of absent heart waves, which is the graphic indication of cardiac asystole. Sustained high amplitude variability indicates healthy heart function. Below is Dardik's illustration.

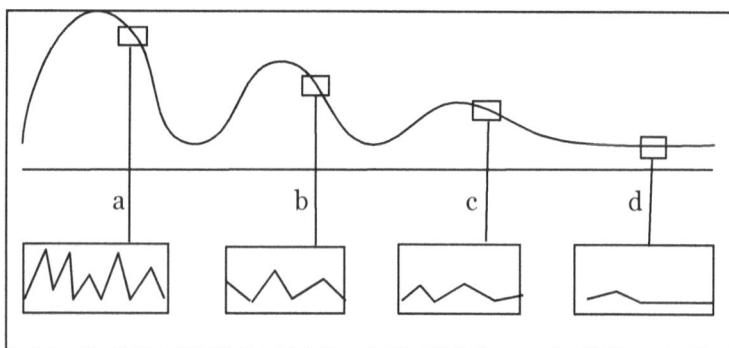

Arrows indicate decreasing amplitude of heart rate variability, but there is also a diminishing chaotic pattern, with increasing regularity within each wave from a to d, at which point chaos and frequency are totally lost, as shown in the magnified images of blocks taken from waves a to d.

There is no question that concepts of non-linear dynamics are essential for unraveling some of the obscurities and complexities that characterize human anatomical (gross and ultra-structural) and physiological form and function, as illustrated above, and as physicians often observe in the illnesses of their patients that follow

232

unpredicted and inexplicable courses. Another example of the utility of the fractal concept in medicine can be gleaned from genetics in regard to the polymeric structure of DNA, which enables it to encode inordinate amounts of information by virtue of its repeated sequences of monomer units called nucleotides: Adenine, Cytosine, Guanine, and Thymine, connected by Hydrogen bonds. Each one consists of a 5-carbon sugar, deoxyribose, to which are attached a nitrogen containing base and a phosphate group, as illustrated below. (The human genome has over 3 billion bases compared to less than 7 million for bacteria and less than 250,000 for viruses). Some scientists believe that the entire fractal pattern of instructions in the gene directs the various functions of the cell including the orderly formation of human tissues and organs. Also, that the 95% portion of the gene that is styled "junk" DNA plays an active role in directing cell function, together with the 5% that is known to be active.

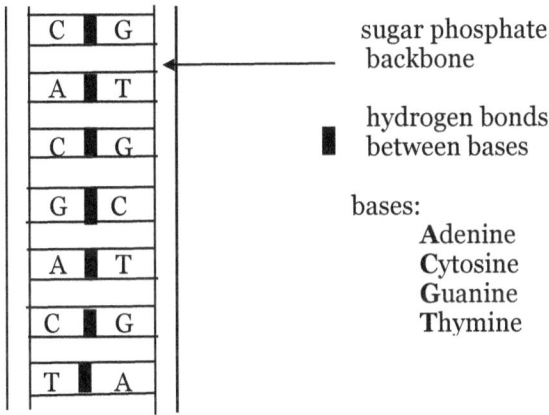

| C | G | | sugar phosphate |
| A | T | | backbone |
| C | G | | hydrogen bonds between bases |
| G | C | | bases: |
| A | T | | Adenine |
| C | G | | Cytosine |
| T | A | | Guanine Thymine |

Structure of DNA

Ultimately, non-linear dynamics and chaos will prove their usefulness to the physician in the physiological monitoring of her patients. She will be able to spot trends toward lower variability in systems that should normally be chaotic without believing that less chaos spells better homeostasis, as in the relative pathologic periodicity of lethal arrhythmias compared to the chaotic dynamics of the healthy heart with its sinus arrhythmia. She will be able to study the simpler components of motion in patients with gait disorders like Parkinson's disease and thereby assess the severity of their affliction with a view to its amelioration with treatments like injections of Botox. With information gathered from airway cast studies of normal and asthmatic patients in different stages of the disease, she will be able to analyze the correlation between the fractal dimensions of asthmatic airways and the pathological effects and severity of asthma as reflected in the loss of structural complexity resulting from remodelling of the airways shown by their digitized images. In addition, researchers have demonstrated a fractal difference between normal and cancerous breast tissue—a useful adjunct to mammography, which may sometimes not pick up the minor differences in X-ray attenuation between normal and cancerous glandular tissue.

In its totality, the foregoing discussion has implications for the following concepts:

(1) Normality, by challenging homeostasis as the hallmark of the ideal state that is claimed for the well-being of all living organisms, since homeostasis entails the inherent propensity to equilibrium and invariability, which is contrary to what chaos represents.

(2) Causality and induction, by divorcing them from mechanistic determinism and placing them in virtually the same class as quantum behaviour and freewill, which are undetermined.

(3) Disease, by introducing undiscoverable variables to the many unknown but discoverable ones that make every clinical case of disease $x$ different from other cases of the same disease and thereby confounding medical diagnosis.

(4) Systemism, holism, and reductionism by the participation of non-linear indeterminable factors in the framing of these concepts as a necessary consequence of the fact that chaos permeates all known natural processes.

Examples of the effects of non-linear dynamics in the philosophical concepts that undergird clinical medicine and its supporting basic sciences are as many as the concepts themselves, because chaos is itself a ubiquitous concept, and the above list can be extended to satisfy many more existing and imaginable medical and scientific concepts.

The role of fractals and non-linear dynamics in medicine will continue to unfold with time, human endeavour, and the growth of knowledge.

234

# 14

# Epistemology

## Definition

Epistemology (Episteme=knowledge+logos=discourse) is the branch of philosophy that investigates the nature and origins of human knowledge about the truths of reality and our methods of acquiring that knowledge, whether by perception, reason, or both. We have already discussed most aspects of epistemology in chapters 2 and 3. Ryle described two ways of knowing, viz: knowing how, which involves a display of skills in performances such as riding a monocycle or driving a car, and knowing that, which involves knowledge of facts—that such and such is the case. Bertrand Russell's knowledge by acquaintance and knowledge by description can be reduced to knowing that, since his two modalities of knowledge acquisition also inform us that such and such is the case.

Ordinarily, we perceive things, but we come to know what they are only after we have interpreted our perceptions with the aid of reason, because the senses alone are inadequate to guarantee and place in proper context what we perceive. The possibility of error lurks in every act of perception; e.g., when a stick that is half immersed in water is looked at from the side, it appears bent. On seeing this phenomenon we use evidence from common experience, but essentially from optics, to correct for the illusion inherent in this unedited act of perception. Besides also relying on past experience and the culled evidence of several observers, we rely on the use of reason, whereby conceptualization, classification, and identification are executed with the aid of deductive and inductive logic to determine if our perceptions are truthful or false representations of reality, and whether they will promote our goals in life. This chapter will deal only with the category of knowledge called propositional knowledge; i.e., knowledge that $x$ is the case, as opposed to knowing how to do $x$, or knowing $x$, as in being acquainted with $x$.

## Propositional knowledge

When we claim to know something truly we are claiming that

we understand and can explain its occult causes on the basis of arguments using propositions known to us to be true. These may include causal explanations like those of the role of tumour suppressor genes in carcinogenesis, scientific laws like Fick's law of diffusion relating to transport of oxygen through the lungs to the mitochondria, and inference to the best explanation (abduction) when other accounts of events do not suffice to explain them. We see, therefore, that knowledge of medical facts is not different from knowledge of scientific and other facts about the world.

In the specific case of medical knowledge, Epistemology enquires into how we come to know the nature of disease, and that our patients are afflicted with the many defined diseases; how we acquire that knowledge, and what our criteria are for justifying it in the same way as we do with other information presented to us. With this information in hand, we can gain a better understanding of the nature of Medicine as a science-based endeavour and as a system that is based on the hypothetical-deductive method that employs both empirical observation and rational deduction to establish its diagnostic and therapeutic parameters. From this understanding, the proponents of Evidence-Based Medicine claim that their discipline has evolved as a universalized, logical approach to the accumulation and systematization of knowledge and understanding of the mechanisms and treatments of disease, as opposed to practices that rely on the personal experiences of doctors and traditional methods of treatment, some of which are merely anecdotal.

As we have already observed, the hypothetical-deductive method, based on the existence of the external world as it presents itself to us, without our mental distortion of it, begins with observation of phenomena, followed by postulation of a unifying and explanatory hypothesis, and then the deduction of universal laws to account for subsequent, similar phenomena to confirm the hypothesis, or dissimilar ones to force a modification of the hypothesis or qualify the phenomena for recognized exceptional status. The method is still subject to error, but it uses this experience to build a functional system of beliefs, which approaches certainty without achieving that ideal. That means even scientific knowledge is not yet justified, true belief, in keeping with the definition of knowledge as justified, true belief. New ways of knowing in Medicine will be discussed in sections (g) to (k).

Epistemology is rightly a sequel to ontology, because we have to be presented with reality before we can investigate it and know the truth about its nature, otherwise we may be tempted to create our own ontological systems to suit our epistemic and other pursuits, and thereby lose track of the authentic trail to follow during our quest for the grounds of our beliefs in the objective truths of medical science

236

and practice. The history of Medicine is not short on seekers after fame and funds who have cheated by cooking results of their research to justify their personal versions of the truth. Epistemology, coupled with logical analysis, alerts us to the invalid methods used by charlatans to arrive at apparently true, but in reality false conclusions in the quest for knowledge. If q is not entailed by p, or if not-p is presented to us in the guise of p, we are able to apply these two disciplines to label these assertions as untrue logically and epistemologically by virtue of their inconsistencies.

## Epistemic Skepticism

Epistemic skepticism refers to fundamental doubts about knowledge of existents and the sources of that knowledge, as opposed to ontological skepticism, which concerns itself with doubts about what actually exists. Although we expect medical facts to conform with logical principles, the situation is confounded by the many variables that enter into medical equations: the environment, the emotions and the human element of caring, and a wide range of what passes as normal or abnormal in anatomy and physiology; in short, all the "original conditions" that we saw in the chaos situation. In the end, our medical judgments may lack objectivity in the same way that our judgments of reality begin (and sometimes end) with a lack of objectivity. Consider what happens when someone is looking at a tree: first she assumes its objectivity or its existence apart from her; then she assumes the absence of illusion and the veracity of her perception, based on the integrity of the pathway of light rays from the tree to her eyes, and then by way of her normally formed and functioning optic lens, media, retinal rods and cones, fovea centralis, optic nerve head, optic tracts and radiation, to her visual cortex. Then she embraces the mystery of interpreting and converting these purely physical phenomena into the mental experience of seeing a tree, an experience that is more than the sum of the physical processes involved in the experience of perceiving the tree. This entire process pushes truth beyond her grasp, if anywhere along this pathway error should unwittingly creep in. We seek knowledge of truth, because if we don't know what truth is, to begin with, then we cannot tell whether a statement is true or not, and we cannot build truthful structures of knowledge on the information at hand.

Truth, what is, becomes illusive if we cannot be sure of the other part in the equation: truth=correspondence with what is in the world, and if we can never rule out the ever lurking possibility of error about what is in the world as we perceive it. Admittedly, we are not always deceived by our senses, because we are already acquainted with what is genuine by virtue of the experience garnered from using the same senses collectively in the past, and we are thus able to use this knowledge to recognize and correct for the mirage that looks like a pool of water on the road, the straight stick which appears bent when partly immersed in water, and the malingering and hysterical paralysis that could easily pass for organic disease. We are able to discriminate between authentic and assumed paralysis on the basis of our knowledge of when to exercise selective skepticism in regard to what we perceive against the background of our total experience of different facets of reality.

# Existence and Truth

Descartes could not rule out the possibility of epistemic doubt about his sensory perceptions and intellectual beliefs in the existence of the outside world until he was convinced by his own thoughts of his existence as a thinking thing. If he was thinking, or doubting, or aware that he had a problem with the existence of the outside world at that time, then it would be silly to say that he did not exist while he was having these thoughts. If he is thinking, or thinks that he is thinking, then he must exist; he can't think that he is thinking when he does not exist. From this standpoint he proceeded to build an ontology based on reason only. The dictum under which he built this ontology was: "I think, hence I am."[1] (*cogito, ergo sum*). That means he derives his existence as mind from his perception of the external world, including his body which is not identical with, but external to, the thoughts in his mind, because if he were not thus thinking of externality as the object of his thoughts, he would not exist. But to be able to say I think, I must already exist, otherwise who is the "I" that thinks? Existentialists therefore think that the dictum should rather read: I am, therefore I think, since if I already am, then I don't need thinking to prove that I exist, because that will amount to putting the cart before the horse. David Hume had the same problem with denying the self as a mysterious internal "I" that thinks, over and above the sensory impressions that people have of themselves. But if thinking is occurring, then it must be doing so in a subject of thought who already exists. Therefore thinking does not determine existence of the person *de novo*; it follows from his prior existence.

Ordinarily we take truth to be correspondence between the world and our representation of it, although this is a long shot, because nothing can be like anything other than itself; so our representations can never be like what they are supposed to represent except in a vague sense. Therefore our notion of correspondence is imprecise, and our notion of truth is also imprecise; and yet truth does not have degrees. Something is either true or false, not partly so; hence the belief in some quarters that we can never know reality but only the phenomenon that presents itself as a representation of that reality, because every time we try to derive truth from our beliefs about reality we have to compare them with what is inaccessible. We have to prove that what we call reality is truly so to be able to compare our beliefs with what is true; but we enter a regress in trying to prove the $truth_1$ of our beliefs against our notion of $reality_1$ and the $truth_2$ of our notion of $reality_1$ against $reality_2$ (a reality which is really our notion of that reality), and the $truth_3$ of $reality_2$ against $reality_3$, and so on, and so forth. Therefore it means that some of our statements are said to be true, not because they claim to represent reality, but simply because they cohere with our system of beliefs; and this in spite of the problem that mere coherence within our belief systems might still not be a veritable representation of this elusive reality. We could be misrepresenting the truth consistently.

If the foregoing argument sounds abstruse, there is also the familiarity of most people with optical and other sensory illusions, which throws doubt on our ability to vouch for truth as correspondence with this misperceived reality that looks authentic. Taking all these problems into account, we might have to resort to relying on the truth of our medical beliefs, not as what we know with certainty, but as what is pragmatic or works best for us. Nevertheless, those beliefs will not be proved true because they work, but rather they will work because they are true and they vaguely represent what obtains in the world, besides also cohering with the rest of our other tested, rational beliefs about the world. Our actions are guided by our beliefs, and if we are to act appropriately, our beliefs must be true and based on the best evidence, because a concatenation of false and uncertain beliefs will only result in another series of harms and disasters. Meanwhile, truth is proving to be elusive; so we have to find solace in the thought that the beliefs that have successfully brought us this far in our journey through time can be relied upon with confidence, and other beliefs that cohere with them are most likely true in our domain of existence. We reached this same conclusion when we considered our formation of the concept of a rose as the sum of the different concepts of different people. The lack of precision in these two concepts only goes to show that we operate mainly in the realm of probability, not that of indubitability and certainty; but it should not be allowed to paralyze our thinking.

Why the chase after truth? Because when we have found truth, we can hope to discover knowledge as defined in the equation: knowledge=true belief with non-accidental justification. Truth, belief, and justification are individually necessary and jointly sufficient conditions for a claim to knowledge. If they are met, then the claimant has knowledge. The obvious question is: can knowledge answer to a different definition? People may entertain false beliefs believing them to be true, or they may feel certain about something that turns out to be the contrary. These states will never amount to knowledge; only true beliefs will, but they must be justified by adequate, rational, unassailable, and conclusive evidence that confers the right to equate belief in them with knowledge. Inconclusive evidence based on insufficient information, arrived at by accident or epistemic luck and by invalid deduction, or that is derived from doubtful authorities, untrustworthy sources, or personal intuition can only yield probable evidence on which a medical epistemology cannot be constructed. If this line of reasoning is followed, the entire web of medical knowledge will be based on a shaky foundation, and it will constitute a hazard to the health of patients, thus going against the Hippocratic principle of not doing harm to them: "I will keep them from harm and injustice."[2] (The dictum, which is sometimes stated as: *primum non nocere*=first, do no harm, is taken from the original statement by Hippocrates, which reads as follows, "The physician must be able to tell the antecedents, know the present, and foretell the future – must mediate these things, and have two special objects in view with regard to disease, namely, to do good or to do no harm"[3]).

We should distinguish between the criteria of truth as the jointly necessary and sufficient conditions that we use for determining the truth or falsity of a statement and the definition of truth as the jointly necessary and sufficient conditions under which a statement is true. The statement that Jim's leg is fractured is true if, and only if, Jim's leg is fractured (definition); but the criterion for establishing the truth of that statement is clinical and an x-ray picture of it. A statement may be true even if we have no way of determining that it is or is not true. It may be true that Jim's leg is fractured even if we have no way of determining that it is; i.e., that the assertion corresponds with the facts. Such a state of affairs does not serve a useful purpose, because the tautological definition of a fractured leg as a leg whose state corresponds with that of a broken leg does not provide us with the means we need to determine that Jim's leg is fractured, so that we can do something about it. We do not bestow any favour on him by theorizing about the conditions under which it is or is not true that his leg is broken. The pragmatic procedure is to confirm the fracture and repair it.

# Probability and Doubt

According to Bayes's theorem, the probability of the truth of any hypothesis is increased by how likely that probability makes the given evidence. If we have overwhelming statistical evidence that erythrogenic toxin from a certain serotype of group A Streptococcus causes Scarlet fever and we have a case of streptococcal throat infection with that serotype, we can infer with a high degree of probability that the person thus affected will have that toxin and will, therefore, develop Scarlet fever. On the other hand, if we have a patient with nondescript pharyngitis, which could be viral, the statistical probability of scarlet fever resulting from such infection will diminish drastically until the culprit serotype can be cultured from his throat to raise that probability. Criticisms have been levelled against this formula to the effect that beliefs cannot be measured in numbers, and that a theory of knowledge cannot be shaped around conditional probability, i.e., how probable a proposition would be if some evidence is accepted. Some people maintain that the probability of the truth of a medical statement to count as knowledge will depend on the strength of evidence backing it and not on the context which shapes the standard to be met for beliefs. According to them, higher epistemic standards than are found in expressions of doubt make it difficult for our beliefs to count as knowledge; e.g., I cannot always rely on vision to be an authentic source of knowledge, because sometimes I see things like mirages that are not veridical. But the standard that obtains most of the time is the lower one that says I know that I have eyes, because I can see, and that is the function of eyes that the nose or the ears can never fulfill, even if I sometimes have to correct for optical illusions. I know that this patient is likely to develop Scarlet fever, because his throat culture is growing group A Streptococcus, which, in this context, has a statistically high causal association with Scarlet fever that viral pharyngitis does not have.

This contextualist perspective is complicated by the Cartesian doubt mentioned above, as illustrated in this borrowed example:

(1) I may know that I have hands on the common sense view.
(2) I don't know that I have hands if I don't know that I am not a brain in a vat.
(3) I don't know that I am not a brain-in-a-vat.

Each one of these statements by itself makes good sense, but together they are inconsistent; so one of them must be eliminated by lowering the standards of doubt by way of appealing to common sense and experience. By so doing we will have to retain 1 and eliminate 3, because all the empirical evidence points to the fact that 3 is not true, and it contradicts and is not compatible with 2; but 2 is true as a logical proposition, and it is compatible with 1. Therefore 1 remains as the only viable choice, and I must assert the correctness of the common sense view that I know that I have hands, since the conclusion of 2 and 3 is that I don't have hands, which is absurd, because I am looking at my hands and using them appropriately all the time. If I didn't have hands I would know it.

241

# Gathering clinical evidence

The advent of evidence-based gathering of knowledge is believed to have produced desirable and adequate tenets in the epistemology of medical science, the establishing of clinical practice guidelines, and the clinical management of patients. This practice contrasts with the traditional or foundationalist methods that rely heavily on didactic instruction coupled with common sense and clinical experience that was mostly accumulated randomly. At the same time, the method of Randomized Controlled Trials (RCT) is used to collect statistical data from large numbers of randomized cohorts consisting of sick people matched for all possible variables: diagnosis, gender, age, socio-economic status, against control groups when placebo is used for comparison with active treatment. They and the trialists are blinded to the selected drugs and to which arm of the trial is receiving which treatment (double blinded) to eliminate possible biases. Half of the subjects are given one treatment, and subjects from the other half are given the other treatment. From this unbiased information, their responses to active treatments (drugs, etc.), different doses or frequency of administration of the same treatment, or different kinds of treatment for the same clinical condition are recorded; e.g., oral versus patch or lower versus higher doses of medications to induce cessation of smoking. Their responses to treatment are compared to determine, as far as it is possible, a cause and effect relationship between treatment and response and the cost effectiveness of the treatment. The trials may be further extended and refined by reversing the treatment modalities. The information thus gathered is then recommended for use by doctors to manage new cases of disease in accordance with the golden standards set by these trials. Comparing active treatment and placebo groups proves nothing, unless the trial is repeated with the groups switched around; but it is also an ethical challenge because it subjects sick people to zero treatment with the use of placebo or to inferior treatment when preliminary evidence concerning the new drug indicates its superior effects. The cost and time consumption of these trials has also spurred efforts to find other ways of matching patients with treatments, spawning such procedures as EBM, Personalized and Precision Medicine, and P4,— discussed below.

Since no two persons react identically to the same disease-causing microbes, or to the same therapeutic measures, this is where the principle of reflective equilibrium can be usefully applied. This is a method originally used in justifying the rules of inductive logic by bringing them into concordance with logically credible particular instances that cannot be sacrificed for the general theory, and brought into prominence by John Rawls through his application of it in ethics and political philosophy. It states that the end-point of a deliberative process in which we reflect on and revise our beliefs about an area of inquiry should be regulated by testing the particular case against method resembles the hypothetical-deductive method by testing new individual cases against previously established general principles, revising either one or both if inconsistency is found, until a stable end-point is reached where general principles and particular cases conform to the evidence. In Medicine empirical observations serve as basic tenets on which practice is founded and by which it is guided. If theories are not consistent with our observations, we have to revise our theories or reject them altogether, provided that our observations have been confirmed to be accurate and as truly reflective of reality as possibly permitted by epistemic skepticism. As Kuhn pointed out, this entails an essential shift in paradigms from the previously accepted methods of medical practice to new methods, a distinction that is claimed by many for the new practice of EBM backed by RCTs and now extended by P4 and the rest of the newest procedures. They all recognize the problem of deriving information about causation, clinical course, and treatment of disease from a variety of single, isolated cases in the face of the multiple factors that compose any single pathological event, and from multiple person trials whose results are applied to individual persons or populations with their own idiosyncrasies that may differ radically from those gleaned from the test group, sometimes with deleterious effects.

# Evidence-Based Medicine

The scientific methods discussed above are also at the root of the practice of Evidence-Based Medicine, a discipline whose aim is to regularize the guidelines, establish standards of effective treatment, and unify the methods of clinical practice while eliminating the use of unproven treatments. Proponents of this discipline claim that it will eliminate the use of unproved remedies and existing variations in clinical practice caused by lack of uniform guidelines and rectify the deficiency caused by absence of any guidelines or presence of poorly evidenced practice guidelines. To achieve this end, and basing its authority on RTCs and their meta-analyses that are sometimes no more reliable than other methods of research, it adopts or rejects hypotheses about etiology, clinical presentation, investigation, and management of illness in keeping with the Popperian hypothetical-deductive method as more research data relating cause to effect are collected. Accordingly, traditional methods of treatment, some of which were anecdotal and sometimes dogmatic, which were not previously subjected to rigorous systematic examination in randomized controlled trials in the manner outlined above, are now

243

being replaced by those that have undergone these trials, evaluation of laboratory and other tests, and controlled tests of practicality and optimum utility for patient care as in Translational Medicine (below).

The basic focus of EBM differs little from, and is not incommensurable with, that of traditional medicine. In that respect, it is not a paradigm shift as defined by Kuhn. However, both seek to establish evidence-supported treatment efficacies for the medical problems of patients, but from different starting points. EBM does it from the controlled and allegedly bias-free population standpoint, although it still depends on the same methods of information gathering as traditional medicine. Nevertheless, it now wants to discredit the latter as prone to bias because of its dependence on an accumulation of varied clinical encounters and experiences by the individual clinician to formulate a rule of thumb in the management of future cases. The truth is that the experienced physician has the unwritten skills and know-how that the inexperienced but suave EBM researcher might not yet possess, since the latter is guided mainly or solely by scientific facts of knowing what and knowing why as laid down in the theoretical texts and rules of operation on which the experienced physician now places less dependence for handling new and complex situations; EBM and its rules do not now direct every step in her clinical decision making. Her diagnostic and management abilities have been sharpened by years of practical experience involving skills that are not found in the texts, but which may sometimes be guided by intuition, which to say the least, may sometimes be erroneous but cannot be supplanted by EBM. Her auditory diagnosis of a mitral stenosis murmur is more than the act of conforming to the textual account of it, although it is essential as one of the basic facts that EBM uses for laying down criteria for the ultimate management of clinical states which entail that kind of murmur. Without this ability to diagnose the murmur, in the first place, the clinician cannot employ these criteria meaningfully, and can, therefore, not manage the case of mitral stenosis as recommended in the protocol. Even phonographic representations of the murmur, which are used to diagnose the murmur have their basis in its preceding direct auditory perception.

The kind of clinical knowledge possessed by the seasoned physician facilitates the gathering of different shades of presentation of a wide variety of clinical conditions for the use of EBM in its effort to assist physicians in the management of these conditions, because it is never the case that any two patients require the same specific treatment for what is diagnosed as the same condition qualitatively (symptoms and signs) and quantitatively (laboratory measurements). Their individual pathophysiologies and responses to treatment are necessarily as different as the phenomenologies that they experience with the same disease, and only experience acquired through

244

knowledge derived from personal contact can alert the physician to these facts. The textbooks and EBM theory will not teach her those skills, nor will they show her how to impart them to her students, because these are the skills that she knows and exercises without knowing or being able to recount how she does it. They accrue with experience. The students, in turn, have to acquire their own such experience from contact with patients in clinical situations for which textbooks and EBM theory and utilitarian rules are no substitutes.

EBM has to acknowledge that we always predict the responses of the patient to infection, trauma, and treatment based on our understanding and application of the basic sciences to the functions of a specific human body and person with unique characteristics that we can never fully comprehend, versus the entire population group that it uses for its clinical trials in establishing the efficacy and safety of methods of treatment—compounding the problem of one with many. In this holistic approach EBM also misses out on the needs of patients with rare diseases that are not included in the data subjected to statistical analysis for everyday use. Nevertheless, basic science lends credibility to the results of its clinical trials, which in turn place the cause and effect relationships among disease etiology, course, and response to treatment on the more secure footing of *propter hoc* than the isolated *post hoc* sequence that could possibly obscure this causal relationship in individual cases, thus giving EBM an edge over traditional methods of data gathering, diagnosis, and treatment.

EBM therefore claims superior facilitation of a more precise definition of the problems of many patients on the strength of empirical information gathered biostatistically from research into thousands of similar problems. In addition, it also claims observance of scientific rules of evidence whereby physicians can better rationalize and validate clinical, investigative, research, and treatment methods to also deal with presentations of illnesses that do not follow the standard textbook pattern. This practice allows them to make treatment decisions that are more balanced by being beyond the limitations of personal and small circle experience, and it also permits them to include the patient's own values and emotional burden in their plans for deserved, compassionate, rational, and artful care in his illness. To achieve these ends, some practitioners of EBM follow four steps in decision making: "formulate a clear clinical question from a patient's problem; search the literature for relevant clinical articles; evaluate (critically appraise) the evidence for its validity and usefulness; implement useful findings in clinical practice."[4] Other practitioners of EBM adopt a five-pronged approach by including the patient's values and goals in the process: "defining the case-based question, . . . integrating clinical expertise and patient values in the context of the evidence"[5] The methods just outlined are essentially the same ones that all conscientious physicians have been using for many years with only their knowledge of basic science and clinical experience of which methods of diagnosis and treatment have worked in the past, as well as the expertise that they have acquired from this accumulated knowledge to guide them, although they did not have a fancy label for them, which only goes to show how much EBM depends on these foundational determinants.

Against this background, the criticism that information gathered from population-based clinical trials and EBM is not geared to individual patient management might as well be levelled against the traditional practice of medicine that is also based on information gathered from the general population for formulating its principles. In both instances information from sources is not applied blindly, but always in light of the particular patient's special situation and with the aid of the physician's accumulated experience. A physician who has learnt to associate certain types of chest pain with heart attacks in older patients will not quickly jump to the conclusion that a twelve year old athlete with chest pain is having a heart attack, even if his symptoms are identical with those of a heart attack. He will consider first the valid statistical evidence for muscle strain in this type of patient while keeping the possibility of heart attack in mind. In the case of an infant with gastro-esophageal reflux, he will cautiously recommend placing the baby on his abdomen, which is the ideal position in this situation, without losing sight of good research evidence for placing babies on their backs as a precaution against sudden infant death (syndrome), which is statistically associated with the practice of letting babies sleep on their abdomens. At every step of the way, it is the combination of careful individual and population observational data that directs the doctor's approach to clinical problems both in standard medical practice and in EBM practice.

As with most ventures that entail utility-based stratification, EBM is subject to caveats in the accessibility of its tenets to all and sundry who are inclined to undermine and restrict doctor-patient autonomy. Non-medical personnel of HMOs and other health care money making machines that favour the accumulation of corporate profits (like some so-called fiscally conservative governments that do not hesitate to slash health care spending for unconvincing reasons) are using it to disqualify treatments that EBM does not place at the top of the list of recommended treatments for a particular medical or surgical condition. The doctor is thereby prohibited from using her clinical judgment, experience, and expertise to manage patients by executives who do not have a clue about the principles and practice of medicine and surgery, but feel empowered to dictate to her how she will treat them. She is thereby also treated like a commercial vendor of the health care commodity, for which she also bears the responsibility by acting like a commercial trader of her skills who does not deserve implicit trust from her patients as their advocate—such is the power of the dollar. This is a highly regrettable state of affairs which calls for firm and concerted opposition from the rest of the medical profession against those of their colleagues and others who engage in these marketplace practices by capitalizing unethically on the high esteem that is generally accorded to physicians by the public, because it erodes that esteem, demeans all of them, and most importantly, it is extremely hurtful to the patient.

Ultimately, however, doctors who value the effective use of medical literature also rely on the updated oracular opinions of experts and authorities, some of whom may still be regarded as leaders in their fields of honest research, while others may be tainted by their cozy association with commercial concerns that influence the direction and results of their research. So, we and they cannot belittle the opinions of these pundits, especially since EBM has so far not proved that it is the best qualitative method of patient management, even to the exclusion of traditional methods of management. The best approach to settling the disputed superiority of EBM over traditional methods is to adopt the principle of reflective equilibrium. In this case we are seeking internal consistency in and mutual support among our beliefs, and we are striving toward a state of ultimate balance that is finally achieved after a process of give and take between the general principles of EBM as they apply to the wider population and the specific situations encountered in individual patient care provided by those physicians who do not practice EBM.

The application of this principle involves the critical review of many individual cases, formulating general principles of diagnosis and treatment as guided by EBM, and deducing new evidence-based practice guidelines, which are then applied to all new cases. If the particular circumstances of the new case don't fit in with the practice guidelines, modifications are made to the guidelines to accommodate any future similar cases, and to avoid harming the patient, his case is categorized as an exception that is eligible for special treatment. The practice of making people's clinical conditions conform to a specialist's theories about that condition is to be deplored. It is dishonest and hurtful in some cases, like one well documented case of a specialist forcing the subject of ambiguous genitalia to live a life that clashed with his physiological drives, simply because he had chosen his gender for him and would not admit to his mistake, even when good psychological evidence was staring him in the face, lest admission of error should offend his ego by scuttling his theories about assigning gender to this kind of patient.

In fine, the acquisition of knowledge in clinical medicine relies on honest biomedical research with no economic strings attached by pharmaceutical and other parties with vested interests. Equally essential are quality assurance studies meant to ensure high performance standards; meticulous peer reviews aimed specifically at avoiding the perpetuation of mistakes; RCTs to eliminate as mush bias as possible in test subject selection plus collection, interpretation, and application of data; epidemiological studies; systematic reviews; continuing medical education courses operated at arm's length from the influence of the pharmaceutical industry; practice guidelines designed to incorporate meta-analysis of only truthful data that will be made possible by the foregoing practices; and the wealth of anecdotal reports representing the judicious clinical observations of practicing physicians, as well as information gathered from patients about their values, ethical and phenomenological. If the doctor is to live up to her allegiance to the Hippocratic oath and to her designation as a teacher (docere=to teach) and agent of healing of her patients, and if she wants to be in possession of the most up to date information about their illnesses and how to manage them successfully, with the minimum of harm, if any, then she will avail herself of some or all of these disciplines and others that serve the same end, and she will temper these efforts with compassion for and empathy with her patients.

## Translational medicine

Translational medicine is a new approach to an old problem of gaining knowledge about disease mechanisms and treatments. Its aim is to unify divergent aspects of biomedical research and to transfer the resulting knowledge "from bench to bedside" and vice versa to facilitate the development of new treatments, e. g., Harry Dietz and his colleagues at Johns Hopkins University found that Losartan can prevent the development of aortic aneurysms in mice with induced Marfan syndrome. Its trial use in children with this syndrome has been found to inhibit development of aortic abnormalities, and it is now being used in combination with β-blockers to treat patients with the Syndrome. Translations have also occurred in the introduction of glutamic acid decarboxylase (GAD), islet antigen 2 (IA-2), islet autoantibody (IAA), and zinc transporter 8 autoantibody( ZnT8Ab) in early diagnosis of type 1 diabetes, differentiating it from type 2 and other types of diabetes. Also, intermittent blood glucose self-monitoring is being superseded by continuous monitoring devices, and the management of Type 1 Diabetes to reduce long term complications with control of $HbA_1c$, pre-prandial and post-prandial blood glucose levels to near normal levels using multiple injections of insulin versus the old method of using two long acting insulins in the morning and evening is being refined with the glucose pump for better precision and reliability.

In the form of Regenerative Medicine, Translational Medicine provides another example of how knowledge of genetics is being used to correct genetic aberrations before they cause disabling disease and to restore normal function free from prolonged use of drugs. It uses stem cells of different origins for repairing, replacing, or regenerating affected tissues; e.g., by injection into coronary arteries to repair damaged myocardium, by direct injection into retinal defects, spinal cord injuries, and injured or diseased joints and ligaments, into nerves for myelin repair in Multiple Sclerosis, and by transplantation.

Because it is a co-operative effort among scientists, clinicians, and pharmaceutical industry that is also individualized by virtue of its limitation to small groups of similarly afflicted persons, and to those with discernible genetic abnormalities and predisposition to certain diseases (e.g., leukaemia in Down Syndrome patients), treatment can also be individualized or zeroed in to such groups in a cost-effective manner, as reflected in the next discussion on Personalized (Precision) Medicine. Eventually, changes are made in medical practice and public health policies and practices that benefit individuals and communities. This transfer from bench to bedside reduces costs incurred in prolonged preclinical testing, facilitates accelerated validation of new products, and helps to prolong life by tailoring treatment to specific patient needs.

# Personalized Medicine (Precision medicine)

So far, we have been treating patients by the hit and miss of one mould fits all on the basis of group pointers that we have modified to suit individuals, fully aware that the experience of disease and the response to treatment vary with each individual's genetic profile and with the different stages of his life. Failure to heed these differences can lead to failures in treatment. Patient-centred medicine employs the patient's personal (genetic) information and communicated experiences to his caring doctor who is fully aware that the same cancer presents and behaves differently in Jim and Jill; they respond differently to their illnesses and to the same treatment regimens, or else they require different modalities of treatment or different doses for the same weight. Now we are able to personalize management of disease on the biophysical and psychosocial interventional levels with the aid of the patient's genetic information. The basic mechanism of the patient's disease is revealed and projections can be made about his body's responses to medications via pharmacogenomics, based on his own pharmacokinetics or drug metabolism and physiological drug responses (pharmacodynamics). We can choose the best treatment for his disease, and as the expression goes, provide "the right patient with the right drug at the right dose at the right time" with minimal deleterious side effects and maximum therapeutic benefit. With the use of genomics, we are able to prevent catastrophes such as fatal bleeding in Aspirin users, since some patients who need Aspirin for preventing coronary artery thrombosis sustain gastric hemorrhage from ingesting it, and it does not prevent heart attacks universally. Other safer drugs can be used in its place.

Precision medicine has already facilitated re-classification of diseases based on genetic markers and fine tuned selection of tests to confirm those diseases and differentiate them from others with similar clinical (phenotypic) presentations but different genetic (genotypic) basis. It also expands the selection of treatments to match particular genetic constitutions as is already the case with blood transfusions—no one transfuses B negative blood into an A negative patient. Its predictive aspect is further reinforced by such pursuits as proteomics (study of the ever changing proteome, the whole set of proteins produced by the genome—all the genes in a human being—at a particular time), and cytomics (investigation of the biochemical features of the cytome or cellular system). All these measures are meant to amass helpful information about the patient to predict health problems, thus facilitating their prevention and prevention of their sometimes devastating effects. Regardless of costs and logistics of information gathering and storage to respect patient confidentiality, the method offers a holistic approach to patient care as I have tried to indicate in chapter 21 on Darwinism.

Another combination of the pharmacogenetics of personalized and translational medicine is theranostic nanomedicine—a system of accurate diagnosis and therapeutics with nano-particles that selectively accumulate in diseased tissue. It enables identification of patients who are likely to derive benefit from or be harmed by specific medications; e.g., children with leukemia and the methyltransferase genotype in whom 6-mercaptopurine is metabolized to a toxic product. The nano-particles act by delivering a safe and effective therapeutic action at the specific site of disease, while sparing normal tissues and undergoing biodegradation into nontoxic byproducts.

# P4 Medicine

P4 medical practice is *p*redictive, *p*reventative, *p*ersonalized and *p*articipatory. It is an integrated system meant to detect markers of diseases in their earliest stages before the appearance of symptoms, thus enabling physicians to categorize and diagnose the diseases accurately, and to treat patients effectively, if they cannot prevent the diseases from developing further. In this latter respect it provides the necessary tools for promoting individual wellness, and hence the wellness of the larger community. "By locating the networks that are perturbed by the disease state, drugs will be designed to perturb these networks in the opposite direction, thus promoting health."[6]

P4 medicine is personalized, since its concern is the genetic make-up of the individual, which will guide appropriate, optimal treatments for him. It is also *participatory* in that one of the driving forces behind it will be the participation of patients sharing their experiences through networks of communication with physicians. Because test groups for the efficacy and other effects of drugs will be smaller than those in EBM and RCTs, it is expected to improve health care while also reducing universally escalating health care costs. The basic method of investigation and assessment of wellness in P4 medicine is blood testing, which also allows simultaneous testing of the functions of multiple organ systems to chemicals and microbes, and their responses the effects of these agents.

Recent developments of "organs on chips" fabricated from stem cells promise to individualize drug treatments and reduce the size of clinical trials to patients with specific genetic and biological profiles. Bret Stetka reports in *Medscape* June 17, 2016 that researchers are "using induced pluripotent stem cell technology to reprogram cells from individual patients into specific organ tissues that are then implanted into chips" to build models of such conditions as amyotrophic lateral sclerosis and the blood-brain barrier to test its selective permeability to drugs in specific patients and thereby help to avoid their toxic effects from increased permeability.

# Narrative Medicine

Narrative Medicine transforms the sick body of biophysical medicine into a person who has a story to tell to a listening ear about his illness and how it has affected his life and what it has come to mean. The patient is trying to get the doctor to see things from his point of view, not by reciting his symptoms and answering stock questions from the doctor, but by revealing himself in his existential predicament and in his cry for help. Narrative medicine facilitates communication and understanding between patient and doctor, if each is receptive to the other. I know that parents (in general and later in paediatric practice) appreciated my advice on disease prevention for their children through immunization, because I told them sad stories about my experiences with infants afflicted with diphtheria nursed in iron lungs (old time respirators), other infants with tetanus having tonic seizures and *risus sardonicus*, others with limp limbs from polio, and still others with seizures and subconjunctival hemorrhages from pertussis, while I let them express their fears and concerns freely. This way we were able to reach amicable conclusions on courses of action that were beneficial to those in our care, because of relationships of trust, understanding, feeling, respect, and fitting empathy.

# 15

# Ontology and Existentialism

## Ontology

### Language and Meaning

Ontology (onto=being) the science of being, is a conceptual scheme that enquires into the nature of reality: what exists, such as material objects; what is, but is not a material object, such as processes and events that happen to those objects; abstract objects like numbers and sounds; qualities like colours, and the properties and relations of all these entities. An entity is defined by Rom Harre as "that to which a unique reference can be made (without implying that it is) an individual . . . or an attested member of an ontological class."[1] Ontology assumes that these existents are not Kantian noumena, which are beyond human knowledge, but knowable phenomena whose inter-relations can also be known. The belief is that if they were beyond our knowledge, reality would be mentally inaccessible and therefore meaningless. All of these entities are supposed to exist in their own ways, in their own domains where it makes more sense to refer to them coherently as stable entities than as merely fortuitously comprescent qualities, or qualities that accidentally happen to occur together in the same location, which they are not. On the other hand, things and objects, as constituents of this ontology, exist only and always as spatiotemporally locatable particulars with predictably consistent physical properties.

Ontology therefore enquires into "the categories of things that exist or may exist in some domain . . . a catalog of types of things that are assumed to exist in a domain of interest D from the perspective of a person who uses a language L for the purpose of talking about D."[2] The categories of these entities and their parts must, however, reflect the consensus of what constitutes the structure of the world, not what exists only as logical and linguistic possibilities, and they must dissuade us from deriving ontology solely from particular uses of language. Hence when Quine says that to be is to be the value of a bound variable, he is saying that the subject of a subject-predicate sentence used within a particular "conceptual scheme" must necessarily refer to what is singled out by that subject term in that domain. That means the subject of this sentence names a process, a macroscopic or (ultra)microscopic identifiable particular, or only what the language of ontology says that there is. The question therefore arises whether every designating medical term names a definite entity, or it is only a verbal means of referring to various categories of medical concepts within the arbitrary domain selected for that purpose; and that includes reference to disease as an entity that afflicts people, or as a conceptual state of the human organism resulting from physical or psychic causes.

The terms in which we understand and speak of what exists do not affect the fact of its existence. Thus we can refer to a brown structure with a flat top resting on four struts (legs) at its corners as a table for pragmatic purposes, while the atomic scientists refers to it as simply a particularly shaped cohesion of colourless atoms and other particles (its how or manner of being) acting in concordance with the laws of physics without denying its status (its what or being). Even if these particles are not directly observable, the fact that predictions can be made about the properties of their spatial cohesion in the shape of a table assures the scientist of their existence, in much the same way as the existence of quanta can be deduced from observations of other actual and consistently predictable behavioural patterns of matter. Tables are identifiable as spatiotemporally locatable objects (necessary condition of their existence) and as foci of predictable characteristics (sufficient condition of their existence). On the contrary, disease is not spatiotemporally locatable, but it is unmistakably and consistently recognizable as an adverse state in the life of the person who is afflicted with it, and its nominal existence in that guise is as certain as the existence of the one whom it afflicts.

## Existence

The verb "to be" takes anything or nothing as dialectical subject, and anything mentionable has existence within the universe of discourse, or the context of the story in which it is named as dialectical subject, e.g., 'unicorn' in "a unicorn is fictitious", although it may not be a spatiotemporally locatable ontological particular. It exists conceptually only, in a particular domain, as the referent of the subject term of the subject-predicate sentence in which it occurs. In the diagram below, a sentence represented by constituents 1-4, viz., "a unicorn is 1) a creature; 2) horse-like; 3) with a single horn; 4) in the middle of its forehead" may refer to what there is in its story-relative context, but it does not commit to the existence of unicorns: "to be" is not logically or ontologically equivalent to "to exist", and "there is" does not entail "there exists". So 5) will be empty.

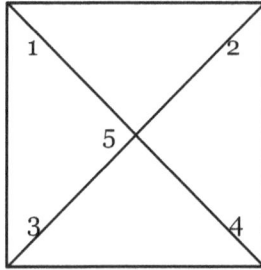

(The descriptive phrases represented by the four lines converge on unicorn, at a point in the centre of the diagram that does not contain anything like what they together describe).

The statement makes sense; it has meaning and story-relative reference, but not ontological reference, because such an animal does not exist and it cannot be pointed out, although comprehensible statements can be made about it as something. That means the concept embracing it is empty or vacuous; i.e., the centre or reference point of the descriptions represented by the four lines is empty. Logic will not tell us this fact; only empirical evidence will make us aware of it and confirm it. Some philosophers, like Meinong[3], believed that the statement about the unicorn refers to something that exists, because impossible and imaginary objects exist in their own way as much as real objects also exist in their own way. Perhaps, then, and contrary to expressed opinion, diseases also exist as objects in their own way.

On the other hand, saying that unicorns do not exist does not commit one to positing and simultaneously denying the being of unicorns. The statement is about 'unicorns' as its grammatical subject, but not as identifiable or ontological particulars whose existence is simultaneously and paradoxically denied; it only confirms that the predicate 'is a unicorn' does not have any application in a statement like 'x is a unicorn' i.e., there is no x. Use of a predicate does not imply that there is something of which it can truly be predicated. So in the original logical formula, only certain values of x entail the existence of x or an ontology of xs; but saying that x exists does not achieve that goal, because 'exists' does not add anything new to the concept "x"; it is not an extra descriptive property of x like 'horse-like' or 'single-horned'. It only means that for all values of x, 'x is a unicorn' is not true. Ordinarily we can infer the existence of some things from their re-identifiable effects, as is the case with an entity named disease, but not with entities named unicorns. So we cannot deny the existence of disease, because it

254

"fulfils our set of stable expectations for things of that kind"[3], although we have no way of identifying it ostensively; but we can deny unicorns a claim to existence on the same grounds.

'Exist' should not, therefore, be regarded as the qualification or predication of an object or a process, because to claim existence is not to confer attributes. The statement 'flowers exist' is not a subject-predicate statement, and 'exist' does not predicate anything of flowers. On the other hand, the statement 'flowers are pretty' is a subject-predicate statement, because the concept "pretty" predicates something of flowers. To exist is to be, not to possess the quality of existence in addition to being; and conceptual possibility suggests, but does not prove, ontological being; e.g., unicorn. Can we say that there is an identifiable and specific entity named disease $A$, with properties $x$, $y$, and $z$, which are instantiated in patient Joe, or that there is a disease $A$, which is the conceptual denotation or diagnostic label of a constellation of signs and symptoms $x$, $y$, and $z$, which is not instantiated in any person at this time, but which owes its potential ontological status to the manifestation of the constellation $x$, $y$, and $z$ in any human being? The answer to this question will be decided by the kind of ontology to which we believe that disease belongs; but as a conceptual denotation, disease can hardly be said to enjoy the same ontological status as its actual occurrence as a state in patients in whom its reality is realized; no one can lay her hands on it.

Against the popular concept of existence, McGinn[4] argues that 'exist' is a predicate, although not of the same caliber as a perceptible property of objects like colour, because it singles out existents from fictitious entities referred to by the grammatical subjects of our assertions; e.g., horses exist versus unicorns do not exist. So when we think of objects as existing, we are really thinking about instantiations of compresent properties at a particular location in an existent object. That means for an object or disease to exist, some property that it instantiates should also exist simultaneously; but we know that disease can be pinned down only by its manifestations and not by its objective radiation of properties as can a flower. He maintains that the concept of existence is not identical to the concept of property instantiation, but to the additional property of objects that constitutes their being. II discuss existence and being at length in my book *Things, Objects, and Persons*.[5]

The word 'is' can be used to signify existence (There is a man named Skip), identity (Skip is the butcher), generic implication (a cat is a feline), or predication (disease $A$ is contagious). The 'is' in the latter case is a predicative copula joining the subject 'disease $A$' with the predicate 'contagious'; it is not an existential quantifier, $(\exists x)$, asserting the existence of disease. In symbolic logical parlance the formula: $(\exists x)(Fx \& Gx \& Hx)$ means there is, or there exists an x, such that this x has the characteristics F, G, and H, or x is airborne, contagious, and deadly. The statement asserts ontological existence by virtue of the existential quantifier (∃x), unlike the expression Fx&Gx&Hx, which means only that x is airborne, contagious, and deadly without specifying whether x is conceptual in the context of a medical statement or an actual existent. In the case where Fx means 'disease A (x) is airborne (F)', the subject expression of the subject-predicate statement does not refer to an identifiable particular, unless such a particular exists as indicated by the existential quantifier (∃x). It merely states that if there is such a disease, then it is airborne. So Fx does not mean (∃x)(Fx) or there is a disease A, such that it is airborne; it only presupposes the existence of the disease, and could be mistaken, because a name does not refer, since it has no meaning. Even if it did refer to something, philosophical conclusions could still not be drawn about the nature of that something: whether it is real or imaginary. Reference does not bestow reality on any referent.

## Ontology of disease

From the preceding discussion, it follows that disease is only a concept that is instantiated in sick or diseased persons without any claim to existence as a spatiotemporal particular. And yet, the fact that diseases can be listed with existents that are individuated by their consistently re-identifiable and compresent constellations of symptoms, signs, and etiologies, and laboratory tests that are causally related and hence facilitate prediction of the course of the disease (its natural history) at each stage would suggest that it is a real (ontological) existent. This is not the case with conglomerates of events within a particular conceptual scheme whose descriptions do not entail a demonstrable causal relationship among the components of the constellation; they cannot be accorded an ontological status. We are therefore persuaded to conclude that diseases exist by virtue of being consistently re-identifiable abstract entities defined by the principles that determine the system to which they belong, even though they do not populate the world like spatiotemporal particulars. The extent of this ontological divide is apparent in the cluster of identifying features regularly observed in the syndrome as opposed to the disease by virtue of the clinical consistency and precise definition of cases of the disease with respect to their macro- and micro-pathology, treatment, and postulated etiology that is lacking in the inconsistent assembly of features seen in the syndrome.

An ontological enquiry will provide a full description of the nature and kinds of existents and beings as they appear to us under ideal conditions, and it will also allow for our interpretations of those of their appearances that are not as clear cut. It will delineate

256

categories existing within a selected perimeter of reality, their inter-relationships, and their relationship to the rest of reality. In the case of disease it will permit dissection and description of its basic, intermediate, and ultimate constituents, as well as its holistic implications, and it will place them appropriately in the context of other entities within the biomedical domain of the deranged health of persons. This kind of interpretation is necessary, because disease does not have pragmatic meaning outside the body of its victim; it causes immediate and remote changes in the life of persons within their particular habitat by imposing on them temporary or permanent disabilities and maladaptations to their environment, and this is what constitutes its holistic existence.

Disease also owes its existence as an ontological category to the joint contribution of its constituents, which, in turn, belong to the different ontological categories of pathology, abnormal biochemistry, and erratic genetics, and which also exist within the domain of the disease as its sub-categories. Without the being of these subordinate categories, there would be no being of the disease categories $A$ with its constellations of symptoms $x$, $y$, and $z$ and signs $a$, $b$, and $c$, and $A_1$ with its constellations of symptoms $x_1$, $y_1$, and $z_1$ and signs $a_1$, $b_1$, and $c_1$. Diseases $A$ and $A_1$ may be similar, but they are not identical. Their underlying aberrational states make them ontologically different.

## Ontological variations, tropes, types, tokens

Ontologists use the term 'tropes' to represent entities on par with diseases in this context. Their tropes are instances of properties or relations conceived as abstract particulars. For example, when they say that objects are bundles of compresent tropes or qualities, they mean that their bundles of qualities are their ways of being; e.g., the roundness, hardness, and whiteness of a golf ball are its ways of being. Similarly, symptoms $x$, $y$, and $z$ and signs $a$, $b$, and $c$ are the ways of being of disease $A$, which is different from the being of diseases $A_1$ and $B$ with their own different clinical manifestations. The different instantiations of the symptoms and signs of diseases $A$, $A_1$, and $B$ confer a differential existential status on $A$, $A_1$, and $B$ within the conceptual scheme that allows for an ontology of disease, as we have already discussed. As more information becomes available about the role of other entities in the generation of disease, so will the ontological categories be extended to include and name these entities and ultimately relate them all to the rest of the reality in which disease exists in a holistic fashion in their own ontological realms.

These variations in manifestation of a disease process from states considered to be typical only go to show that multiple variables make it impossible to designate any single set of manifestations as typical of any disease—the dose of infecting microbe, the immune status of the victim and his nutrition, environmental and psychological factors, all exert varying degrees of influence on the nature, presentation, and course of the same disease in its many forms of presentation. *Forme frustes* thus also qualify for ontological status by virtue of their inherent differences with the typical case and their ability to exist apart from the full- blown syndrome.

A further demonstration of this concept is provided by differences in the etiology of the disease genus Diabetes mellitus. The disease is an ontological genus, because it exists in recognizable form, while type 1 and type 2 Diabetes exist as mutually exclusive species of the disease genus Diabetes mellitus. Their ultimate pathological effects are the same, viz., tissue damage from excessive circulating blood glucose; hence they are species of the same genus. But their etiologies are different. Type 1 Diabetes is identified in its pre symptomatic stage by at least 2 islet autoantibodies to insulin, GAD65, IA-2, and/or ZnT8, and later by T-lymphocyte immune-mediated destruction of the ICA69 protein of the islet cells. Among the postulated culprits is $\beta$-casomorphine-7, a cleavage product of the $A_1$ and B fractions of cow's milk $\beta$-casein, which causes disease in genetically susceptible subjects who carry HLA DR3 and HLA DR4 antigens. Type 2 Diabetes, on the other hand, is related to excessive body weight, insulin resistance, and multiple genetic factors, and that makes it a different ontological entity with a different clinical course, treatment, and prognosis. Along the same lines, the condition known as Chemical Diabetes constitutes another (silent or subclinical) type of Diabetes mellitus, and hence another ontological species, as does gestational Diabetes. (Liam Davenport lists five types of Diabetes mellitus in "Medscape Medical News" of March 1, 2018).

The occurrences of Diabetes in individual persons are the tokens of the four types. The tokens are ontologically dependent on the patient (person) without whose existence, they would not have occurred. This kind of dependence is both a counterfactual and an existential dependence, meaning that, *ceteris paribus*, if the patient did not exist, his disease token would not have existed, and the disease token necessarily exists only as long as the patient exists, because it depends on him; it ends when he does. The disease type continues to exist in its conceptual framework as the genus embracing all the species of Diabetes that we have mentioned and those still to be identified. It owes its extended being initially to the existence of the tropes, and subsequently to their potential existence. Tropes can be regarded as closed concepts in their limitation to individuals, and types as open-to-membersip concepts.

It appears that the area of ontological proliferation is one area where Ockham's razor or principle of ontological economy does not apply, viz., entities are not to be multiplied beyond necessity. Perhaps this is not the area where entities can be thus limited, especially since Ockham's razor does not draw limits of its own application. One can envisage the birth of more species and sub-species as the science of medicine advances, calling for a more comprehensive classification of the resulting multiplicity of associated clinical conditions.

Based on their classification, therefore, the proliferation of biomedical entities with ontological status seems to be endless. As long as a part of one genus or whole system constitutes a qualitatively different entity in its own right, it seems there is no reason why it cannot become an ontological entity or species and be referred to as such to individuate it and distinguish it from other species. For example, the cells that constitute the bony tissues of the human body persist through the time frame of the endurance of that body, but they have specific functions, which are neither the functions of the body *per se* nor those of the other types of body cells like blood cells. Therefore they are existents in their own right, and they enjoy their own ontological status in the service of their specific tissue and in the service of that tissue in the body and ultimately in the person. This movement from basic cell structure to specialization in the formation of different organ tissues like liver, spleen, muscle, bone, and on to the formation of body systems like cardiovascular, neurological, endocrine, and respiratory necessitates a hierarchical categorization that conforms with a difference in ontological status as determined by differences in form, function, and dysfunction (disease).

## Granularity and Mereotopology

Going in the opposite direction, further down the ladder, to consider the constituents of the cell, we can apply similar arguments to cell constituents like mitochondria, nuclei, chromosomes, and Golgi apparatus. These different ontological levels are distinguished by what is called granularity, defined by Barry Smith as "a means of representing the hierarchy among elements of bodily systems; such a hierarchy is arranged according to the functions of the respective elements."[6] Granularity therefore applies at every level from the whole body to its protons and electrons with all the different but integrated functions that determine their place in the body's granular hierarchy. Functions at the lower end of the granularity scale are considered less critical to the survival of the living body, because other elements belonging to the same system can assume the function of the disabled element in that system. For example, loss of a kidney cell and its function is less detrimental than the similar loss of a nephron, and this in turn is less detrimental than the loss of a kidney, although all of them can be compensated for by increased function and size of the healthy kidney. But loss of both kidneys spells the end of the line. Such a loss, if not compensated for by dialysis alone or followed by renal transplantation, will affect the acid-base balance of the body, implicating the respiratory system in a transient compensatory mechanism, which can only result in the failure of the cardiovascular and neurological systems and the eventual demise of the person when the compensation fails.

Hence the systems form a mereotopological unit: a unit of discreet and overlapping parts in relationships of connections across boundaries, all contributing to the survival of the living body. The terminology derives from mereology (meros=part), a theory of the relations of parts to the whole and to one another within that whole, and topology (topos=place), a theory of qualitative spatial relations, e.g., continuity and contiguity. Maureen Donnelly cites the example of a parasite within the lumen of the intestine (topology) in which the connection is spatiotemporal but not a sharing of common parts as is the case with the intestinal wall and its lumen (connection relation) and the infected cells which are a part of this wall (parthood relation). Thus "mereotopology is a natural basis for spatial reasoning in medicine"[7], because for bacteria to spread and cause their damage, their vehicle has to be able to carry them to any and every part of the body, in the same way that drugs and nutrients depend on continuity of tissues and a vehicle that can traverse all the tissues where their intended beneficial effects have to be exerted. Without the continuity within and the connection among body parts, nutritious and therapeutic substances would remain localized and bring less benefit to the person. The trade-off for this disadvantage is that the same limitation would save the entire organism from harmful and toxic substances by restricting their spread through the body. We will encounter this same kind of trade-off in chapter 21 (Darwinism).

From the above considerations, we conclude that disease exists only as an identifiable, nameable, and classifiable conglomerate of signs and symptoms related to underlying pathophysiological processes in the person, but not as a concrete existent with spatio-temporal location other than the location of the diseased person.

# Existentialism

## Existential Being-in-the-world

Existentialism is the discipline that studies persons primarily as existents, beings with a place in the world, and secondarily as thinking subjects of existence and initiators of action, meaning, and value in the world. For existentialism what matters is life as it is more than life as we would like it to be. Since the person is the centre and origin of existential thought, without persons and their use of language to express ideas, there cannot be meaning; without persons to appreciate what exists, there cannot be value; and without persons to create an interactive environment by which they are also shaped, there cannot be a world for them to share with others and in which to live out their hopes, wishes, and dreams.

Our bodies mark our presence in the world of relationships with which we interact in both an ordered state of good health and a disordered state of illness; and our existence, which consists of standing out from the world of objects, is essentially being-in-the-world-with-others with whom our only means of communication is through verbal and body language as extensions of ourselves. We are thus able to transcend our bodies, although we still remain tied to them, because we cannot act and react, or participate in the world with other people without them. We communicate with them through the use of language; language reveals to us what someone is thinking and experiencing, because we have no other way of knowing what goes on in his mind. It also signals to us that someone has not been able to overcome her existential predicament when she is found to be projecting her distorted thoughts onto the world and other people by her deviant behaviour, e.g., a liar calling others liars. Furthermore, it conveys our genuine or spurious feelings and emotions, and tells the other whether we relate to them as "I-You" with our whole being, i.e., treating them as persons on par with us, or as "I-it" with an attitude of detachment, i.e., treating them as mere objects, and only as means to our ends and not also as ends in themselves. Physicians are often guilty of assuming "I-it" relationships with their patients.

## Authenticity and Freewill

Existentialists maintain that a human being and her existence or being-in-the-world are best understood through her authenticity than by the truths of natural science or moral enquiry. They believe that each person should freely choose her own values and set her own standards of value without persistent coercion by others. They do not, like realists, believe in the objectivity of intrinsic values, universal guidelines of conduct and choice, or the making of moral decisions with objectivity and rationality. Their outlook is subjective, and by authenticity they mean commitment to being oneself, moulding oneself in the image of one's own choice, taking responsibility for who one is, and not letting others dictate who one should be and what one should be duty bound to do to conform to their standards. The existentialist thus realizes and actualizes herself in any given set of conditions. But we know that no one can escape the sometimes adverse influence of all her circumstances, including other persons; therefore no one ever exercises complete choice on what she will do, or may do, and what she will be. On this reasoning, inauthenticity is the opposite existential possibility of Dasein by which it conceals itself from its responsibilities by assuming a non-human factuality.

Furthermore, since some human beings cannot always be trusted to conduct themselves appropriately, they will need restraint in their actions. This conflict with circumstances, coupled with the stress, anxiety, and existential despair brought on by the further realization of the emptiness of her life and lack of freedom of choice, is what fills the person with *nausea,* a term used by Sartre to denote her helplessness to take full control of her circumstances by herself in the absence of universal guidelines of action. Both sources of frustration are often overcome by what Sartre calls *bad faith*[7], a species of self-deception where the person regards her behaviour (towards patients and colleagues) as determined by the sometimes greed-driven, maneuvering role that she has elected to play in life, and in which she claims that her circumstances have divested her of freedom of choice and the attendant responsibility for her actions: "We shall willingly grant that bad faith is a lie to oneself, on condition that we distinguish lying to oneself from lying in general."[8] The result is existential guilt, which resides in the fact that limitations are placed on her willingness and ability to carry out her obligations and responsibilities to others for whom she should be caring, and which manifest as a state of conflict between selfless service to patients and her misplaced concern about her own being as being-in-the-world-alone and making decisions to promote her own welfare at the expense of the welfare of her patients with whom she is in the world.

Nevertheless, Sartre maintains, a human being should still be able to project herself into the world by making responsible decisions to define herself as a being whose existence precedes her essence: the "one being in whom existence precedes essence, a being who exists before he can be defined by any concept, and that this being is man, or, as Heidegger says, human reality."[9]. In her role as an existent who stands out from the world of objects and is capable of self-assertion, she has freedom of choice to become passionately involved in the world as subject. She is not forced to be a detached observer or a mere object among other objects that lack subjectivity. She also carries the responsibility of making choices and decisions that reflect on others by setting generally acceptable moral standards of behaviour. Hence the admonition by Kant, a non-existentialist, to "act as though the maxim of your action were by your will to become a universal law of nature."[10] So the person who is involved in making choices should also choose the duty to be available to others in their particular predicaments, not for her own material benefit, but because it is the right thing to do; and for physicians this availability is best expressed in the modern version of the Hippocratic Oath:

262

"I will apply, for the benefit of the sick, all measures which are required, avoiding those twin traps of over-treatment and therapeutic nihilism. I will remember that there is art to medicine as well as science, and that warmth, sympathy, and understanding may outweigh the surgeon's knife or the chemist's drug."[11]

## Existential predicament

The existentialist Robert Olson maintains that "The life of every man, whether he explicitly recognizes it or not, is marked by irreparable losses. Frustration, insecurity, and painful striving are the inseparable lot of mankind."[12] The irreparable losses that people are said to suffer are due to frustration in the conflict between their perpetual quest for worldly possessions and their inability either to acquire them or to be content without them, insecurity in holding on to their possessions in the face of adverse circumstances, which can decimate them, and painful striving resulting from the insatiable desire for more possessions. These losses constitute their inevitable existential predicaments of anguish and suffering, which come with simply being in the world; they deprive their lives of value, sometimes resulting in depression, but also at times paradoxically prompting them to seek to generate and pursue worthwhile values that are vital to their survival. The mental health expert has a role to play in recognizing these snares and ensuring that her patient is not victimized unduly by the inauthentic life that he has already assumed, which amounts to living the lie that replaces the truth that goes with an authentic life. Failure on the part of the doctor to appreciate these influences on her patient's life will escalate his mental imbalance.

The person who becomes frustrated and depressed by her failures in life can take comfort from the existential viewpoint that she can never reach an end point in her quest for idealization, because it is not possible for her to accomplish her projected quintessence during her existence, since her existence precedes her essence, unlike artifacts that are made according to specification— their essence precedes their existence. She does not come perfect or near perfect with only the last few finishing touches left to realize her essence as a person so that she can go through life as *the* perfect being, because in this context perfection is a terminal event. So she first exists, and then she chooses to project herself into the future to attain the achievable part of the perfect being that she eventually wants and hopes to become by pursuing successive realistic goals.

Striving to achieve the unachievable may stimulate the stable mind to reach respectable heights, but it will only result in the disintegration of the unstable mind that has not learned how to interpret reality in all its subtle dimensions. Therefore a person will remain "incomplete and unfulfilled" up to the end of her life, because she is always trying to find new ways of coping with life's vicissitudes, including the threat of ill health, which ultimately triumphs over her by terminating her existence. In her efforts to complete and fulfill her being, she sometimes entertains outlandish and conflicting beliefs about reality, assumes illogical and conflicting roles, and then succumbs to a split of her personality to fulfill those roles, ending up in clinical states like schizophrenia. Alternatively, she constructs her own reality to suit her maladapted state, and then she becomes paranoid, hearing diverse commands of how to fulfill herself and how to defend herself from those whose actions appear to be meant to hurt her and thwart her ambitions. She may also combine these two abnormal states into a paranoid-schizophrenic state. The doctor's role is clearly carved out for her here.

## Transcendence

Transcendence means that no one exists for herself; each one is always striving to reach beyond herself and her spatiotemporal boundaries into the world around her, including relationships with other persons. Like the anxiety induced by the dread of the non-being or nothingness caused by death, suffering is one of the fundamental aspects of our existence in-situation; it constitutes a part of our facticity (the facts of our being or those qualities that predicate our persons in the context of our environment and our reactions to it), but it should not be our destined lot. Hence the need for relief of anguish affecting body and mind, which medical practice can offer to some extent by adopting a transcendent approach that goes beyond facticity and the assumed predictability of human behaviour by the behavioural scientist who reasons inductively and therefore mechanistically. This attitude does not eliminate the importance of inductive reasoning in our dealings with patients, because without this assumption of the predictability of future events, cohesion would be lost among the categories of etiology, clinical presentation, treatment, and prognosis, and diseases would be seen to be behaving in a haphazard fashion, making nonsense of the science and practice of medicine as a whole, and preventive medicine in particular. We have to admit, though, that mechanistic prediction is still an important part of humane medicine.

The body as a psychosomatic unit is the visible form in which persons are in the world and through which they interact with it. When the body suffers trauma, disease, other illness, or abuse, the person suffers as the transcendent but inherent constituent of that body. As a subject of experience of both internal and external stimuli, she transcends her body; but as the object of physical forces she is that body, and what happens to the body happens to her as an active

264

participant in her world. When her toe is crushed, it is she who feels the pain in her toe, and when her blood glucose metabolism is deranged, it is she who has to be restored to normal health not only by restoring her blood glucose to normal values, but also by restoring her disabled personhood, transcending her anatomy and chemistry.

Physicians can unwittingly become trapped in the undesirable practice of losing sight of this transcendent part of their patients and fall into the habit of using them as mechanical cash cows in their revolving door practices, as objects for sexual exploitation, and as guinea pigs for personal profit and prestige, especially in the research field. These are the unsavoury habits whose contrary is embraced by Martin Buber's reference to the primary words of existence in social and communal relationship, which are the basic word pair 'I-You', and in which 'I' (the doctor) is not the dominant word:

> The basic word I-You can only be spoken with one's whole being.
> The basic word I-It can never be spoken with one's whole being.[13]

The separate, non-basic words of individuality 'I' and 'You' can only be derived from this pair, meaning that although self-recognition entails simultaneous recognition of the other as not-I, but you, I am not more privileged than she is, unless I regard her as 'it', in which case I have unjustifiably self-assigned precedence over her. From this kind of relationship, therefore, it is not difficult to appreciate the message of the Hippocratic Oath exhorting physicians to be less pre-occupied with themselves, but to come out and become more available to their patients—a call for transcendence: "I will remember that I remain a member of society, with special obligations to all my fellow human beings, those sound of mind and body as well as the infirm."[14]

The scandalous media scenes, allegedly involving physicians, where patients are left to die unattended in emergency waiting rooms are certainly not part of this virtue of transcendence, or even of the doctor's solemn undertaking under The Oath. The profession seems to be losing its moral fibre in this and several other ways, falling victim to the decadence of lack of concern for the other that is threatening to devour the society from which she cannot totally divest herself, but in which she can still wield an influence to ensure that these deplorable acts do not happen within her field of service and operation. Now, more than ever before, the physician is required to subject herself to serious self-examination to re-affirm her commitment to the promotion of human dignity and respect in the execution of her duty to those who are her actual and potential patients, and to rise above this pervading fray of self-indulgence and marginalization of the other that has usurped the characterization of her society.

# 16

# Phenomenology

## Essence of Phenomenology

Phenomenology (*phainomenon*=appearance) is "a science of experience"[1], the reasoned inquiry into the essences of appearances as they are presented to consciousness. It entails the understanding of phenomena without preconceptions about them. Consequently, Phenomenology studies the meanings of the detailed descriptions of people's subjective or first person conscious experiences of these objective phenomena or appearances in their purity. Their study of these pure first person experiences of reality is made possible by bracketing or holding in abeyance every aspect of their prior scientific and other knowledge of reality. The essences of their experiences can thus be conveyed to their consciousnesses untainted by inferences, interpretations, and biases, imposed on them by presuppositions derived from previous scientific or other acquired third person conceptualizations of reality, i.e., what they have learnt from other sources and from their own unfiltered impressions. These phenomena and appearances cannot yet be referred to as real or illusory; they are simply first given in perception or intuition, to be evaluated subsequently. Accordingly, only pure experiences can fully engage the consciousnesses of those engaged in the process, providing the untainted basis that they need for the construction of their ultimate knowledge of the world. Phenomenology does not assume that there are so-called noumena (unknowable elements of reality) behind the phenomena that people experience, nor does it discriminate between reality and illusion, or mind and matter. In that sense, the lived and experienced world of Phenomenology is as prior to the derivative world of science as the ill body-as-lived and experienced is to the patient's well biological body. Although the holding in abeyance or bracketing of these tainting influences about the existence and nature of the external world sounds naïve, it is useful for relating appearances to experiences. For instance, the scientific basis of medical practice leads us to believe that a patient who is afflicted with pneumonia will suffer from shortness of breath, some pleuritic pain, and general malaise as a result of the infection. It will not tell us anything about his fearful thoughts of the implications

266

of his illness, and it will not give us a handle on the holistic approach that is necessary for responding to the patient's concerns by treating him also as a person who is more than a complex of physical problems. Descriptions of the pure lived experiences of illness provided by individual patients can thus be viewed from a particularized, unbiased, and untainted perspective, without the doctor's added perspective on it as the diseased state of a dysfunctional biological organism, and apart from the perspective of the all encompassing descriptions of the textbooks, which describe everyone but no one in particular. Phenomenology focusses on this patient that the doctor has to deal with to the exclusion of all others, and on what he is going through in his illness, not on the ideal clinical state of ideal organs described in textbooks.

# Essence in Phenomenology

The existence of what is presented is not the primary concern of phenomenology; only its appearance, or the phenomenon, and its experience as presented to consciousness counts. We could say that in phenomenological terms essence precedes existence, as opposed to Existentialism whose concern is with mode of existence or being-in-the-world-with-others. In Sartrean terms, it is existence that precedes essence. In direct contradistinction, Ontology's sole concern is with being or existence *per se;* essence is of secondary concern to the ontologist as evidenced by his talk about unicorns and golden mountains, which are non-existents while they can be named as the grammatical subjects of propositions. Phenomenology maintains that we cannot ever know anything beyond what is presented to our consciousnesses, whether it is by way of perception or by intuition. Therefore any other knowledge that we claim about what is in the world is derived deductively or inductively from this kind of perceptual and intuitive knowledge, and to assume that this perceptual or intuitive knowledge is determined by external objects and phenomena is to put the cart before the horse. This attitude arises from the assumption that we can't have perceptions unless something is causing those perceptions. So when we explain the perceptions in terms of this something while we also explain this something in terms of the perceptions that it causes in us we are arguing in perfect circles. Hence, when we have described the phenomenon as it appears, without adding to this description any of our biases resulting from emotion, desire, or anticipation, we have described its essence, and there is nothing more to be said about it that will not distort its nature.

So, any person's description of what it is to suffer from depression or pneumonia is unique to that person's lived experience

267

of the illness and is as much as he can say about how he experiences his new lived body and mind in its disordered state and his uncertain future as the changed person that he is now. The doctor does not have first hand knowledge of that particular patient's experience, although she might have a broad descriptive concept of it from learning about it from many other similar descriptions of the experiences of other patients, and perhaps also from her own experience with that illness or one of a similar kind, if she has suffered from it. It therefore behooves the doctor to re-orient herself from her position as a remote outsider to a position as close as possible to that of the patient, as insider, to be able to acquire a perspective that is close enough to that of her patient so that she can be able to empathize fully with him and embark on the right course of helping to heal her patient's "biological and lived' body and mind together. This she can do only by listening and responding appropriately to the patient's description of the how and not the what of his feelings of alienation from the world of healthy people and from self due to his unique experiences with his particular illness. The similarity in the natural constitution of patient and doctor makes it easy for the doctor to appreciate her patient's feelings, unlike the impossible situation of someone trying to imagine what it is like to be a bat and see the world from a bat's perspective. It can't be done; no one can assume a bat's perspective on life and the world, because there are practically no areas of correspondence between the two perspectives; but anyone can try to see things from the perspective of someone else with whom she shares her physical nature and many common worldviews.

## Phenomenological Being-in-the-world

In maintaining that phenomena are all that is given in perception and introspection, and therefore all that we can ever know about the essence of what is thus presented, Phenomenology is describing one of our ways of being-in-the-world called intentionality or the directedness of our experiences toward things in the world. Whatever the given is, whether it be persons, objects, occurrences apart from us, or feelings and thoughts within us, it always consists of phenomena, which constitute the objects of consciousness. That means consciousness is not a passive process, a mere presence of the object of perception to the mind, but an active mental process of reaching out to the object of thought, from which it is self-evident that without consciousness as one of our ways of being-in-the-world, we cannot be in meaningful contact with our world of persons, either perceptually or cognitively.

The other way of being-in-the-world is through our bodies, either as lived-bodies or subjects or as mere bodies-in-the-world or objects. According to Sartre, someone's body is being-for-others who are subjects, because it is a mere object of their gaze, which does not necessarily have powers of communication; but it is also an object of gaze for its owner in his dual role as subject and object, as we saw in the discussion on disability, as disabled parts of the body become objects of attention apart from the whole body and the person whose parts they are. For Foucault, however, the gaze is the means of communication between doctors and patients. The physician's clinical or observing gaze, which is directed at her patient, is important not only for seeing those bodily signs that suggest disease, but also for picking up evidence of the suffering that the patient as person is experiencing and having to endure. At the same time, patients invite the physician's gaze to explore their bodies (and souls), because it initiates the chain of events through which they hope to obtain relief from their suffering. The gaze is sometimes the only means that some patients have to convey their feelings of distress, and the doctor must learn to recognize it and act on it.

Ordinarily, we do not differentiate ourselves from our bodies (our selves), but we become conscious of our bodies as objects-for-us when we have to perform certain tasks like trying to reach for objects beyond our range and when parts of our bodies will not carry out certain of their designated functions, e.g., climbing stairs with a fractured first metatarsal or a paralyzed leg, or chewing meat on a sore tooth. The obtrusion of the body thus possessed of illness or infirmity dominates our entire being, demanding our full, constant, and sympathetic attention until we can be relieved of those disrupting burdens in our lives, because the diseased body disrupts the integrity of the self and becomes its sole focus of attention. Such are the occasions on which we resort to curative treatments like having a tooth extraction, an appendectomy, or specific treatment for any other acute disease that consumes our attention. In such cases, we become pure objects for the other as we surrender our bodies-as-lived to the doctor who is charged with the responsibility of restoring those Cartesian physical bodies to normal by eliminating the physical and biochemical factors that have deranged their integrated functioning.

Sometimes, however, acute conditions progress to the chronic state without allowing the patient the opportunity to make the requisite acute changes and adjustments to his life-styles that will render his body only a temporary burden on him. Chronic illness and the scars of some acute illnesses often inflict damage to body and spirit that demand lifetime adjustments on his part if he is to be able to carry out many of the physical and mental functions of daily living. We become conscious of our bodies-as-lived as objects to us in much the same way as they are objects to other people, and also in much the same way as their bodies are objects to us. If the doctor cannot divorce her attitude from this alienating frame of mind, she also cannot overcome the barrier erected by the impenetrability of the other person's mind. As a result she gives up on trying to come to grips with her patient's personal problems, which require her to act out the patient's role from her empathic understanding of his story. She thus continues to treat him as an object whose verbal and physical behaviour only need to be fitted into existing moulds of mental or physical disease from which a therapeutic venture can be undertaken. Such interpretation of human behaviour is not altogether unnatural in the context of the inherent privacy of our thoughts, feelings, and emotions. It happens, for example, when we say or even only act as if we are in pain; other people can only read the expression of pain in our faces—the grimaces, the groans, and even the tears—from which they conclude that we must be in pain. They compare our third person expressions with their first person experiences, and from this analogy they draw their conclusions much as we draw our conclusions about their presumed experience of pain from comparing our behavioural and verbal experiences with theirs. These modes of being and the expressions of the contact of their owners with the world create room for the sick patient's body to be the object of his doctor's detached scientific investigation during his experience of his "body-as-painfully-lived" and of its physical and socio-cultural limitations as his new way of being-in-the-world.

# Thematization

How a patient regards the effects of his illness is not limited to the experiences of pain and discomfort only, but it also includes his fears, hopes, misconceptions, inconveniences, sense of vulnerability, and the humiliating loss of the power of control over his own life. He now has to have other people do for him the things that he could previously do for himself, and he now has to adjust to a different and sometimes lower quality of life imposed by his range of limitations. He acquires a new attitude to his whole life, which is now dominated by dependency and sometimes hopelessness while the world goes on without him, leaving him behind in his inability to keep pace. The doctor's perspective is different however; she sees the illness as an abstract dysfunctional state with symptoms and signs that amount to a particular quantifiable disease with a definite categorial label that typifies it as a case of left lower lobe pneumonia or duodenal ulcer. This attitude is somewhat removed from the patient's existential predicament. Each one thematizes the same condition differently. The doctor also sees it as a challenge to her diagnostic and therapeutic skills, and once she has scaled this hurdle, she is satisfied that she has done everything for this patient, and she is ready to

270

move on to the challenge presented by the next patient with his lymphoma. Her first patient might, however, still be feeling ill, in spite of satisfying certain criteria of normality relating to the clinical signs and laboratory data of his illness. His fears, hopes, and misconceptions may still remain unresolved, so that he is still not healed, although he has been cured of his disease.

In effect, therefore, the patient and his doctor approach this common meeting ground of the patient's illness from opposite poles and with conflicting worldviews; they are said to thematize in different ways what has become their common problem by virtue of the doctor's acceptance of the responsibility to help the patient overcome his malady and the patient's dependence on the doctor for providing the relief that he needs. But that is no reason for failing to blend their approaches to achieve complete conquest of this common challenge, because remaining securely mired in their divergent positions will result in failure to allay the patient's total concerns about his illness. The patient does not have the expertise to align himself with the doctor's learned perspective on diseases and their treatments, including all the known and anticipated complications of treatment with which the doctor is familiar, but which present an added threat to the patient and his survival when they arise in the midst of his already threatened state of being. In this predicament the patient has to rely on the doctor to make comprehensive decisions about the treatment of his disease and the restoration of his whole person to a satisfactory functional state. The doctor can and should step down from her reductionist, academic pedestal to the holistic level of the common human experience that she shares with the patient in the ordinary affairs of life, and from this standpoint share a re-assuring compromise perspective with him in the resolution of his multi-pronged predicament.

For the patient, illness is a new and deranged manner or state of his existence or being-in-the-world. Nevertheless, he still regards himself subjectively as being-in-himself and not objectively as being-for-others or -himself, meaning that he is his body, not that his body belongs to him in the same way as the clothes that he wears belong to him and cover him. The skin that covers his body is still a part of his body, but the clothes are apart from his body; they can be stripped at any time without causing him the grievous harm that stripping his skin would cause. His new state is one that threatens his present existence and casts a dark cloud over his future; he cannot plan for his own and his family's future while he is in the grip of a potentially devastating situation. He cannot do the things that he was used to doing everyday for his comfort, amusement, and survival; and much as he would like to, he cannot dump his broken body, because he is this body that has now turned against him and his purpose in life.

Nevertheless, he still depends on that body as the pivotal point around which the rest of the world that he knows revolves and as his means of making contact with that world. If it fails him, he has no other part of himself to appeal to. His 'soul' or spirit will not be of any avail to him in these circumstances.

# Doctor's duty

The doctor who fails to take this reality into account is operating within a restrictive and nomothetic environment, which mistakenly regards the lived-body as a Cartesian machine, and which only sees mechanical tools as the only instruments that can be used to repair it. She has failed to accommodate the extended paradigm of illness as more than just pathology causing disease, because she has not been receptive to the implications of constructivism and systemism as previously discussed. She should think of her patient as more than the living example of the autopsy specimen of renal amyloidosis that she saw last month, or of the abnormal renal biopsy and renal function test results that she received from the laboratory for that patient. She should think of applying the art of healing to her patient in addition to curing his disease, restoring his integrity and helping to adjust his shattered dreams in addition to providing him with that artificial hip joint. If she cannot do these things, then she should direct him to those who can offer such help. Failure to become involved with more than the patient's disease will result in failure to "remember that there is art to medicine as well as science, and that warmth, sympathy, and understanding may outweigh the surgeon's knife or the chemist's drug."[2]

Nowadays, fewer patients suffer from the infectious diseases that plagued earlier populations, because some of them can be prevented with immunizations and most of them can be fairly quickly cured with antibiotics, antivirals, and antifungals. But the new and persisting burden comes from chronic disorders like cancers, mental disorders, and illnesses related to the modern-day life-style and its stresses (emphysema and lung cancer related to smoking, hypertension and heart attacks related to stress, diabetes related to obesity and inactivity), which call for a revision in the doctor's relationship with her patient. These are situations that require a social approach and the education of the patient on how to live with his chronic problem. The doctor who cannot make her patient realize that she cannot offer a cure for some of these conditions but only relief, and who thereby gives her patient a false hope of cure, is also engendering an unnecessary feeling of despair and abandonment in her patient who now thinks that he and his illness have become a burden on his doctor. If her practice consists only of prescribing drugs and therapies for the body, she will also find fault with herself for failing to provide a cure for the patient's bodily derangement when empathy, care, and support are required to supplement drugs prescribed for relief—a gesture that will be appreciated by the patient in his existential predicament, as in the case of my ex-patient Mr. M.

The problem is, of course, that the doctor starts off by attempting a cure even when the circumstances of her patient's illness clearly do not favour such an approach. Her job is to save lives, and some patients and relatives expect her to live up to that requirement. Failure to do so upsets everyone, but it renders the doctor culpable for creating high expectations to begin with, and failing to substitute her healing skills (if she has any) to comfort and support her patient. She will always be at fault for the failure of the permitted or promoted unrealistic expectations of her patient to materialize, because she caused them by making his disease into something that can always be cured with the help of science; but she can still be exonerated because she does not personally have the magical ability to cure all diseases scientifically where the tools are lacking. On the other hand, she can never be exonerated from blame for failing to promote the healing of the person that is her patient, because healing does not call for the possession of magical potions or the creation of unrealistic expectations; it calls only for commitment, the commitment that will negate opinions coming from within their own profession and from the community recommending that some doctors be treated with no more trust than that reposed in some, if not most, used car salesmen, and in some unscrupulous commodity dealers encountered in the common marketplace.

> it is clear that healing requires an understanding of illness-as-lived. The phenomenological analysis has revealed that suffering is always personal. It relates explicitly to the particular patient's life situation and to the meaning and significance which he or she attributes to the experience of illness . . . suffering may only be relieved if explicit attention is paid to the meaning that illness has for a particular patient within the context of a unique life world.[3]

## The listening ear

At the other end of the scale is the type of case like that of the father who told me that he had taken his daughter to several paediatricians to seek help for her constipation problem. He remarked that no one had ever cared to listen to his account of the agony that his child and family had to go through whenever the child made an attempt to pass a stool. I heard what he said, but more than that I listened to him and empathized, and with his cooperation we resolved his daughter's problem with a few dietary adjustments and elimination of the many laxatives that he alleged she had been using for two years. On the other hand a mother complained to me that the doctors who had failed to treat her child for milk allergy told her that she was "crazy", making up the symptoms that she attributed to the child. After the cow milk and its products in her diet were replaced with non-dairy foods, the child's problem was put to rest. Her mother was elated with the outcome, and she also felt vindicated. Another mother recounted how she trusted the diagnosis of this stranger more than she trusted her child's paediatrician's previously given diagnosis, because the stranger had taken the trouble to listen to her detailed account of the child's illness before undertaking a thorough examination, and doing selected tests to arrive at that diagnosis—she came from a revolving door type of practice. Children also evince this attitude of trust when they tell the story of their minor injury to an adult. The adult rubs or blows on the site of injury and assures the child that all will be well, and the child leaves satisfied, forgets about the injury, and goes back to play, because her concern has received a sympathetic hearing and her trust in the adult's word has been boosted. Physicians should be sensitive to the needs of patients, be there for them, and relate to them in ways that engender their trust. They should stop behaving like disdainful super human beings sitting on a pedestal that is out of the lowly reach of their patients. They are not gods on Mt. Olympus; they are mere mortal human beings like their patients, despite all their grandiosity.

In this matter-of-fact milieu, it is not difficult to envisage why doctor-patient relationships are limited to cold encounters between physicians and the living embodiments of pneumonia, diabetes, fractured femur, and other pathologies with which they are confronted. Somewhere in this scientific jungle the patient got lost, and despite his best efforts to direct attention to himself as the living subject and bearer of these conditions, and also as the legitimate primary focus of the doctor's attention, he always finds himself treated as an addendum to the disorder or disease that claims all of the doctor's attention. Correct diagnosis, prediction of its course, and successful treatment of the patient's disease are the doctor's goals; failure in these endeavours is unacceptable to all concerned, because it will be costly to the survival and socio-economic welfare of the patient, and sometimes also to the protection of the community on which his disease impinges. It is always regarded as more dismal than failure to communicate and empathize with the patient at the same time, as if the cure and empathy are incompatible, and as if disease can be cured in the abstract without the patient whose disease it is. In fact, the patient's limited role, beyond that of serving as a vehicle to provide answers to questions that will lead to the diagnosis of disease, is well defined by its absence in the training and the qualifying examinations of doctors.

Patients are people, yes, but they are regarded primarily as foci of abnormal form and function that call for restoration to normality, whatever normality is, given the information that they are able to provide about their maladies and the doctor's conclusions about what is wrong with them. This unfortunate attitude still prevails in medical practice as it did many years ago, especially in the surgical specialties where the appendectomy and the knee replacement constitute the only areas of interest and the total points of contact between doctor and patient. The rest of the patient with his existential conflicts, personality concerns for survival, and how he relates his infirmity to his being barely enter into the picture. Doctor and patient have widely divergent attitudes to the same problem: subjective for the patient, objective for the doctor, and they don't overlap.

As a result of this attitude of mind, medical students do not always have adequate training in learning to appreciate the worldviews of their patients and in paying heed to these as they acquire the mechanical skills of assessing their patients' clinical status and planning their course of disease-focused treatment, rather than their comprehensive management. The result is that the concept of the dynamic of the doctor-patient relationship becomes so unbalanced that their patients derive only limited existential experience from their encounter with their doctors who, in turn, learn little about the values entailed in the humane aspect of medicine from the same cold and sometimes insensitive encounters. These attitudes of lack of balance between the physical-mechanistic and the psychological-axiological aspects of patient care are carried into their subsequent contacts with their own patients, resulting in the commonly made observation by patients that some doctors have a poor or no bedside manner. Such demeaning observations erode the sentiment of trust and confidence that patients should entertain in their personal physicians if they are to be healed of their maladies, and they reduce the doctors to stethoscope swinging, prescription writing pill peddlers who do nothing to provide the emotional support that patients need to heal their mental anguish.

In this kind of atmosphere it is not possible for the doctor to mend her patient's fractured self beyond simply mending his fractured femur, to restore his shaken personal integrity and his ability to continue to be a social link in his community, or to help in healing him. Curing only his pneumonia as shown by the favourable X ray changes and other technological parameters that treat the patient as a test tube specimen for which empathy is not necessary reflects lack of humaneness. Can the doctor reshape her patient's shattered and crippled future image of himself as he relates it to the detached attention of someone who deals with him as if he is a dysfunctional automobile engine that is devoid of the suffering that the patient has to endure from his dysfunctional body and soul? My own illnesses made me understand what patients suffer bodily and mentally, and I would guess that many other doctors have had similar experiences.

## Personal reflections.

At different times, I sustained a fractured 5th metatarsal and a lumbar spine strain, which made me understand what patients suffer bodily and mentally from the limitations imposed on them by the effects of illness. The intense unprovoked pain and the pain from even the slightest pressure or movement were agonizing for me, because I was personally unfamiliar with such degrees of pain, although I had seen other people in severe pain before. These experiences changed my entire being-in-the-world by highlighting my disordered state, with limited mobility, dependency, and loss of ability to control my body and to participate in and manipulate my environment to my advantage. In the meantime the rest of the world went merrily on without me, with the notable exception of my wife who empathized with me all the time, even though she did not feel the actual pains that I was feeling and could not do anything to abolish the pains. Then I suffered the extremely excruciating and crushing pain of a heart attack (coronary thrombosis with myocardial ischemia and infarction), and that brought home to me my vulnerability and utter powerlessness to be the master of my fate and the captain of my soul. After my recovery from the acute event, I needed more than medicines to restore my integrity: the same family support as before; cardiac rehabilitation in the consoling company of other victims of heart attack, which made me realize that I was not alone in that predicament; and the will to change my plans for the future, realizing that my life has been forced into a new direction, even if my laboratory reports were back to normal and even better than before my heart attack—my lipid levels have never been so good since I started monitoring them before the heart attack.

These lessons from the other side, the patient's side, enhanced the practical lessons that I had learnt from my patients very early in my medical career: to temper the scientific treatment of my patients' illnesses with corresponding effusions of empathy for their situations. This was not a difficult feat to accomplish; the politics of the time and place had reduced their human status to the deplorable state where their lives did not count for much in the eyes of the entire group from which the selectively ruthless rulers were extracted and in the whole scheme of things. So much for that painful fact. In my own general practice, I did not have much of a choice in the matter of how I would regard my patients and their plight, because many of them were poor, desperately sick sole bread winners who lived in single room dwellings with three or more other members of their families, without furniture or appliances, few utensils for the little food that they could afford, and practically no source of solace in their existential predicament. Most of the time I treated them for token fees, my purpose being to restore the patient to good physical health so that he could return to work and continue to earn his meagre wages to support his family by providing even less than the barest necessities of life. In those days we also dispensed medicines, because medical insurance was reserved for the select (by race, or what I prefer to call 'tribe', since there is only one human race and its many tribes); so the poor patient, always the sickest, because also the weakest from lack of good nutrition, received the most expensive medicines from me for nothing; but he never knew it. It would have been unfair to expect him to be able to bear the financial and emotional costs of his predicament at the same time, because he did not have the means to do either. Those were acts of beneficence that went unsung from the mountain tops for all to hear about or see, and they were performed ungrudgingly to the patient, but not without extreme rancour to the originators, perpetrators and perpetuators of the iniquitous system under which my people were suffering.

When I started my family practice of medicine I ran a home office, but I also ran mobile clinics in three outposts that I visited once a week with a car full of a wide assortment of drugs for treating the common diseases that I had learnt to anticipate. My car also served as an ambulance to transport patients to a hospital where I was denied admitting privileges because I and my kind would have had to give a "black" doctor's orders to "white" nurses. Once, and only once, I was allowed to leave orders for one of my patients, because the doctor on duty had gone home and would not come back in to see the patient that night! House calls were still a big part of the doctor's daily routine, and I would sometimes drop in on my elderly patients to offer them moral support in their chronic states of illness, after I had helped them to survive the acute episodes of disease. Their chronic conditions, with all their ramifications, could not be treated with drugs but only with the empathy and emotional support that they had come to expect from their health care provider and which they deeply appreciated. Centres for rehabilitation were not even on the cards for my "racially" discriminated population of patients.

One night I was called to see one of my elderly patients Mr. M. After the usual exchange of greetings, he told me that he did not have me brought in to attend to his medical complaints, but he wanted to thank me personally for having taken good care of him and his health problems over the years. He felt that he had now reached a point in his life when he did not need the services of a doctor any more. We talked for a while, as we had often done before, and then I bade him farewell, my mind still reeling from the unique experience and its implications. That same night he passed on. Why did he have to call me in to have this last conversation with me? Was he trying to teach me something that he must have sensed I was not taught at medical school? I realized afterwards that that was what he taught me. I had spent years attempting to keep him from falling prey to the enemy of the life that we both valued but which for him had outlived its value. Now he was teaching me how to let go without feeling like a loser in the battle to rescue him from his and my ultimate fate and not to engage in futile medical acrobatics when the candle of life has run out. I attended his funeral, like the many previous funerals of my deceased patients that I had attended in my capacity as their neighbour and additional source of support to their families and dependents in the darkest hours of their lives, but this time my outlook was different; I was not there to effect closure to a battle that the patient and I had together lost, but to celebrate the climax of a life fully lived and relinquished without regret. It was also a life that had left behind a lesson for me—that there comes a time in our existence when the inevitable must be accepted by doctor and patient alike, when the most honourable action is not to be found in the futile heroics to which doctors sometimes resort to try to prolong life in their misplaced effort to prove their mini-omnipotence, or to save face, but in seeing what looks like failure on their part as necessary closing of chapters in the biographies of their patients.

From this experience I learnt that beneath the outward appearance of the body of every sick patient there is a person with an encompassing illness who can distinguish between his body that needs the wonders of modern medicine, and his person that calls for transcendent care. I became more convinced than ever before that my function was more than that of dispensing medicines and collecting fees. I began to pay even more attention to my patients as persons with medical and personal needs; not as cases of pneumonia, or as some living creatures carrying a disease label of which I was called upon to divest them. I came to realize more and more that a person is more than a sick body waiting for the exercise of my exclusive, highly technical skills to restore his ailing body to good health; he is a body-mind expecting me to help in healing his total illness of person as compounded of body and mind, more than simply cure his bodily disease. I became more aware that patients can be cured without being healed, healed without being cured, or healed and cured at the same time, depending on their diseases and illnesses, on their healthy relationships with their doctors that outlive their own failing health, and on how much emotional support their doctors will give them.

# 17

# Ethics

## Nature of Ethics

Ethics is the second-order philosophical discipline that studies the theories and principles underlying the standards that we use to make our first-order judgments on the merits of our actions and the concepts that guide those actions, whether those actions are right or wrong in the circumstances of their execution; e.g., why do some people say abortion is wrong or immoral while others do not think so? What criteria do they employ for arriving at those conclusions? Should these criteria be governed by theories or principles, or should they depend on discretion and good sense? Discretion and good sense are the pillars of casuistic ethical judgments in which each considered case receives specific assessment on the basis of prevailing intuitive ethical principles and presumptions, extrapolation from paradigm case judgments by means of analogical reasoning, and an all round knowledge of human affairs. The casuist does not pretend to offer universal principles by which to judge all cases of morality like the blanket admonition of Kant against all killing; she only lays claim to what she considers to be the best action to take in a particular case after careful consideration of all its surrounding circumstances of motive and intent, while remaining fully aware that individual cases do not have to conform with specific rules.

Hence, when she decides on the moral propriety or otherwise of therapeutic abortion in a case of rape, her reasons will necessarily differ from those applied to a case of aborting a fetus for personal convenience, such as a pregnancy that foils plans for an exotic tour. Such decisions, which reflect regard for personal interests in keeping with phenomenological principles, still do not alter the fact that the procedure and its outcome are the same in both these cases. The difference stems from the motivation behind each case, which is necessarily different for every case that has to be decided. But this position raises questions: why should the act be wrong in one case and not in the other? What gives one motivation priority over another?

To answer such questions, we must clarify the descriptive meanings, applications, and normative implications of words like 'abortion', 'right', and 'wrong' as they are used in those specific contexts, and as that use affects every concerned person. We must further justify our reasons for approval of the rightness or disapproval of the wrongness of those actions and prescribe future norms or standards to be used in assessing actions of a similar nature. The quest for these criteria is expressed in second-order considerations relating to the language used to talk about these first-order concepts and actions as contained in the ethical theories designated as consequentialism, deontology, intuitionism, and virtue and the ethics of caring. All these theories are driven by the ultimate search for universal values, because without an appreciation of the nature of values, we cannot decide on the value or disvalue of actions. Besides, these values should be universal, not parochial, because some local practices may be offensive to the general sense of morality of the majority of people and would consequently not qualify for universal adoption. The immediate problem is, however, that we need criteria to guarantee the correctness of each of these theories and to justify the propriety of their use as standards of assessment.

# Ethical theories

Ethical theories attempt to provide these criteria, and we will now briefly outline the content of these theories.

## Consequentialism

Consequentialism is an ethical theory that labels acts as morally right only if they produce the best overall results; the ratio of benefits to costs of the consequences of the action should be high. This definition of consequentialism includes situations where the best results are for everyone (Utilitarianism), or solely for the long term good and happiness of the agent (Egoism) in the hope that other people will behave similarly and thereby also spread happiness, provided that their interests do not conflict. Where interests conflict, however, as in the abortion and euthanasia controversies, egoism appears to encourage polarization in the promotion of self-interest. The definition also includes the confinement of happiness exclusively to a select group (Relativism), or to people other than the agent (Altruism). Furthermore, best results means the superlative degree of good, and that calls for a clear delineation of what constitutes "the good" if our aim is to produce the greatest good for the greatest number. Does it refer to happiness, pleasure, absence of pain, liberty, free will, rights, justice, friendship, or well-being? Does it matter if the happiness is of high or low quality? Mill answered this last question this way: "It is better to be a human being dissatisfied than a pig satisfied; better to be Socrates dissatisfied than a fool satisfied. And if the fool, or the pig, are of a different opinion, it is because they only know their own side of the question."[1]

Other questions whose answers require a lengthy discourse that we cannot now undertake spring to mind immediately. Is happiness the ultimate good-in-itself for whose achievement all these others are instrumental goods? How should these be measured, compared with one another, and pitted against their individual costs to determine a benefit to cost ratio in the absence of established standards? How should people's individual acts be judged against equally valid consequentialist rules that judge the rightness or wrongness of these actions, not according to their consequences *per se* in individual situations, but in accordance with how their consequences are consistent with general rules governing the performance of those actions? Is it possible to frame rules to cover every possible situation? Do rules hold *prima facie* (explained below)? Do extenuating circumstances help to exonerate one who breaks these apparently inflexible rules on purely moral grounds? How can anyone know all the consequences, immediate and remote, of each of all his acts and, therefore, their ultimate rightness or wrongness? When is one permitted to sacrifice the happiness of certain individuals for that of the rest of the people? Do intrinsically good actions issuing from good intentions and resulting in the greatest happiness for the greatest number of people count? What happens when competing interests and commitments conflict in producing good results, which one wins the choice? How are the differently ranking values of different people accommodated within the ambit of the happiness principle when some people derive pleasure from base pursuits that are without value? Again, as Mill said,

> Men lose their high aspirations as they lose, their intellectual tastes, because they have not time or opportunity for indulging them; and they addict themselves to inferior pleasures, not because they deliberately prefer them, but because they are either the only ones to which they have access or the only ones which they are any longer capable of enjoying.[2]

These are engaging philosophical questions for the reader to pursue.

## Deontology

Deontological (deon=duty) theories examine the role of the concept of duty or obligation in ethics to the exclusion of the consequences of people's actions. The rationale for deontology is that unforeseeable and hence potentially unknowable consequences cannot be used as the basis for the determination of what is the right

281

or wrong action to take at this time. The duty entailed is not the hypothetical kind that says: if you want result $y$ you should perform act $x$. It is rather one that says: you ought to do $x$, because it is the inherently right thing to do, in keeping with the principle that we should treat people as equals, recognize their rights and the reciprocal duties that these rights entail, promote their good, and avoid harming them and using them arbitrarily only as means to our selfish ends. According to Kant's Categorical Imperative of duty, which bases the moral worth of an action on duty versus the intended results of the hypothetical imperative, we should be guided by the dictum: "Act so that you treat humanity, whether in your own person or in that of another, always as an end and never as a means only."[3]

Kant bases the possibility of universalization of acceptable behaviour, in the face of a variety of personal and group opinions on all matters, on the human faculty of reason as the rational basis for action to determine if that action is right and therefore moral. He maintains that all reasonable people will inevitably reach the same conclusion on all matters under their consideration, not by appealing to a principle, tradition, or law, but by the use of intuition—being aware of the right thing to do; and therein lies the Achilles heel in his theory, which he rescues with his Categorical Imperative of morality, "[presenting] an action as of itself objectively necessary, without regard to any other end."[4] It is this imperative that makes euthanasia and suicide wrong, in his opinion, because the person is here used by others and by himself as the means to these ends, even if he may be reaching these decisions autonomously. Besides, these selfish and inconsistent appearing acts violate the sanctity of the life of the rational, intrinsically worthy being that the person is, and they fail the test for serving as universal laws of nature. This is the same innate sanctity that is denied the poor people of the world and that warmongers also violate with impunity all the time, because they make the sanctity and ultimate value of these lives circumstantial.

According to Kant, we should, as morally autonomous agents, always act as if we intend our actions to form the inflexible basis of all moral laws. He makes no allowance for lying to protect the interests and welfare of the person lied to, arguing that to permit one lie is to recommend lying as a universal principle for all of humanity to adopt, and that is to stifle and supplant truth telling and the good will that undergirds it as the norm. So in a lying society there would be no advantage to be gained from lying, since there is no truth to take advantage of, and lying cannot be made a universal rule, since it is the exception to truth telling, which is the norm. Truth telling is like a genuine coin while lying is like a counterfeit coin; the two can never exchange roles.

That means a doctor can never lie to her patient in a well meant

282

attempt to shield him from devastating news about his health with the intention of breaking the bad news when she believes that circumstances are just right for doing so. That kind of parentalism does not wash with Kant, nor does the parentalism of making treatment decisions for patients who lack the means to make those decisions for reasons of mental or physical infirmity, immaturity, and ignorance, because they undermine the patient's autonomy. The problem with this stand is obvious; untold suffering will result from leaving treatment choices and decisions to these relatively rudderless kinds of patients, which is what this kind of autonomy entails, when even competent patients need the expert direction and advice of their health professionals to make appropriate treatment decisions.

In his insistence on respect for autonomy and the ethic of the categorical imperative of duty, Kant did not recognize William Ross's *prima facie* or conditional duties "which an act has, in virtue of being of a certain kind . . . which would be a duty proper if it were not at the same time of another kind which is morally significant."[5] These are conditional duties that at first sight appear to be binding but are not, because they allow exceptions and can be overridden by other more pressing ones that present better moral reasons for abandoning the original duties in favour of the latter. For example, it is a truism that no duty or rule is absolute or without exception, when all the circumstances involved in its execution are taken into consideration, e.g., a lie that saves a life may be *prima facie* wrong, but it is morally justified by the life that it saves. This holds despite what Kant and some consequentialists (rule utilitarians who base the rightness of actions on their conformity with prescribed rules that are circularly derived from best consequences) maintain.

The question, however, is that of knowing which one of the conflicting *prima facie* duties or rules to obey, whether by intuition or by pure reason. It is clear, though, that *prima facie* duties that override others and thereby become actual duties should impartially promote the good of the parties concerned and be achievable without unduly sacrificing the happiness of any one individual for that of the other people involved. That fact alone imparts a consequentialist flavour to Ross's deontological position. In his opinion, the rightness of an action has nothing to do with its utilitarian goodness, only its balance of *prima facie* goodness over its *prima facie* badness as given to us by intuition, after taking all things into consideration: motives, consequences, and wide variations in intuitive capacities. According to Ross, good motives, good consequences, and intuitions that impart a good feeling make otherwise wrong actions right. Not many people would agree with that mixed principle that would, for instance, approve of murder for exercising the good motive of eliminating a thief and thus promoting the happiness of a community at large.

# Principlism

For principlists, the categorical imperative does not apply, and there is no dogmatic following of rules of conduct as guides to action, the latter being considered by some to be the weak point of this ethic. The gist of moral life for principlism is following guiding principles faithfully, so that if a doctor respects a patient's autonomy, she will be frank with him about his illness and its implications and allow him to choose between treatments and no treatment. She will not lie to him or hide the truth from him as has notoriously happened to those from whom their diagnosis of and treatment for syphilis was withheld. Sometimes the temptation to be less than forthright with a patient may be occasioned by the desire to protect him from emotional trauma. In such cases of conflict, the doctor should not be bound by the absolute rule of "thou shalt not lie", but she should be able to resort to the four principles of medical ethics, viz., autonomy, beneficence, justice, and non-maleficence (discussed below) to arrive at what Rawls calls a reflective equilibrium. That means weighing the pros and cons of all the directives derived from these principles, matching them for coherence and mutual support and explication, and arriving at a balanced conclusion that is not tilted unfairly towards any one prevailing principle. It also means that the prima facie force of all principles should, as far as it is possible, be equal as determined by the circumstances of the person concerned, and everyone of the principles should be modifiable or defeasible in the face of more ethically demanding circumstances.

Besides the criticism of luck of guidance on action, principlism has been criticized for lack of a standard to resolve conflicts among principles, because they are not derived from a unified moral theory, and individual principles are also subject to cultural relativity. They vary from one cultural society to another.

# Intuitionism

Intuitionism does not depend on consequences or obligation to determine if an act is right or wrong. It holds that this knowledge is intuitive or self-evident, not derived inductively or inferred deductively from prior facts, but known as readily as the axioms of Euclidean Geometry, *a priori* or without prior proof. Rational people know that it is *prima facie* right to help those who are in trouble and wrong to kill others (without cause), and no proof is required. These moral beliefs are believed to be as true as that when someone seems to see a book on the table she does see a book on the table, unless she or other people can produce evidence that she is having an optical illusion or some other perceptual aberration. By and large, we rely on our eyes to see things, because they are our only means of seeing, and we believe what our eyes represent to us, unless we have good reason to doubt our one or more links in our visual mechanisms. Similar arguments apply to the things we hear and touch, and ultimately to our intuitions about ethical norms.

The problem with intuitions, however, is that every person has a perspective on everything and intuitions vary concomitantly; so no one intuition on any one subject can be held to be the right one. That means we have to rely on the consensus of a network of the most rational and credible intuitions as the bases of our moral beliefs and judgments; but it also means having $criteria_1$ for selecting those intuitions, $criteria_2$ for selecting these $criteria_1$, $criteria_3$ for selecting $criteria_2$, and so on *ad infinitum*. Besides, since skepticism about the reliability of intuitions as the basic tenets of beliefs simultaneously disqualifies axioms as the basic starting points of Euclidean Geometry, causing the whole structure to tumble without them, the same element of skepticism with regard to the reliability of intuitions will produce the same result, because there is nothing against which to check the veracity of both basic tenets; they are the rock bottom or fundamental truths of the two systems. If they fall, the systems fall. But that still does not render intuitions totally reliable, although they are excellent guides from which to launch rational ventures.

Without starting points like axioms and intuitions, geometry and moral thinking would not be able to get under way. In the case of moral beliefs, it is significant that they do not follow logically from factual statements, but only from statements with a moral content. Can medical practice rely on intuition? Yes. It is conceivable that in the beginning it was intuition that led to the appreciation and documentation of certain symptom clusters as characteristic of particular disorders, and that the systematization of these basic facts led to the construction of the ever increasingly complex structure of the science and art of Medicine. Evidence-Based Medicine notwithstanding, hunches or intuitions still play a role in leading physicians to seek out information that will help them focus on a possible diagnosis that may not seem to follow readily from the facts at hand. Every physician has had a brain wave at some time or another that led him to the right diagnosis, much like the experience that Archimedes had when the idea of the equality of the weight of displaced fluid and the object immersed in it dawned on him, and he exclaimed: *Eureka!* (I have found it).

## Virtue

Virtue theorists place less emphasis on consequences and obligations; they focus instead on the cultivation of excellence in performance, which springs from excellence of character. Only the person who is habitually benevolent and just will of necessity act in accordance with the principles of beneficence and justice. She does not have to try to produce the best consequences or try to fulfill her obligations, nor does she have to wait for that spark of intuition to guide her action to do the right thing. The right action will always be second nature to her, as long as she does not desist from constantly striving for excellence of character. She can be relied upon to act with virtuous consistency in a wide variety of situations over a long period. She performs benevolent actions not only to achieve beneficence, but also because benevolent actions have an intrinsic value that does not depend on external factors for their worth; they provide their own justification quite apart from theories of beneficence. On the other hand, her actions could still be guided by beneficent principles, even if she lacks benevolence, while her sporadic efforts and outbursts of presumed virtuous activity, which are prompted solely by special circumstances of advantage to her, could never amount to excellence of character, since they are not the mark of a virtuous person. The morality of the person cannot be separated from the morality of her actions and the manner of their execution; hence virtuous persons can be trusted to act both dutifully and virtuously and to do the right thing, and that applies to the doctor's attitude toward her patient.

The striking feature about the list of virtues proposed by various persons is that they all appear to have been developed intuitively; e.g., the cardinal virtues: prudence, justice, temperance, courage, and the other virtues of generosity, honesty, tolerance, etc. So have the opposite traits or vices that should be avoided: rashness, injustice, unrestraint, cowardice, and the other vices of stinginess, dishonesty, intolerance, etc. Most actions should be geared more to steering the middle path than to being carried to excess, such as avoiding cowardice and foolhardiness by being courageous, avoiding permissiveness and intolerance by being tolerant. Another feature of these virtues is that they are supposed not to be *prima facie*, because the daily conduct of life depends on their consistency; e.g., we have to trust that the doctor is always telling us the truth about our diagnoses and prognoses, and that she always prescribes curative medication and not poison. And yet occasions will arise when some of them will have to be violated, like when the same doctor has to temporarily withhold the devastating diagnosis of advanced cancer from a patient until he has had time to take that long anticipated holiday which will add much quality to his remaining months of life.

These and many other excellencies in the character of a physician surpass by far any efforts on her part to observe laws and regulations that help to keep her out of troubles like litigation. The doctor who is virtuous in her practice has no fear of being sued for negligence,

286

dishonesty, and insensitive exploitation of her patients, because her actions are virtuous. If she makes honest mistakes and admits her mistakes, she will go a longer way than her colleague who tries to lie her way out of her callous mistakes and cover up her shortcomings. Her qualities of virtue and patience allow her to place her caring skills at the disposal of all categories of patients, including those with AIDS-related illnesses and other diseases from which the business minded doctor will shy away, or those from communities that are habitually singled out for maltreatment by her cultural group.

## Caring

Caring is the attitude that connects doctor and patient closer than all the theoretical aspects of their relationship mentioned above, and certainly very much closer than all the technology that often intervenes between them to make the patient a cog in the wheel of that technology. It is not theory that is governed by rules, but practice that replaces I-It with I-You and thereby helps to generate the needed flow of empathy from doctor to patient as she responsibly strives to promote his health, compassionately protect him from harm, and maintain a good relationship with him by respecting his autonomy. In many cases this empathy has not been existent and in other cases where it might have been dormant or nascent, it that has been eclipsed or even displaced by the biomedical character of medicine where the doctor does only what is expected of her to mend the patient's deranged function of body or mind under the auspices of reductionism. Caring calls on the doctor to go the extra mile to connect with the patient emotionally, place herself in his position or, as the saying goes, walk a mile in his shoes, in order to appreciate what and how the patient is feeling under the weight of his often alienating affliction—I have already recounted personal experiences in this regard. Also, as I indicated in other sections of this book where I discussed being-in-the-world alone or with others, it is the doctor's prerogative to bring the patient in from the cold and make him feel that his moral value as another human being is equal to that of any other person, despite the inevitable but inconsequential constitutional differences that exist among all people. If a certain amount of altruism is required to achieve this end, then so be it; altruism does not entail self-neglect or extreme supererogation, it only entails feeling for the other in a positive way. Doctors should not entertain negative feelings toward their patients, or treat them grudgingly or callously and feel snug about it, because that kind of behaviour offends against common decency and medical ethics.

We can see, therefore, that unlike the maximally-good-results-oriented ethics of consequentialism and the duty-motivated ethics of deontology, the ethics of caring emphasizes human connectedness and interpersonal relationships of understanding, compassion, and

287

empathy, leaving no room for the domineering attitude of human separateness often seen in rights-based adversarial contacts, which aim only at duty, cold justice, and "fairness" for participants. Its proponents contrast its contextual approach with the abstract approach of duty and justice whose aims are laudable only to the degree that they can be fulfilled with genuine fairness and concern. However, if they are not tempered with concern, then those at the receiving end of this exchange could get emotionally hurt by the impersonal and dispassionate distribution of their abstract rights by aloof public officials and physicians. Conformity with the principles of duty and justice does not necessarily imply performance of a moral action, nor do morally right actions always conform with dutiful and just principles. Emotionally detached, indifferent persons can distribute ostensible entitlements to citizens and patients without caring a hoot about their welfare, while sparing no effort to engage in self-worship and enhancement of their own images for their benevolent deeds. Underlying these transactions is the attitude that a just distribution of goods is considered right and proper if it is done as per stipulated guidelines, whether or not it is done with compassion and the necessary follow up to ensure that recipients derive benefit from this gesture, as long as a duty has been fulfilled; and no one can complain of being short-changed as a result. The only problem with this kind of robotic dispensation is that it is only duty to oneself and justice of a kind that is not steered by empathy, caring, and concern.

A moral theory of care should refer to the nature of the action that makes it right and to the character of the person performing the action in a manner that reflects an attitude of extreme concern and the willingness to help alleviate situations of distress and dependency on others for the needed relief. It should have universalizability, and be known as such intrinsically and via its doer, otherwise it becomes a virtue that renders the category of "care" superfluous, reducing it to a qualifying term describing an attitude that might not even be accompanied by benevolence, love, or concern. On the other hand, the doctor does not always have to act on the basis of rules or a principle, seeing other people in the abstract or as lying outside the limits of two standard deviations on the bell curve and therefore excludable, but she should act purely from inclusive caring, empathy, compassion, and an even distribution of justice directed with genuine concern at each individual in his specific predicament of vulnerability.

# Medical Ethics

Whilst recognizing traditional ethical disciplines and theories that provide the overall moral bases for our actions, clinical medicine further recognizes four guiding principles of moral obligation that cut across the different theories and that, taken together, are believed to provide a sound basis for the tacit contractual and moral relationship of mutual respect for each other's values between doctor and patient, regardless of their other affiliations, whether they are deontological or utilitarian, theist or atheist, liberal or conservative, etc. These are the principles of autonomy, beneficence, non-maleficence, and justice that have been found to be adequate guides to how the doctor should conduct herself in her dealings with all patients, if she is to accord them the respect that is due to them as moral beings.

Meanwhile, the evolution of biophysical medicine and its ethics has resulted in the discovery of new diseases and the development of new technologies, increasing the complexity of ethical problems with which doctors have to deal and the ways in which medical care is distributed, evaluated, and applied among different patients with their many novel afflictions. As we have already noted, new controversial concerns now relate to 1) the status of human embryos, which have been endowed with personhood and inviolable personal life by certain groups; 2) the ethical propriety of harvesting embryonic stem cells as tissue repair units to treat neurological and other clinical conditions in which existing well differentiated tissues cells have lost their repair and regenerative capacities; and 3) the elimination of embryos burdened with uncorrectable genetic anomalies that will constitute postnatal sources of persisting emotional and economic hardship for afflicted persons, their caretakers, and society. The same society has generated dilemmas that include the use of genetic techniques and ultrasound to identify and slate for elimination fetuses with anencephaly, some trisomies, and some chromosomal deletion syndromes. All these technological advances are dogged by controversy stemming from the need for framing ethical theories in ways that consider the morality of responses to the problems that they generate, versus ensuring that those responses accord with standard ethical theories like the categorical imperative of Kant, which is interpreted rigidly by some people as prohibition of any kind of killing, including that of fetuses, regardless of how they came into being—voluntarily or by coercion.

How we resolve the conflicts between opposing moral convictions as to the personhood or otherwise of these embryos and fetuses may not be achievable by employing the four principles alone. It might, however, be rendered so by overriding them with the use of pragmatic considerations relating to all the parties affected by the final attitude of the adjudicators concerned with deciding the fate of the embryos and fetuses (casuistry). What is required in these situations is the sheer common sense that would be applied if there were no conflict-prone ethical theories such as Deontology and Utilitarianism, which did not descend to us on a platter but were framed by people in response to poring over similar problems. The prescription that right actions are those that conform to a certain pattern or formula does not always make them right or square up with how people act and act rightly and morally in the particular circumstances confronting them without having to justify their actions by invoking an ethical theory; e.g., when to abort a fetus or assist in a case of euthanasia. There will always be many situations in which clashes occur with regard to the moral correctness of the choices made by any one or more of the diverse interested parties, such as the patient and his wishes, the physician and her duties, the medical community and its standards of practice, and the social community and its laws and economic constraints; but there will also be only one or two incontestably right actions in any particular situation, and the challenge is how to pinpoint those actions. Those are the moral dilemmas facing the doctor in her daily encounter with her patients and their problems.

## Autonomy

Autonomy refers to a patient's capacity for rational self-governance and the right to hold clearly understood personal views, values, and beliefs that determine his free and competent choice and initiation of actions in his own interests without pressure or threat from anyone. It also entails his right to determine how far others can intrude on his body and person, how much, if any, of his personal information they can divulge, and how much he can rely on them to deal with him truthfully and not lie to him about his medical condition. Physicians are sometimes prompted to lie to patients by the desire to indirectly protect patients from emotional trauma, not realizing that they are thereby directly compelling them to make decisions and take actions that they would not otherwise make and take if they knew the truth, even if the truth is likely to hurt them. A cognitively competent patient who is protected from the shock of knowing about the rapidly terminal nature of his illness is thereby deprived of the opportunity to wrap up his affairs in a timely manner, and that is not fair to him and to those that he will leave behind to carry out that task. They all deserve time to prepare for the parting.

Since autonomy entails self-determination, self-expression, and the liberty to choose one's actions, it has to deal with the problem of persons who are too young and immature to have the capacity to exercise these properties reasonably, if at all, and who must therefore depend on surrogates to make decisions for them in a manner that also accommodates their dignity and their autonomy to the degree that it can be ascertained. Autonomy also extends to the duty owed by the physician to her older patient not to pander to his irrational

290

wants or yield to his self-destructive ignorance. Like other rights, his is a right that does not end with making the doctor yield to his desires indiscriminately, but entails the exercise of reason by him while it demands acknowledgment, respect, and facilitation of his interests by everyone else. So as long as these interests do not infringe on their own rights or cause them harm in any way; e.g., deprive them of their legitimate entitlement to resources, the doctor can entertain them. If she claims harm for any reason, she will bear the onus to prove that the patient lacks the capacity to claim autonomy. The doctor is obliged to yield to her patient's autonomy when she cannot demonstrate that the exercise of that autonomy is inconsistent with his character by reason of some incompetence on his part, such as his inability to understand his clinical situation and its implications. In that case, she has to assume that her patient is acting rationally, and she can only resort to beneficence as her guide to dealing with him, failing which she has to balance her own autonomy against his in good faith and with the best interests of the patient, not society, as her priority. If his rationality is still in question, his personality as consistently expressed in his previously expressed values should, as much as is possible, be reflected in all decisions made on his behalf; but his present uncertain values should not be ignored, because they are an essential component of his entire personality.

Autonomy is best expressed in the principle and practice of informed consent, which entails authorization by the patient for the doctor to treat him in specified ways, provided that the doctor has disclosed all the relevant information relating to that mode of treatment, including short and long term complications and their gravity; and provided that the patient has the cognitive competence to understand every facet of what he is voluntarily consenting to. That means the doctor should communicate at a level that is commensurate with the patient's comprehension of the facts and resist the temptation to bedazzle, coerce, or manipulate the patient into consenting to what he does not understand; it is always possible to simplify technical information to accommodate the patient's level of non-technical understanding and maintain respect for his autonomy. No patient should ever be deceived or dismissed because he does not understand medical jargon and technicalities. Doctors should not rely on giving patients consent forms to read and sign without first explaining all the facts to them and entertaining their questions, or try to make them sign blank forms (details, which may be upsetting, to be written in later) as one doctor tried to do to someone I know. Even information about their conditions that would otherwise traumatize them if it is presented callously can be communicated with compassion that will soften any hard blow coming to them.

Sometimes doctors may withhold information from patients,

because they rightly believe that their extensive knowledge of medical facts, compared to the patient's relative lack of knowledge of those facts, entitles them to convey to him only that selection that he can handle comfortably in his present condition. To do so, however, is to deprive the patient of the full exercise of his autonomy by providing him with inadequate information by which to make decisions that are compatible with his interests and goals. The doctor may be exercising the utmost beneficence in so doing, but if she is concerned only with the pathophysiological aspects of the patient's problem and its objective implications to the exclusion of his socio-cultural concerns and preferences, she ends up undermining his autonomy. She does likewise if she violates the confidentiality that the patient can count on when he confides his most intimate concerns in her for the better management of his illness. Without this bond of confidentiality, the patient will lose his trust in the doctor, and he will withhold from her the most private and sensitive but critical information that is required for the diagnosis and treatment of his illness in the fear that she will make it common knowledge. The ethical doctor respects the patient's confidentiality and every aspect of his autonomy.

## Beneficence

Beneficence entails promoting and doing what is best for the patient after balancing the benefits and burdens of treatment; i.e., risks and costs against the patient's goals as advocated by utilitarian or consequentialist theory. Beneficence is the natural precursor to benevolence, and thus it dictates that the doctor's sole function should be to promote the patient's good whilst respecting his autonomy. Therefore she should always aim for concordance between these two concepts. Sometimes, however, her exercise of beneficence may clash with the patient's autonomy, but if the patient's interest and welfare are her foremost considerations, she will find it easy to override the patient's autonomy when there is convincing proof that his actions and desires are motivated by incompetence of a kind that is detrimental to his health and welfare. The same prerogative will rest with the doctor when the exercise of her patient's autonomy involves risk to the interests of other persons, or when acts of beneficence may be looked upon as contravening established ethical or legal principles, as in the controversial example of euthanasia in its many faces. When a doctor encounters statutory laws, she is forced to terminate her quest for ethical problems *statim;* the laws do not allow her the opportunity to apply ethical principles to the situation at hand. Her only obligation at this point is to obey the law.

It is noteworthy that the exercise of beneficence is wholly dependent on the doctor's moral temperament, because no law mandates beneficence, and the only culpability that accompanies failure to exercise it is moral, rather than criminal. Therefore doctors should be sensitive to both the goals and the plight of their patients and respond with beneficence to threats on their health and welfare to prevent unnecessary disability and loss of life. They need not go overboard to try to provide unachievable "super health" when the situation only calls for them to promote conditions whose additive effect is to create a better quality of life for their patients, especially if they are the only ones who can prevent these ills from befalling the patients and can do so without significant risk and other burdens to patients and themselves. Beneficence is not supererogation; it does not call for sacrifices. It only demands honesty, compassion, and humility on the part of the very important and infallible person that the doctor thinks she is, and it entails always being ready to ascertain from the patient what his goals in life are, what makes his life valuable, versus assuming that only restoration of his physiological function by his all-knowing doctor will suffice to make him happy.

On the other hand, beneficence necessarily entails a degree of paternalism, because the doctor acts without the permission of her patient and against some of his autonomously made indiscretions and irrational choices with the sole intention of protecting him from harm and furthering his best interests in a truly utilitarian way. The assumption is that if the patient were cognitively competent like the one mentioned in the opening paragraph of the section on autonomy, he would make different choices that are consistent with what his doctor is recommending for him and with what most rational people would have wanted. Hence the truly beneficent action may at times justify flouting deontology by lying to the incompetent patient to protect him from inflicting harm on himself and others physically and emotionally. This is particularly so in suicidal patients and those in states of depression who cannot adequately assess the import of the information that they are given. In that case it is better to gloss over or postpone complex information until the patient has been rendered stable and competent enough to handle it. The opposing libertarian view will always promote patient autonomy over physician beneficence, even if the patient's resulting choice is contrary to the best medical advice and will not further his best interests.

## Non-maleficence

Non-maleficence refers to the obligatory duty of the doctor not to harm her patient or wittingly expose him to harm. Although it is a negative duty by not specifying what the doctor should do, it is a categorical prohibition whose infringement invites culpability beyond moral censure, and it often results in legal action for damages against her, because of its positive stance against doing harm. Beneficence cannot issue a similar prohibition, because it depends on the goodwill of the doctor, which cannot be mandated. However, since the absence of curbs on human behaviour can sometimes result in unpalatable actions, codes of ethics and the imposition of standards of performance set by medical and other regulatory bodies are in place to ensure that actions that stifle the patient's welfare or violate his rights to medical treatment and other entitlements are expressly forbidden when common sense is not sufficient to act as a guide for some doctors to respect the rights of patients and promote their good.

For the doctor, the tricky situation arises when she has to decide between the two choices of letting the patient suffer until he dies, which amounts to harming the patient passively, and respecting the patient's discretionary autonomy by assisting him to escape in a dignified manner, i.e., with intact self-control, self-determination, and self-esteem, from prolonged terminal suffering and incurring the wrath of those who consider this act to be also one of harming the patient. These dilemmas will be discussed under euthanasia.

## Justice

Justice refers to the treatment of like cases alike and unlike cases differently, on a universally acceptable basis, according to the extent of their moral likeness or unlikeness, without diminishing the moral worth and dignity of anyone whose lot is worse off through no fault of his own. It entails the arguable distribution of medical treatments and benefits according to fixed criteria of desert, financial ability of a particular society, respective contributions of its members, versus the challenge posed by the entitlement of all members of the society to egalitarian consideration for health care that is constituted by their common humanity and basic need. The criteria underlying these principles should be rooted in promoting the equal distribution of maximum health benefits to the greatest number of citizens as a duty of the state. The same principles should also deprecate the biases of those who favour minimal access to health care for patients without financial means and maximally prioritized access for patients who can pay for more than the basic health care services. This is an alternative that falls short of the principle of beneficence and should not be encouraged anywhere at any time, because it excludes the most vulnerable citizens from health care.

Rawls went beyond these principles of justice in his treatise on *Justice as Fairness* by taking into account the existing inequalities in society that render unfair the use of the criterion of merit for the distribution of benefits, because of differences in head start and history of acquisition of benefits that have been perpetuated through the years, bestowing an advantage on select sectors of society. This principle of justice as fairness urges the doctor to treat all her patients on the same level, whether they are rich or poor, and recognize that poverty and disease are bed-fellows. Furthermore, the doctor should realize that the poor do not have the means to pay for preventive measures against ill health and they can also not afford to pay for life saving procedures against conditions that they did not self-inflict; e.g., hyaline membrane disease, congenital heart disease, cancers, etc. These are responsibilities that should be assumed by the state, and the doctor should be an advocate, no less an agitator, for that to happen. The doctor's cognizance of these facts, her advocacy, and her genuine determination to correct for them in treating her patients will persuade society not to discriminate against them or punish them for their lack of means to pay for costly and exclusive treatment like those who have the money, in the same way that it should not discriminate against all people on the basis of gender, sexual orientation, colour, religion, and other inexcusable biases. Society sometimes shirks its obligation to the many who are deprived of basic health care services by its use of funds diverted from them to experimental, unproved, and sometimes unnecessary fad treatments that suit the doctor's choice or are undertaken in response to undue pressure from patients and their relatives in their push to procure what they think is the best treatment, but which often proves to be futile in the prevailing circumstances.

A society with a moral conscience will not encourage a two-tiered system of medical care; it will rather double its efforts to ensure that all citizens receive the treatment that they need and deserve in good time, instead of wasting money on bureaucratic pursuits that benefit a few chosen individuals at the expense of the health care of the masses of citizens. Doctors should not encourage or condone it, unless they do not care about the moral discredit that they are earning for themselves by so doing. Their duty is to remove impediments to the equitable health care of their patients and to relieve them of the burdens that the state unfairly and callously casts on them and their health needs.

# Applied Ethics

# 18

## Right to life

## Nature of rights

The practice of Medicine entails recurring contact with situations that involve patients who are in pain and suffering from their illness, many of them terminal. Being autonomous individuals, they want to exercise their incontestable right to self-determination. But in all cases the state intervenes to preserve what it calls the sanctity of human life, ostensibly on moral grounds, but really on religious grounds based on so-called natural law which, it is claimed, comes from God. (I discuss rights and religion in *The Human Agent*). For this discussion, we will consider the bearing of morality on the right to life only from the secular point of view as the one that we can understand, since religion says we do not have the capacity to unravel transcendent matters. We can argue about whether the killing in voluntary euthanasia is morally wrong and thus culpable because God says so, or else morally right because it was sanctioned by an autonomous human being in his own interests, which in his sane mind he knows better than any of us (keep God out of it).

When we make decisions in medicine we should consider the principles of autonomy, beneficence, non-maleficence, justice, and rights. The first four principles were discussed in chapter 17; here we will discuss rights. Rights come in many forms, but underlying all of them are personal entitlements to actions and states of being that we are free to exercise without prohibition except by duty to ourselves and to others. Most of our (negative) rights entail a moral duty on the part of other people to respect those rights and not to interfere unjustly with our exercise of them. In other words, we are mostly immune from the self-assumed but unlawful attempts of ordinary citizens to try to direct our lives in accordance with their own morals outside the stipulations of binding civic laws. The right to life is a fundamental, unassailable right of self-determination according to which each person, as a sentient being and as the sole owner of his body, is freely entitled to do with it as he pleases by living his life his own way, including terminating it, without interference from other persons. The only proviso is that he does not jeopardize or encroach on the right and freedom of others to do the same or otherwise with their lives or impose an unjustifiably onerous burden on them to allow him to exercise his rights. Therefore each rational person has the reciprocal right to sustain or to terminate his life if its quality is grossly inadequate and of no value to him alone, and no one should assume the prerogative of preventing him from doing so, or of urging him to do so if he is not thus inclined, because he alone knows best what his life means to him and what intrinsic value it has for him.

The intrinsic right to terminate one's own life sometimes encounters an impediment to its execution when the person does not possess the substantive means or the capability to accomplish his intended result. In that case, he has to sacrifice his autonomy and depend on others, generally his physician, to execute that function for him. She then finds herself in the untenable position of complying with two opposing obligations: one with the welfare of her patient by respecting his autonomy and his rights, and the other with her general undertaking and public expectation to save lives, not to terminate them. She also has to wrestle with the concepts of rights as goals or side constraints. If rights are inviolable goals, then they will compel her to comply with her patient's wishes; but if they are side constraints that can be violated for the attainment of certain ethical goals, then they will allow her the option of subjugating those wishes to the trumping effect of some other moral considerations that obtain in her own sense of values and the principles on which those values are based, and in the values and principles of the society to which she and the patient belong. Some philosophers take the view that we all have non-imperative, positive duties to help other persons and binding, negative duties to refrain from hurting them, and when the two conflict the negative duties trump the positive duties for the sake of realizing the maximum amount of good from our actions. Such is the case when the life of a fetus doomed to die in utero is sacrificed to save the life of its mother, instead of letting both of them die and thereby cause greater overall loss, which goes against utilitarianism.

Situations like those outlined above have persuaded some people to regard the holder of rights as exercising controlling authority over those who are duty bound to act only in certain ways to respect his rights, because rights entail obligations on the part of those over whom those rights are held, while others argue that rights serve merely to protect and promote the interests of the holder of those rights against the selfish majority, where these interests exist. The holder of moral rights may not use this prerogative to make others perform acts that will allow him the exercise of his rights while depriving them of their liberty and rights, unless the state compels them to do so; and he cannot compel them to provide health care for him, except by agreement or contract; but the state, as his guardian and the guarantor of his right to life, is morally bound to provide that care. He can, however, forbid them from harming him on moral grounds with the blessing of the state, which is also morally and legally bound to ensure his exercise of all his basic human rights.

Although the state may try to shirk its responsibility to the person, the right to life is still morally binding on it to provide him with the barest minimum means for survival, even if nothing else, from resources that most states squander thoughtlessly on ventures that not only deprive citizens of their life sustaining entitlements as members of those states, but also severely limit extension of those same benefits to citizens of less privileged countries and states—behaviour that amounts to a violation of the fundamental human rights of all those concerned. For physicians, however, their code of ethics obligates them to provide health care, unless the patient refuses it or demands an inappropriate mode of treatment. The physician then has the moral obligation to enlighten the patient on his situation and attempt to reach a consensus with him before she decides not to treat the patient. Notwithstanding the achievement of such compromises, controversy still dogs the duty that is due to the terminal patient who requests treatment to shorten his life of suffering promptly when preserving and prolonging it only worsens his ordeal, as we will see when we discuss euthanasia in chapter 20.

One would like to think that on any controversial issue accommodation and compromise are always possible, and that no one should be coerced into doing what other people consider to be the right thing to do, even if it is to his advantage, or else forcibly prevented from doing what they consider to be wrong, even if it is to his detriment—both paternalistic actions that violate the patient's autonomy and place him at the disadvantage of being forced to make choices that he would otherwise not have had to make, if that option had not been imposed on him. As Mill says, "The only purpose for which power can be rightfully exercised over any member of a civilized community, against his will, is to prevent harm to others. His own good, either physical or moral, is not a sufficient warrant."[1] David Velleman notes that if, when faced with the choice of having his fruitless life terminated or allowed to ebb slowly to its close, the patient fails to choose the former, which everyone else thinks is the logical choice, he will have to justify his action to those who think that his life is no longer worth living and who have allowed him the choice that he has ignored, failing which he will now feel obligated to end his life against his better judgment. The right for someone to decide when to die, as controversial as it is, will then have become the obligation to die. "The worry that a right to die would become an obligation to die is of a piece with other worries about euthanasia, not in itself, but as a problematic option for the patient."[2]

When it is applied to human fetuses, the right to life assumes a further controversial character relating to whether the fetus is or is not a person. By definition, only persons as innocent human beings have a right to life, but human organisms, as mere genotypic and phenotypic members of the species *Homo sapiens*, do not have that right, because their lack of self-consciousness ($P$ predicates) qualifies them for only the material component ($M$ predicates) of personhood ($M$ and $P$ predicates) in the Strawsonian sense. Some other people maintain, however, that the fetus is a person even as it still lacks $P$ predicates, and it therefore has an intrinsic right to life that it acquired either at the start of pregnancy or somewhere (at some unknown point) along its course, and which it carries forward into the infant, youth, and adult stages.

Contrary to this position, and as a prelude to his thesis justifying abortion and infanticide, Michael Tooley argues that if fetuses, embryos, and zygotes are innocent beings, then on the premise that such beings have a right to life, it becomes *prima facie* seriously wrong to kill them as in abortion, except when they are products of rape, incest under coercion, or fetuses with severe brain damage or anencephaly: "Therefore abortion is at least prima facie very seriously wrong, unless—as, for example, in the case of rape, or many cases of incest—one did not intentionally, and without coercion, run the risk of becoming pregnant."[3] His version of the right to life does not, however, morally have to extend to beings who are merely members of the human species, but only to those organisms that have certain psychological capacities and potentialities that depend on the appropriate neurological organization of their brains. These are features that confer the capacity for thought and self-consciousness, properties that are in turn basic to the acquisition of the right to life. Their lack in the fertilized human egg, the anencephalic, and the severely brain damaged human organism disqualifies these three from claiming the right to life. He is silent on the status of unconscious, sleeping, or anesthetized persons to whom we do not deny a right to life while they are in those states, on the grounds that their right to life precedes the temporary suspension of their self-consciousness during sleep and induced states of unconsciousness.

Philosophically, decisions$_3$ about rights and their exercise hinge around the third order category of who makes decisions$_2$ about who makes decisions$_1$ about the issues, as discussed by Frederick Schauer: "If first-order decisionmaking is about what should happen, and second-order decisionmaking is about who should decide what should happen, then third-order decisionmaking is about who should decide who decides."[4]. The first category concerns the answer to the first-order decision of what should happen, i.e., the patient should die, or the mother should abort the fetus. Next is the second-order decision making of who should make the first order decision in accordance with the circumstances of the case (material and non-material costs and benefits), its legality, and the ethical affiliations of the parties concerned in the making of this decision. The rights of the individual are here pitted against the rights of society as exercised through its appointed or elected officials to advise on and uphold the practices and laws that are relevant to that decision making process, e.g., hospital ethics committees and the lower courts of law. Third order decision making involves the highest courts of law as the ones which answer to the question of who decides who decides or, in Platonic terminology, who will be the guardian of the guardians. So the first order autonomous decision by the patient or the prospective mother is only the beginning of a process that often has to make its way through the next two levels (of parentalism) before her right to self-determination can be respected or her wishes and intentions fulfilled. Her rights to herself have been severely restricted by this hierarchy of deciders. The same scenario will apply *mutatis mutandis* to euthanasia where the patient is seeking relief from the suffering caused by the physical and mental pain of terminal disease.

Ultimately, rational decision-making will depend on balancing the utility (values) derived from cost-benefit trade-offs of the consequence of a chosen action, and the probability of occurrence of that consequence. Proponents of this view calculate the expected utility as follows,

$$\frac{\text{probability} \times \text{utility}}{\text{consequences of the action.}}$$

## Right to live or die

The question to answer here relates to the necessary and sufficient properties that any biological being must posses to have an undisputed right to life, as proposed by Mary Anne Warren on page 310. The right to life, while entailing the right not to be killed unjustly, also entails its converse: the right to die or have his life terminated when it has irrevocably lost all its goodness and value only for the person concerned, because he is suffering terminally with no chance of relief in sight. The right to life does not force him to live or compel others to keep him alive, and it certainly does not coercively prohibit him from ceasing to live; it only allows him the privilege to live or die as he chooses. If he waives that right by deciding to die, he permits others also to waive their obligation not to kill him, regardless of what they might regard as the goodness of life and his intrinsic human value, which he has now given up. Goodness of life includes living a life of quality that is free of overwhelming and intolerable suffering and pain and is filled with the minimum of happiness in communal relationships with other persons. Loss of this minimum in quality can be considered justification for some persons to express the wish to exercise their right to die. It also imposes a duty on others to provide them with the kind of humane atmosphere in which their personal wishes can be fulfilled. This requirement is in keeping with the liberal tradition's demand that legitimate rights be matched by duties and obligations; and since we concede that the right to die is entailed in the right to life, we cannot, therefore, shirk the moral duty to respect the patient's request for the termination of his life in the same way that we cannot disrespect his right not to be killed. Nevertheless, the unassailability of such action is still the subject of both ethical and legal scrutiny that prevents it from being assumed with impunity, especially since it appears to hold to ransom those on whom the duty falls to execute this right on behalf of the patient, even if they may consider his action to be irrational or to be contrary to their beliefs.

So, if a patient expresses his wish not to be resuscitated when he sustains cardiac or respiratory arrest because he deems his life to have lost the quality that matters to him, but his family and friends rely on him to fill an important position in the lives of all of them as long as he is alive, the doctor is theoretically faced with the dilemma of matching the principles of egoism and utilitarianism from which he is rescued by respect for the patient's autonomy against the wishes of other individuals, including herself. On a practical level, she has to yield to the commanding prerogative of the patient's wishes and grant his wish. The problem is compounded if the patient is a minor whose wishes clash with those of his parent or legal guardian. In that case, everyone should try to place herself in the patient's position to feel what he is feeling—a difficult feat to accomplish, but one that is necessary if his rights are to be respected. Such was the case with a ten year old boy in my practice who told his parents that he did not want any more treatment for his leukemia after he had endured several bouts of intensive chemotherapy. His parents empathized with him, and however much they hated to lose him, they felt that his wish was not unreasonable, and they decided to support his request for termination of treatment. His doctors also readily respected his wish. This case illustrates the harmony that can prevail when moral rights and empathy are allowed to prevail over the upholding of opposing dogmatic societal biases based on fictive religious ideology.

The Kantian principle of treating rational humanity also as an end in itself, and not only as a means to an end, is one such view that places intrinsic human worth above the happiness and interests of society that are served by a person's life, and the happiness and interests of that person in his own life. Kant's principle prohibits us from using any person as a means to ends such as the convenient termination of his life for utilitarian purposes. People should,

304

therefore, be free to exercise their autonomous choice of refusing initiation and continuation of treatment, and doctors should be bound by respect for this autonomy and by both beneficence and non-maleficence to honour the patient's wishes. Similarly, respect for the autonomy of the woman who chooses to abort her fetus for personal reasons, which should not be the concern of other people who are not faced with the burden of her choice, should receive equal consideration. Her autonomous and competent choice to have her pregnancy terminated when that pregnancy has become genuinely burdensome on her for a valid reason deserves to be respected. Those who wish to restrict the principle of freedom to exercise certain rights, like the right to an abortion, or the right to die, owe it to the holders of those rights to show why they should not be allowed to exercise those rights in a free society that respects people's rights to self-determination. If they claim to base their stand on the premise that acts like abortion and euthanasia are acts of killing, and unqualified killing is wrong in principle, then they also have to show why killing with qualifications is not equally wrong in terms of the same principle, or whether a different principle applies in the case of killing with qualification or in withholding life-saving treatment, both of which still result in the death of the person. They should not equivocate on the word 'killing' for their own convenience.

According to the moral symmetry principle, which is the converse of the doctrine of acts and omissions, if it is wrong to initiate treatment or another process that could lead to the death of a patient, then it is equally wrong not to attempt to halt another one that could produce the same "morally significant result"[5] when the intention of the action and inaction is the same; viz., death of the patient. The same argument applies to refraining from acting to save his life. Tooley admits that there are "factors which make it generally the case that killing someone is more seriously wrong than intentionally letting someone die . . . [because] the motive of a person who kills someone is generally more evil than the motive of a person who merely lets someone die."[6] But, since in euthanasia the motive is to spare the patient the agony of a painful, lingering death, not to kill him maliciously, then it is a mistake, he thinks, to believe that there is a significant moral difference between the act of killing (to be interpreted as actively causing the death of someone) and the inaction of letting die. Killing may be rendered more culpable than letting die by the antecedent circumstances than by the action or inaction; from which it may follow that the duty entailed in refraining from killing is more binding than that entailed in refraining from letting die. However, apart from the question of duty, there is no moral difference between active and passive euthanasia; both result in the death of the patient. Nevertheless, according to the adherents of the doctrine of acts and omissions, there is a moral difference between acting to cause a certain result and refraining from acting so as to let it happen. The corollary from this position is that not doing anything to save a salvageable patient may be worse than killing the terminally ill patient for whom there is, after all, no hope, if intervention can save the life of the former.

The doctor always feels that it is her duty to care for patients to the best of her ability from the time of facilitating their entrance into the world to easing their passage out of it, and that includes never withholding treatment where it is needed or failing to withhold it where it is not needed by the patient. In the latter instance the patient's autonomy demands that she forgo her life-saving stance, because the patient voluntarily requests termination of treatment (and life indirectly). The doctor can't unilaterally decide to withdraw supportive or other life-sustaining treatment from a non-terminal patient, allow the patient to die, and then claim not to have killed the patient, in the same way that she may not unilaterally decide to kill the patient who does not want to die. So why should it be offensive for the doctor to oblige the patient who requests active termination of his ebbing life with the full knowledge of what he is asking for?

In considering the right to life, however, we have to ask the crucial question: what kind of life? Life lived attached to machines that sustain vital functions, life that is no better than mere existence by virtue of its detachment from its environment, or life that is characterized by independent activity entailing the realization of desires, wishes, and goals and amounts to a quality-filled existence in all possible respects for the person concerned? Equally important is the answer to the question whether legislating the right to die will also impose an obligation on patients to have their lives terminated when we and they think that their exacting chronic illnesses, dementias, and dependence on life supports beyond the point at which they would naturally have succumbed to disease have made them a prolonged and demoralizing emotional and financial burden on their families and a counterproductive drain on the resources of their communities, thereby casting upon them the onus to justify their continuing existence against all odds, as hinted by Velleman.

To this stand, John Hardwig takes the contrary view that he "may very well some day have a duty to die. . . . [and that] many of us will eventually face precisely this duty."[7], based on the same concept of the inter-relatedness of the lives of individuals in the context of society and the devastating effects of chronic incurable illnesses that sometimes leave their victims in a state of insentience and their families in financial ruin, without minimizing the sobering effects of the support and sacrifices that go into caring for a chronically afflicted family member. He appeals to Kant's dictum of not using other persons as means to one's ends as it applies equally to the sick person whose exclusive care drains his family physically, emotionally, and financially. He uses the same dictum to indict the patient's family for now using the patient's plight to compensate for their neglected duties to him before the onset of the present predicament for all of them. He also resorts to the utilitarian principle (although he denies doing so) of matching the limited benefits and burdens of the individual in his terminal illness on a "slice of time perspective" against the more extensive benefits and burdens of his family on a "life time perspective". From this comparison and contrast he concludes that the sick person's limited happiness resulting from buying a little more time for himself should not trump the extensive happiness of his entire family, some of whose members are being deprived of their future, i.e., security and sometimes careers.

# Ethics of killing

The ultimate question to be answered is whether the universal application of the presumed major premise: "all killing is wrong" can justify conclusions like: "therefore abortion, or euthanasia, is wrong". We will therefore discuss abortion and euthanasia, because the most divisive conflicts surrounding the nature of persons are found at these two extremes of life. Both processes involve the concepts of person, life, and death, which are central to our origins, being, and destiny, as we have already discussed in preceding chapters. We do not experience being dead, which is not an activity but a state—a state of non-existence indistinguishable from being unborn. But we very much experience dying, and just as we try hard to make our acts of living pleasant, we likewise desire that the process of dying, over which we will have no control, will not be frightening, painful, unpleasant, or undignified. Dying, as the final event of living a long or short, happy or miserable life, and as the antithesis of birth in the history of the independently existing human organism and person, can be pleasant or painful, timely or untimely, and tragic or desired. It is central to the character and quality of a person's life as only he can experience it; no one dies someone else's death. Each person should, therefore, be able to anticipate it with equanimity and face it with sufficiently fitting dignity when it comes his way.

Some people maintain that dying in severe, prolonged pain that requires heavy sedation to the point of inducing total oblivion to one's person amounts to a loss of human dignity and does not benefit the person or those around him in any way. With this situation in mind, the person might want his life terminated before he loses his dignity in that fashion, unlike the one who has already lost his autonomy and is now entirely dependent on others to make every decision for him as they would for a toddler who has not yet acquired autonomy. Other people deny that a life without dignity in this manner is worthless or that it gives any one reason to seek to end it for the sake of avoiding being a burden on self and others. They say this imposes an unfair duty to die on persons in this situation, which is inconsistent with the Kantian principle of not using self (committing suicide to escape suffering), using others (to keep oneself alive), or having others use one as means only to their own ends (eliminating him, because he is a financial, resource, or social burden). In effect, they generate a three-pronged impasse that defies solution with this opinion, placing themselves on the horns of a trilemma consisting of suicide, self-preservation, and 'euthanization'. What the dire state of such a patient calls for is rather more engagement and concern for better palliative care than elimination, bearing in mind the fact that sometimes valuable lessons on how to live with adversity can be gleaned from even the most onerous lives.

One should also ask if nurturing a fetus and eventually giving birth to an infant with severe congenital defects that will only result in the foreseen beginning of a life of extended misery is the only choice allowed by some persons, or whether they would grant that it would be better to abort such a fetus than allow it to be born and then kill the infant or child after witnessing its untold suffering from which there is no chance of recovery? The question is not irrelevant because parents have been jailed for empathically terminating the lives of their children who were suffering with no end in sight. On the other hand physicians and other empathizing adults have also been jailed for assisting in the termination of the lives of those who were eking out an existence of lingering, terminal suffering from which their only reprieve was eventual death; e.g., disconnecting a terminal patient from assisted respiration may allow him to linger on for many days when assisted dying would have ended his suffering sooner. A compromise solution to these problems appears to be remote in the face of the extreme polarization that exists between the flexible ones who consider certain concessions as warranted and the inflexible who will not brook any concessions, although they will not hesitate to terminate the lives of their pets if their suffering cannot be relieved.

We have to wonder also about the professed sanctity of human life and the slippery slope argument against euthanasia in light of the following quotation from Peter Singer: "White colonists in Australia would shoot Aborigines for sport, . . . with no discernible effect on the seriousness with which the killing of a white man was regarded. If we can separate such basically similar beings . . . into distinct moral categories, there is surely not going to be much difficulty in marking off severely and irreparably retarded infants from normal human beings."[8] Another convenient equivocation; on 'sanctity' this time.

308

# 19

# Abortion

## Defining the issue

The abortion issue revolves around the following attitudes and beliefs:

1. the presumptive definition of a person as an innocent human being;

2. the possession of inalienable rights to life by all persons and innocent human beings;

3. the person as the being that results from a fertilized human ovum;

4. induced abortion or the permanent termination of the further development of this potential person as the morally wrongful killing of an innocent human being. The legal implications of killing in this sense are secondary to religious dogma of the sanctity of human life. The issue can be encapsulated in the form of a syllogism as outlined before on page 216.

1. The killing of a person or an innocent human being with a right to life is wrong;

2. the human fetus is a person or an innocent human being with a right to life;

3. therefore the killing of a human fetus is wrong;

4. abortion is the killing of a human fetus;

5. therefore abortion is wrong (without reservation).

The argument pits the *prima facie* rights of the fetus against those of the mother, if her condition should require termination of her pregnancy for any reason, such as to save her life in cases of severe toxemia of pregnancy, incest, or rape, thus presenting an ethical dilemma about which rights should take precedence. But opponents of abortion advance this kind of argument to support their point of view. The argument assumes the truth of both the major premise (1) and the minor premise (2) in arriving at the valid conclusion (3); but the validity of the argument does not necessarily make its conclusion true. Premise (1) is true by intuition, but the truth of premise (2) still has to be proved to render the conclusion true, even if it is validly derived. (See chapter 12, Embryos, stem cells, etc). We still have to establish possession of the following traits by fetuses to prove that they are actual (not potential) persons, as per Warren: "consciousness, reasoning, self-motivated activity, capacity to communicate, presence of self-concepts, and self-awareness"[1], but can we? Anti-abortionists will argue that possession of these features is not necessarily the limiting factor; the sanctity of human life is. That is why even if infants, like animals, do not have all of these features, we can still kill animals, but we do not, like Tooley, condone infanticide. So why condone abortion, they say?

The human fetus is technically an organism that belongs to the human species, as defined on page 67, by virtue of its chromosomal constitution of 46XX or 46XY chromosomes, which does not entail innocence or the right to life, or carry any moral commitment. The argument based on the personhood of the fetus must, therefore, derive its moral significance from personhood *per se;* but that link with personhood remains to be proved, because we have no present evidence that fetuses possess consciousness and other hallmarks of personhood stated above. Proponents of abortion, while admitting that the fetus is human as stated above, but not admitting to its possibility of personhood, advance the contrary argument:

> if p, then q;
> not p,
> therefore not-q;
> i.e.,
> if the fetus is a person, then it is wrong to kill it;
> the fetus is not a person,
> therefore it is not wrong to kill it.

This argument is contrary to the logical principles established before, where we saw that to argue that "if p, then q; not p, therefore not-q" can be false. The only time that this premise can be inviolable is if it states that "*if, and only if p*, then q". So what the pro-abortionists are trying to do is to use an invalid argument to prove the invalidity of the argument and conclusion proposed by the anti-abortionist. Both sides are resorting to fallacious logic to prove their cases when they should be trying to establish the truth of their premises, using valid arguments and relying on their soundness to prove their cases.

## Logically incongruous arguments

In arguing that since the fetus is an innocent human being[1], and it is wrong to kill an innocent human being[2], therefore it is wrong to kill a

310

fetus, anti-abortionists (not pro-lifers, implying that pro-abortionists are anti-life), equate human being$_1$ with human being$_2$ by virtue of the use of the common phrase "human being", regardless of the difference in the nature of the referents of human being$_1$ and human being$_2$. The logical step of proving that the two uses of "human being" have the same referent has been omitted, because it is lacking, or else it has been strategically withheld (equivocation); but the unproven impression is created that the 46XX or 46XY organism is the same entity as the self-conscious person who evolves from this primordial complex. So it makes this argument sound like this one: tough meat is better than nothing$_1$; nothing$_2$ is better than wisdom; therefore tough meat is better than wisdom. Clearly, nothing$_1$ (no meat) and nothing$_2$ (no one thing) do not have the same referent, and tough meat is not better than wisdom by any stretch of the imagination. Both arguments for and against abortion thus fail to prove their points; they only help to further obfuscate the issue.

Let us simplify matters by assuming that the fetus is a potential person, just like the zygote and the cloned body cell that also have the potential to develop into fetuses and persons, but are not yet persons. If we then arbitrarily refer to all three entities as persons, we can argue that since it is immoral to kill a person, it is, therefore, equally immoral to kill a fetus, to destroy a cloned cell, or a to pulverize a less than 14 day old zygote, which is a nondescript cluster of cells with the potential to develop into one or more persons (twins, triplets, etc.), or into a chimera or a hydatidiform mole on some embryological claims. Arguing thus, we confuse the rights of actual persons with those of potential persons (zygotes), and by retrospective implication, sperms and ova, which are potential zygotes twice removed from actual persons. One cannot imagine depriving the sperm of its future happiness and goals in life as a fetus, an infant, a child, and an adult, or envisage the personhood identity persisting through all these phases as if one is the same as the other. One can also not imagine the absurdity of trying to euthanize an early fetus-person for the same reasons that real persons would be euthanized, except by analogizing such a process with the euthanization of a permanently unconscious, depersonalized person who is now just a simple human organism like the fetus.

Having previously proved the fallacy of this kind of argument, let's pose another question: if we grant that the fetus became a non-moral human organism when its non-moral chromosomes were constituted, then we have to answer the question at what point in its development it acquired morally significant personhood with self-consciousness and rationality. Was it during the arbitrarily chosen process of fusion of the gametes, reconstitution of the contributory XX and XY parental chromosomes, or later during another arbitrarily

311

chosen process such as implantation or closure of the neural tube? So why not during the morula stage, quickening stage, ex utero viability stage, four months gestation, or eight months of gestation? If not during any of these times, then when, exactly? Even if the onset was gradual and imperceptible, it still must have had a beginning—when? Barry Smith and Berit Brogaard have given their reasons for selecting the 16th day post-conception, after implantation of the zygote as the point at which the future adult (person) comes into being, while claiming no moral implication in their postulation—"For all of these reasons we shall argue that, while human life is present at earlier stages, it is gastrulation which constitutes the threshold event for the beginning to exist of the human individual."[2]

They disqualify the single-cell zygote that exists before this time, because it undergoes fission whereby it ceases to exist as one substance that will result in a human being, since the original single substance has been destroyed and is now replaced by two or more individuals in the form of twins, triplets, etc., arising from separate and uncoordinated cell divisions. But Gregor Damschen and others argue that the starting point is at fertilization, because division of the zygote does not result in the bundle of disconnected cells, as claimed by Smith and Brogaard, but in a "hierarchical structure" with a common covering membrane (zona pellucida) that permits nutrition, growth, and coordinated division of the enclosed cells, with the potential to culminate in one or more human beings from the same zygote without any observable breaks in physical continuity:

> Their interpretative claim that a zygote divides immediately into two substances and therefore ceases to exist is highly implausible by their own standards, and their factual claim that there is no communication between the blastomeres has to be abandoned in light of recent embryological research. . . . the vast majority of humans begin to exist at fertilization . . . at any rate much earlier than sixteen days.[3]

But they do not assure us that this spatiotemporal continuity also amounts to persisting identity; that in the process of development a new entity has not replaced the original entity that started the whole process as in the development of two individuals from one zygote, unless they tell us what consistent functional arrangement of characteristics (in addition to structural ones like a 46XX or 46XY chromosomal constitution) to look for in evaluating this continuity from gametes to zygote to embryo to fetus to neonate without self-consciousness, and finally to infant or child characterized by this property that endows it with personhood.

The dilemma generated by the onset of fetal personhood poses further problems to both sides: The moral difference between killing this same fetus in utero at 26 weeks' gestation and killing it ex utero as a spontaneously born 26-week-gestation premature infant. Anti-abortionists have to answer these questions before they can claim that the fetus is a person, to justify their stand and to defend the sometimes vicious physical methods that they use for trying to force others to adopt their point of view when rational arguments are what is called for. Pro-abortionists also have to justify killing a 26-week-gestation neonate against the charge of infanticide that could be levelled against them, even though the infant does not live inside any one's body to cause the same health problems that fetuses can cause, considering that "except in such cases as the unborn person has a right to demand it . . . nobody is morally required to make large sacrifices . . . for nine months, in order to keep another person alive."[4]

## Fetus as person

Carlos Bedate and Robert Cefalo argue that "it is incorrect to assert that the zygote possesses informative molecules for the future person in its genome"[5], in spite of its human genetic constitution, because the earliest process of spontaneous and undirected cell division to the blastocyst stage is replaced by the later ones of differentiation and organogenesis that depend on the occurrence of "some type of interaction between molecules of the zygote and extra-zygotic molecules"[6], most likely from the mother, which do not appear to be genetically coded but are participants in a process that is sufficient for producing a human being. Therefore "the zygote makes possible the existence of a human being but does not in and of itself possess sufficient information to form it. . . . The zygote does possess sufficient information to produce exclusively human tissue but not to become an individual human being."[7] *A fortiori*, it cannot also be or become a person, because it "does not possess all the necessary, and surely not sufficient, information to become a human person"[8], since the zygote can develop erratically, as we have seen, resulting in a non-human organism (chimera) and a non-person.

Antoine Suarez contradicts the above statements, claiming that "the biological identity of the human embryo . . . depends basically on the information capacity of the embryo itself"[9], because its development is independent of any determining mechanisms that can be presumed to be coming from its carrier. He states categorically that *"the preimplantation embryo is the same individual of the human species (the same human animal) as the adult into whom the preimplantation embryo can in principle develop."*[10] He does not say it is the same *person,* only that it is the same *individual;* but then he later advances a hypothetical argument meant to prove that *"the preimplantation embryo is the same human person* [or *living being*] *as the adult into whom it can in principle develop."*[11]

313

In response, Thomas Bole III observes that the notion of human being is rendered ambiguous by the possibility of one zygote becoming two or more persons (twins, triplets etc.) He queries the contention that only one of these persons is the actual person related to the zygote, leaving the others in limbo and thereby vitiating the claim that the whole zygote is a potential person; and he further maintains that a zygote is a human being with doubtful personhood, because it lacks the ability to psychologically integrate its functions—"A person is a psychologically integrated unity, because it must unify its experiences in morally imputable actions. To say that the zygote is a person requires one to assert that the zygote has the same principle of psychological integration."[12] To answer these questions entails reverting to answering the same questions posed before: what is necessary and what is sufficient for something to count as a person, as opposed to a non-person, and at what point in development did the fertilized egg acquire self-consciousness to become a person?

The importance of answers to these questions resides in the fact that if the fetus does not satisfy the conditions for being a person, then it loses its personhood claims and thus its right to life. If it satisfies those conditions, then it can rightly claim personhood and a right to life. As we have noted, opinions differ widely, but most opinions tend to gravitate conveniently toward conception as the starting point of personhood, even though objective evidence for the onset of personhood at conception or anywhere else along the path of human development is lacking and there is no way of establishing it, thus permitting some arbitrary, extreme religious opinions to assume that zygotes are persons, and also to go on to claim that embryos developed in petri dishes deserve the same right to life as conceptuses and persons, even while they are still ex-utero. By parity of reasoning, as we have already indicated, they might as well go one stage further back and sanctify the sperm and the ovum on the basis of their individual potentialities to form a zygote, which may be of lower moral standing than, but has the potentiality to develop into, a sanctified human being or person.

Fetal personhood has another defender and champion in Carol Tauer who sets out to prove that the experiences of the prenatal human organism are part of the total chain of experiences that culminate in its adult experiences. She argues that "since my total experience is morally significant as the experience of a person in the strict sense, and since prenatal experience affects the development of the characteristics which give moral significance to personhood, therefore prenatal experience is actually personal experience."[13] To the critical question how and from what point in its development the fetus can retain its experiences, she suggests, without proving it, that in what she calls the psychic sense, "the human fetus attains significant personhood by the second half of the first trimester, [but before that, it] is best described simply as a 'potential person'."[14]. She then goes on to define persons as moral agents with moral rights, rationality, self-consciousnes who assume responsibility for their actions. The problem is that by its nature, the fetus does not have these characteristics. It "can have a right to life only if it now possess, or possessed at some time in the past, the capacity to have a desire for continued existence",[15] which only a person can satisfy; but it will not ever have such a desire during fetal life. So the frustration of the attainment of these characteristics and of the capacity to have mental states, experiences, and self-consciousness, by virtue of its basic constitution, violates the right of the fetus to personhood.

It is obvious, therefore, that if the right to life depends on the desire to continue to exist as a person, or on having the desire to be a person, then the rug is pulled from under the fetus which has only the potential to acquire the desire to be a person so that it can fulfill the desire to continue to exist as a person. It is also unfair to exclude it from further potential personhood by stipulating with Tooley that "if something has not been capable of having either of these desires in the past, and is not now capable, then if it is now destroyed, it will never have possessed the capacity in question"[16], because it was never given the chance to reach the future point in development from where it could look back and say: I did possess that capacity in the past, therefore I have a right to further potential personhood. Besides, how many people living today had or ever entertained the concept of a person demanded by this theory in their fetal life? And who knows when a functional brain emerges during fetal development that can cogitate on all these requirements demanded by Tooley?

Furthermore, desires belong to persons; they do not define persons, and if, as Tooley maintains, "one cannot have a desire to continue to exist as a person or a desire to become a person unless one has the concept of a person",[17] which fetuses and infants lack, then the natural order of things will have to change to make this game of life fair, because playing it this way is not playing fairly. For instance, infants on whom the right to life as persons is generally bestowed from the moment after they are born, and not the moment just before, still run afoul of the requirements that would qualify them (and fetuses) for existence as persons, viz., self-consciousness, envisaging a future for themselves, having desires about their future states, etc., by virtue of their intellectual immaturity. Does their immaturity and lack of personhood by this definition therefore justify infanticide? If not, then it also does not justify the killing of fetuses.

Still, some other philosophers respond by maintaining that lack of consciousness and the capacity to feel pain, to reason, to engage in self-motivated activity, to socialize and communicate by any means, and particularly to be self-conscious and thus have an appreciation of its entire life from past (in a 'peach tree'/petri dish) to future (law maker) deprives a fetus of personhood, though not of human being status. The abiding question persists: when do these characteristics appear in the developmental life of a zygote or a cloned cell? We have no way, it seems, of knowing the answer to this question in our present state of knowledge, nor does it appear likely that we will know the relevant facts any time soon; but octaves are rising in the bitter tone of debate on the propriety or impropriety of abortion.

Another concern about these claims is the immediate problem that they create by virtue of the applicability of many of them not only to infants, but also to those with incompetent minds who should be denied personhood on their basis. Consciousness involves personal experience or awareness, both of which presuppose a related mind. Self-awareness or self-consciousness involves subjective experiences, how things seem to a person in the phenomenological sense, whether he is perceiving, thinking, or reasoning, and whether he is aware of the existence of these states within himself as an agent with the ability to act on them. To accomplish all these processes, the human organism would presumably require significant psychological complexity, which would, in any case, not characterize the severely mentally defective person or the anencephalic fetus, thus depriving both of the privilege of personhood.

As we are already aware, many difficulties are caused by the use of the terms 'person' and 'human being' to refer to the same entity in different contexts. The categorization of human beings as genetic (46XY or 46XX) and as moral beings does not make things easier. The claim is made by some that the zygote exists as an individual with the inherent (teleological) power to grow into a fully formed infant by organogenesis, if allowed the right conditions of development, simply because it is a genetic human being. Others claim that those entities that answer to this specific genetic code, like human fetuses, are just that, and do not automatically qualify as moral beings, unless they are potentially capable of engaging in rational thought, which will show that they are also potentially capable of assuming the moral nature of human beings. So what do these concepts of potentiality and actuality mean for the fetus?

# Potentiality / Actuality

The potential, or dispositional, is that which has the constitutional ability or capacity to become its end product, or the actual, with all the capabilities that accompany the actual but not the potential state, only in the right circumstances. The medical student is a potential physician; he is in the stages of becoming an actual physician. But until he becomes an actual physician, he cannot legitimately claim the rights and privileges accorded to physicians. Similarly, the fetus as potential person cannot claim the rights of the actual person that it is in the process of becoming, in spite of arguments that its potentiality has the moral relevance that endows it with an interest in life. Potentiality is not actuality. In Aristotle's philosophy, that which has only the possibility of having form and is in the stages of becoming is potential, and that which has form is actual. Hence he speaks of corresponding causes, viz., material cause —the elements out of which an object is created (potential), and formal cause—the expression of what it is (the actual). For the potential to become the actual, conditions have to be conducive to development from one state to the next. Interruption of this process will abort the realization of the end state and confute the equation: fetus (potential) = person (actual), as we have already seen in the preceding discussions, and as the discussion on chaos theory so clearly illustrated.

Therefore the potential, even though equipped with what is necessary to become the actual, can't be accorded the same status as the actual on the basis of that potentiality, *ceteris paribus* (all things being equal), because the process could still abort spontaneously or follow a route that will result in a deviant end product; e.g., chimera or hydatidiform mole. We could argue that because it is wrong to kill the adult, it does not follow that it is wrong to kill the potential adult, or else that these two potential calamities don't justify the termination of the process of becoming the actual by the act of abortion. To use them thus is to proffer a lame excuse; as lame as the argument that on the scale of human values the fetus occupies a lower rank, and can be eliminated without taking away very much from the totality of those graded human values. So the refuge from the discomfiture confronted in the potentiality/actuality issue may ultimately prove to be in the barrier erected by self-consciousness as the sole criterion of personhood and the right to life. On this criterion, if we embrace it, the anencephalic fetus does not have any chance of converting its potential personhood into the actual, since it starts life in utero without the necessary apparatus for generating consciousness (cerebral cortex), which is the *sine qua non* for personhood, and which it will never acquire in its life. This double lack of necessary and sufficient conditions for personhood leaves the anencephalic fetus in an endlessly hopeless state as to their eventual acquisition of personhood. But it is also a problem for the normal fetus, because it is not self-conscious, even though it has the basic neurological framework for consciousness and self-consciousness.

Many philosophers have taken issue with the moral protection afforded fetuses on the grounds that they are only potential persons.

Some have done so by rightly reducing to absurdity the condemnation of all methods of contraception as mass murder, even though the gametes cannot individually be accorded the same potential status as the zygote resulting from their fusion, for which sanctity and a right to life are claimed, because it is destined to become the whole person. They argue that anything else that is not, or not yet, a person does not have that right, because it has not attained the status of meriting rights. But the arbitrary position that these philosophers take does not prohibit others from claiming rights for potential persons *per se,* based on the premise that a particular 46XX zygote is intrinsically programmed and destined to become a particular person, $x$, and not someone else, $y$, who develops from a different 46XY zygote, or even gorillas $\chi$ and $\lambda$, which develop from zygotes with intrinsic 48XX and 48XY chromosomal constitutions. This intrinsic potentiality is believed by some to confer rights (and protection thereof) on this potential person that any extrinsic potentially does not do to realize his actuality and all the benefits that it entails without losing them to premature termination.

The problem is, of course, to demonstrate that the 46XX or 46XY genetic constitution of a human organism, and not the 48XX or 48XY genetic constitution of the gorilla, guarantees or is a sufficient condition for its moral humanity as expressed in personhood. In the absence of this capability, or the necessity for it, we can only postulate that the community of moral beings excludes those beings that answer solely to the genetic constitution of being human organisms without also answering to the moral requirement of being persons with self-consciousness as their distinguishing characteristic. This means that only actual persons constitute the moral community, which potential persons like 46XX or 46XY fetuses cannot do. The same concept has also been extended to cover reverse cases where persons have lost many of the arbitrary criteria required for personhood, and have thus been "reduced" from the rank of person to that of a mere organism with an ethical value that is different from that of a self-conscious being, even if it may be deemed to be less; e.g., PVS. According to this theory, therefore, at no stage during its kind of life, from conception to birth, is a fetus a person with any rights that supersede the kind of rights wielded by its carrier, its prospective mother. So, even if we equalize their rights, the fetus will always have prima facie rights, and at any time when their rights conflict, the absolute rights of the pregnant mother will always trump those of the fetus in her quest for abortion, except in some religious communities where the leaders have decreed that the mother's right to abortion to save her life for legitimate reasons cannot ever supersede the rights and welfare of the fetus. All things considered, however, the woman's rights to her body should prevail.

318

A case in point is that of the report of a Colombian mother of four with cancer of the uterus who was forbidden by church and state laws to procure an abortion at two months' gestation so that she could receive chemotherapy for her cancer. She had to wait for seven months until after the birth of the baby, while suffering constant, excruciating pain that she described as like having her insides torn by the bite of a dog that would not let go of her. Of course, it was too late to save her life by the end of gestation. The question therefore arises, is it better, moral, and rational to sacrifice the life of a mother and the welfare of her four children to ideology and for the sake of preserving the precarious potential of a two month fetus to develop to maturity? Does this accord with the utilitarian principle of ensuring that the consequence of this relationship should result in the maximum of good (saving of one actual life and four actual dependents) and the minimum of bad (loss of one fetus with only the uncertain potential for becoming like the five lives that it has trumped)? On the other hand, advocates of personal values and rights in the development of the fetus and its right to have its life preserved might argue that the fetus-person should not be subjugated to the rights to life of other persons. With this attitude, they hope to justify the pitting of the uncertain potential survival of this fetus in its continuing precarious circumstances against the actuality of the existing members of the family. The reader can decide on the wisdom (or lack thereof) of this and modern-day illogical and indiscriminate legalized prohibitions.

# Fetal Rights

Some people argue retrogradely from the internationally recognized rights of children to fetal rights, and in their language the killing of a child before birth is as much a violation of its right to life as the killing of the child after birth. This argument is related to the slippery slope argument that lack of respect for the life of a fetus will easily lead to lack of respect for the lives of infants, resulting in their unbridled killing and escalating the already distressing amount of infanticide that is claimed to exist in the world today. They go further to point out that since human beings share a common humanity by virtue of the chromosomal constitution that they acquire at fertilization, they all qualify for inclusion in the human family of persons and none among them can be disqualified as non-persons on any grounds. That means personhood cannot be justifiably withheld from fetuses, anencephalics, and people in PVS.

Since rights have already been discussed, we can only observe that rights entail duties and obligations, and if the fetus has rights, then we have both the duty and the obligation to respect those rights and not violate them by terminating its existence. Against this stand, some other people argue that even if the fetus is a person with a right to life, it can still not claim the right to be a parasite in the body of its carrier. This situation is believed to obtain regardless of how the fetus got there: by planned or unplanned pregnancy. The human product of rape, for instance, does not have fewer rights than that of a desired pregnancy; once they exist, there is no difference between them. So a special argument cannot be mounted against the rights of the product of rape simply because it was forced on its carrier, and similarly for fetuses in other unwanted pregnancies.

Can rights be defined in terms of interests, so that whoever is not capable of formulating and expressing an interest in life can be assumed to lack self-consciousness and the right to life? Does it matter that all fully mentally developed judges on this subject were themselves helpless creatures who also lacked the same faculties that they are placing as barriers to the rights of others to their own lives? Do they want to change the natural process of mental development to suit their theories, so that self-consciousness emerges at the point of conception, or can they accept the reality of the existing process of gradual developmental and adapt their theories to it in the spirit of achieving a state of reflective equilibrium? The people who stress the potentiality of the fetus to satisfy at a future time all the requirements that are now being unfairly imposed on it are right; even in our daily lives we need time to qualify for certain privileges, although some who are infirm of mind or body never reach the goals set for them by those who are capable. Should they be selfishly eliminated, or do they still have a right to life? Where is the existential being-in-the-world-with-others that allows them to occupy their rightful place beside those who claim to be "normal" ones? The problems raised by entrenched positions pro and con abortion never cease to proliferate; but they all need to be answered to eliminate irrationality and bigotry from the debate. There is always room for intelligent compromise.

# Life, Killing/Murdering

The next question is that of the meaning of the word 'kill' and its application to a zygote, a morula, and a cloned cell. To kill is to deprive of life. Can effacing any of these entities be said to constitute depriving them of life? If so, will it amount to depriving a person or a human being of life? Can it further be categorized as murder? Murder is defined as the crime of unlawfully killing a person. The use of the term 'murder' to label the act of killing during the procurement of abortion is rendered unfair by its value-laden connotation, and by the fact that it constitutes a pre-judgment of the issue. It makes the abortionist a criminal, it assumes that the fetus has a right to a life of which it has been deprived, and it therefore lays the abortionist open to criminal charges and the sentence that goes with them if her presumed guilt of murder is confirmed. The antiabortionist has already determined that she is guilty by labeling her action as murder; he only needs to confirm the guilt on his terms, not prove it from fundamental principles, although as the one who is making the case for murder, he owes it to the pro-abortionist to prove that abortion is murder. The onus should not be placed on the latter to prove that abortion is not murder, thereby committing the fallacy of arguing from incomplete evidence while also appealing to ignorance, viz., the pro-abortionist has not produced evidence that abortion is not murder; therefore it is murder, and the anti-abortionist does not need to prove his case—bigoted and unfair, to say the least.

As we will see in the discussion of life in the next paragraph, the fetus satisfies both necessary and sufficient conditions for being alive. So it is not out of order to call abortion killing, although it is so to call it murder. Aaron Ridley makes the point that "killing is justifiable when the person to be killed has forfeited his or her right to remain alive."[18] It would be extremely difficult to prove that any fetus falls into this category, because, whilst having rights presupposes having interests, which in turn seems to presuppose having the desire to exist, which in turn presupposes being a person, fertilized eggs and fetuses do not have any of these interests and desires, so they cannot be persons, and they cannot have the same rights to forfeit as persons who wield these rights.

At this time we need to backtrack to delineate the concept of life that we have been bandying around. The dictionaries define life as: "An organismic state characterized by capacity for metabolism, growth, reaction to stimuli, and reproduction. . . . The sequence of physical and mental experiences that make up the existence of an individual."[19] and "The condition that distinguishes animals and plants from inorganic objects and dead organisms, being manifested by growth through metabolism, reproduction, and the power of adaptation to environment through changes originating internally"[20] [homeostatic changes].

These changes enable organisms to exchange matter and energy with their surroundings in accordance with the Second Law of Thermodynamics, which states that in all energy exchanges, if no energy enters or leaves the system, the potential energy of the state will always be less than that of the initial state. The reason is that potential energy is converted into kinetic energy (energy of activity) with some of it being lost in the form of heat, but never destroyed, as stated in the First Law of Thermodynamics: although it changes from one form to another, energy cannot be created or destroyed; it is always conserved, and thus it remains constant. So to maintain a constant state of total energy, living organisms have to harness some more energy from outside sources; and because they are open systems that exchange energy and matter freely with their environment, they can readily avail themselves of this means of maintaining orderly lives. The fetus satisfies all these conditions; so there is no question about its being on par with other living organisms, although its personhood is still in question.

# Down the slippery slope

As we approach the conclusion of this contentious issue, it behooves us to consider the implications of the slippery slope argument, which warns against initial acceptance of what appear to be less extreme positions at the risk of being forced to accept progressively more extreme positions, if we have not drawn a definite boundary line to limit the extent of escalation. If it is reasonable to "kill" a fetus at five months' gestation, then it is equally reasonable to "kill" it at eight months' gestation, and also at term just before delivery. So one can also kill a neonate, and a one month old, and a five month old infant with impunity and not risk moral wrath. The whole argument rests on the merging of one stage into the next without permitting the drawing of a line of demarcation between adjacent stages, because the changes are so minute as to be almost imperceptible, although the farther removed stages look vastly different from the earlier ones. But this line of argument is soon seen to be false and to fail to justify the premise that allowing $a$ will lead to allowing $b$, $m$, $q$, $t$, and on to $z$. So, if we can falsify the premise that allowing $a$ will eventually lead to allowing $m$, $q$, or $t$ by showing that it is not the case that allowing $m$, $q$, and $t$ will also lead to allowing $z$, then we can also show that allowing the first step in the slippery slope argument does not necessarily entail allowing the subsequent steps. In fact, we can show that it would be absurd to kill a one month old infant for flimsy or no reasons, and hence that it is false to argue that "killing" a fetus will permit the slide into killing infants and children for no reason or for flimsy reasons, since no fetus is killed wantonly. Those who uphold this view will have to produce empirical evidence, not only the potential for these occurrences, to support their claims.

In the end, the tug of war over abortion will not be settled by any attempts to set standards and draw lines of demarcation in an area where states blend imperceptibly into one another by a continuous process of development from zygote to old age. Each side has to be adult and intelligent enough to admit that the other side's arguments have merits and demerits, and that deciding the issue is not a matter of adding points and awarding the decision to the side with the most points, as if they are engaged in a game to see who will cling on to his status of one-upmanship. Deontological and utilitarian questions, and strictly intuitional and socio-cultural issues have to be taken into consideration in making decisions pro and con abortion, and each case has to be considered individually on its merits, because situations are as varied as the number of persons seeking abortions. The decision making process will, therefore, include present duties to actual and potential persons, consideration and fostering of the happiness and other factors that will count in promoting good quality in the lives of actual persons now and potential persons in their time, and much more. A blanket formula is a short-sighted way of dealing with this problem rationally and intelligently; it only widens the schism as each side trumps up more and more reasons to support its outmaneuvering point of view in the struggle for supremacy. Further polarizing the already escalating hostility defeats efforts to synthesize a rational decision on this atomized issue; it does not elevate it to the seemly place where consensus can be achieved through rationality.

Perhaps this is the point at which the casuist can come in to help resolve the impasse by eliminating the dogmatism of each side and substituting thoughtful consideration of individual cases as such from some common ground. Starting from, and remaining mired in, polarized positions that answer only to opposing moral codes lacking foundation in logic is not the answer. The doctor has an important educative and advocacy role to play in helping to resolve this impasse by promoting the desire for intelligent compromise through a dialectical process that recognizes the driving principle behind each position and finally arrives at a workable compromise solution to the problem. The process of compromise will entail judicious use of facts that are relevant to paradigm cases on each side to establish mutually acceptable actions to be taken in the different circumstances of the life of the fetus at different stages, since it has proved difficult to agree on the stage in development at which a fetus can be aborted. From this point, it is possible to deal with novel situations when they arise, because a rational and morally acceptable method will be in place, obviating also the resort to violence to defend positions that rest strictly on unjustified biases. Perhaps the Hegelian method of dialectic will be useful in this case to elevate the level of discourse to a higher plane out of the polarized mundane one where some people seek to make decisions for others how they should live their lives, mainly for religious reasons to which these others do not subscribe.

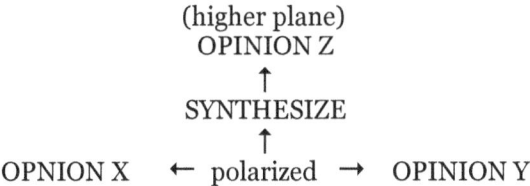

(higher plane)
OPINION Z

↑

SYNTHESIZE

↑

OPNION X  ← polarized →  OPINION Y

## 20

# Euthanasia

## Why euthanasia?

Euthanasia (from *eu*=good, and *thanasia*=death) is the intentional termination of one person's life by another, or her non-intervention to save that life, supposedly in the exclusive interests or for the sole good of the person who dies, although the person will not survive to enjoy the realization of his alleged interests or good, which may be relief of intractable pain or alleviation of suffering. Opinions differ from strict prohibition of all forms of "mercy killing", through condoning it in exceptional cases, to making it freely available to terminally ill and suffering patients who request it for those reasons. In such cases of assisted death (suicide), the doctor only provides the means to procure the drug, but the patient administers it himself in the absence of the doctor, or he may decide not to proceed with the suicide. The whole point of this maneuver is to help the patient to end his suffering, not to kill him because he is poor and cannot afford to pay for medical treatment, not because he is vulnerable, not to offer him the easiest way out of his illness, mental or physical, not to make him feel obligated to die because he is a burden on others financially or emotionally, and not to substitute for palliative care. It has nothing to do with religion or disrespect for the sanctity of human life; it's only concern is compassion while protecting both patient and process from abuse, and safeguarding patient autonomy and rights.

The assumption is that the interests and the good of the patient are supposed to be determined by him and those whom he consults, if he is able to do so; otherwise they will be determined by the attending physician and those associated with her in the care of the patient, which may involve social workers, nurses, other physicians, lawyers, ethicists, and relatives. All of them should be motivated solely by the expressed or assumed interests of the patient. In all cases the intent is to employ acts of merciful killing to administer timely relief from suffering to the incurable and terminally ill patient who has consistently and competently expressed his voluntary desire to have his life ended before its uncertain natural termination. On the other hand, anticipated and temporary suffering from which full recovery is possible do not qualify anyone for euthanasia on utilitarian or other moral grounds, but irreversible unconsciousness, which is loss of personhood that is defined as comprising consciousness, autonomy, rationality, and self-awareness as the prime examples, should, because of the presumed futility of keeping such people alive.

Some people maintain that assisted death subverts the aims of medicine—promoting healing and preserving life—and downgrades the intrinsic value of human life, while others argue that such an element of mercy does not affect the worth of the person *per se*, although it reflects on the worth of the person's life, justifying euthanasia morally, even if it may not always do so legally. They base this opinion on the principles of autonomy and utility, the latter of which provides for the moral correctness of an action that produces the greatest overall amount of good or happiness, as long as it does not violate the rights of any one. The patient's demise is believed to be justified by the balance of the good that it brings to him by relieving him of his prolonged suffering with pain and helplessness, which is a greater evil than his demise. He thus faces a shorter period of pain and suffering than he would if he were left to live longer. His family, those who are nursing him, those who expend resources on him against all odds and thus deprive other persons of the more certain benefit of those resources to them, and all those others who empathize with him in his misery also derive relief from his demise, provided that no one's rights and autonomy are abused in the process, no one is harmed, no one acts selfishly by euthanizing anyone against his will, no one is made to feel guilty for continuing to live and be a burden on others, and everyone is left with the maximum amount of deserved happiness—a paradoxical idea in relation to the one who is euthanized, in spite of the fact that the action is meant to uphold his autonomy.

On the other hand, those who apply the principle of deontology are not concerned with the consequences detailed above; they judge the act of euthanizing someone purely on its merits. They want to know if carrying out the act is the dutiful thing to do to respect and uphold the patient's right not to be killed since rights entail duties, unless he gives up that right and releases them from their duty by requesting to be killed. In this instance, he imposes on them the duty to kill him as a gesture of respect for his right to do with his life as he pleases. It is up to them to respond positively or otherwise as they see fit, but he cannot compel them to honour his request if they do not feel duty-bound to do so for several reasons, including their assessment of his competence to make that grave request or the dictates of their consciences for reasons of rigid Kantian deontology or religious belief: "thou shalt not kill".

It is noteworthy, though, that the originator of the principle of deontology regarded both suicide and euthanasia as expressions of disrespect for the person's autonomy and the use of self or other as instruments or means for ends, and not also as ends in themselves. So if we must compare euthanasia with suicide we should be aware that people as autonomous beings generally have the liberty to terminate their own lives, although society may frown on the practice and posthumously censure them for the action. But if society has not made enough effort to ensure that people are not trapped in situations where they have to resort to that extreme measure to deal with their social problems, it would be unfair to find fault with the victim. However, where fatal illness is involved and the patient and doctor see no hope of relief after all the best efforts to alleviate the situation, it is conceivable that if they are not physically able to kill themselves, patients should be free to request others to do it for them without those others having to incur culpability for acting as their proxies. The doctor does not have to earn the title of angel of death or lose the trust of patients generally for executing an act of mercy on a suffering patient who has requested relief from his ordeal. Furthermore, the lessons learnt from the past misuse of physicians for politically and racially motivated murders are enough to ensure that any tendency to slipping down that slope will be avoided.

The upshot of this argument is that there should be no doubt to the patient about the benefit of this action of euthanizing him, as there is no excuse for irrevocably depriving any person of present and future years of happy living under the mistaken impression that his present and easily manageable temporary suffering is permanent and cannot be sufficiently ameliorated with time. His lack of enjoying a rational existence should be incontestable, and where this fact is in doubt no action should be taken to euthanize him, because it will not be justified even on utilitarian grounds, since we cannot predict whether his life from here on will be worthwhile, on his terms, or valueless. The intention matters more than the consequences of the action or inaction entailed in euthanizing someone. As regards non-intervention, when in doubt, do not let die, because people generally value their lives, even if they may be plagued by illness or disability, and by episodic periods of depression so that they sometimes wish to end their lives as a means to ending their suffering; they are mostly always glad to see the dawn of another day and would not wish to be deprived of that joy unreasonably.

On the Kantian deontology view we are asked to entertain the contention that the preferences of the patient and his attendants should not be the measure of his life, because that amounts to treating him as a means to his and their ends and not also as an end or ultimate value in himself. On this principle, therefore, euthanasia is to be condemned for using a representative of rational humanity only as a means or instrument for the overall happiness of others and himself (even though he will not be there to enjoy that happiness), which is presumed to be the ultimate goal of the entire exercise. But deontology overlooks the fact that illness may reduce a terminal patient's life to a level that is inconsistent with his human dignity while his doctor stands idly by, because all previously effective treatments have failed and palliation is ineffectual. In that case she and others would be justified in ending his life of degradation and pointless suffering, if he so desires, since no one can claim a higher morality for letting him suffer to the end than ending his suffering promptly. To be sure, a patient's right to life includes the right not to be killed wantonly, especially by his doctor, and the duty of others and himself to value and preserve his life; but, as we have noted, the patient who abdicates his right to life by asking to be euthanized withdraws this right and releases others from their obligation not to kill him. Instead, he places them under a *prima facie* obligation to carry out his wishes, unless there is an overriding duty against the execution of those wishes.

Another perspective on euthanasia compares it with suicide. People generally have the liberty to terminate their own lives without posthumously earning moral or legal censure for that action. So it is conceivable that if they are not physically able to kill themselves, they should be free to request others to do it for them without those others having to incur culpability for acting as proxies. This perspective assumes that suicide is always free of legal and moral culpability.

After critically reviewing most of the notable definitions of euthanasia and trying to formulate a comprehensive definition that is not too broad, too narrow, and too vague, Tom Beauchamp and Arnold Davidson presented their definition of euthanasia, which was modified by Carson Strong to convey the following impression of euthanasia: "The death of a human being, A, is an instance of euthanasia if, and only if A's death is intended by at least one other human being, B, where B is either the cause of death or a causally relevant feature of the event resulting in death (whether by action or omission)."[1,2] They listed provisos to this cause and effect relationship as B's belief, with good reason and evidence, of A's actual and potential acute suffering, her intended termination of A's life is motivated solely, but not exclusively, by the desire to halt this suffering and promote his overall benefit and self-determination, and either one of them believes with good reason that A's death can be caused painlessly and without the added suffering that would result from B's non-intervention, unless pain cannot be reasonably avoided.

Some of Strong's reasons for his proposed changes are that since irreversibly unconscious persons do not have interests, causing their death is not euthanasia, because they lack cognitive awareness. Furthermore, it is not possible to euthanize a person after he has died, which is what is being suggested by honouring the interests of a previous person by euthanizing him when he is irreversibly comatose, i.e., dead by definition. Justification for the entire action is based on prior evidence of a net benefit accruing to A, and on the promotion of his interest in self-determination. Again, following Strong, when we consider the case of abortion in this context, it becomes apparent that abortion could possibly be a special case of euthanasia when it is done to spare the fetus a subsequent life of suffering from having severe hydrancephaly or a devastating and uncorrectable congenital defect. He believes that his definition is non-prescriptive and it maintains the distinctions that already exist among the different categories of euthanasia: active/passive and voluntary/involuntary.

## Exclusions

The concept of euthanasia as defined excludes the cases of those patients who suffer from progressively debilitating or disabling diseases that are certain to persist into the foreseeable future and are depriving them of those enjoyable items in life that make its quality optimal for them in that period of their existence. They therefore wish to forestall the escalation of this undesirable state of progressive suffering by the early termination of their lives. So it is questionable if it takes into consideration the ultimate quality of the life lived by physically disabled persons who do not suffer severe pain, but who find their lives burdensome; or even the demented patient who is barely aware of what goes on around him and is eking out a pointless existence. Fortunately, it also excludes the periodically depressed patient who temporarily loses his competence at making rational decisions and who might change his mind about his death wish when his mood is better and he realizes the irrevocable implications of terminating his life. It does, however, allow the patient to express the values that guide his life and to exercise a right over his life, particularly the right to die with dignity. This situation is not very different from the one where we respect the requests of terminal and other patients that we desist from active interventional treatment of their illnesses, deliberately allowing nature to take its course.

A quotation from a paper prepared by a commission on these matters epitomizes the concerns succinctly:

> Law must also recognize . . . the principle of personal autonomy and self-determination, the right of every human being to have his [her] wishes respected in decisions involving his own body. . . . that every human being is, in principle, master of his own destiny. He may, of course, for moral or religious reasons, impose restrictions or limits on his own right of self-determination. However, these limits must not be imposed on him by the law except in cases where the exercise of this right is likely to affect public order or the rights of others.[3]

Mill expressed the same sentiment in different words when he stated that we are sovereign over our bodies and minds, and "The only freedom which deserves the name, is that of pursuing our own good in our own way, so long as we do not attempt to deprive others of theirs, or impede their efforts to obtain it",4 then we are amenable to society. However, "When someone chooses to live or die it doesn't harm any one else's rights."5

Following directly from this stand is the recognition of the patient's right to refuse treatment and care or to receive only the care that is necessary to relieve pain and suffering endured unnecessarily by terminally sick patients, which adversely affects their human dignity. In this condition they require doses of analgesics that often contribute to the shortening of their lives—a subtle form of euthanasia that stems from observing the principle of the autonomy of the patient. In this regard, the question arises as to whether any person is free to get anything that he asks for, with no proviso or exception, or whether that liberty should be tempered with reason. If the former is the case then, people who are not candidates for "mercy killing" could also ask to be put out because they are depressed or have other temporary, onerous problems in their lives that still do not justify their termination, and that would be indisputably wrong.

People endure pain and suffering from many other causes than terminal illness, and it is not always that they can get the care that they deserve for alleviating or ending their suffering of poverty, neglect, persecution, oppression, exploitation, etc., and who is to say they do not deserve equal consideration with the terminally ill? Besides, people's autonomy is violated often by others who kill them for no reason, as in murder and war; so why can they not make their own decisions about whether they live or die, if others like heads of state, who send others to war to die for their personal glories, can assign themselves the right to make those decisions for them willy-nilly? However, not every case of suffering is a signal for immediate termination; perseverance sometimes pays off with lasting relief.

# Categories of euthanasia

The categories of euthanasia listed below are neither exhaustive nor exclusive; e.g., a patient who is severely depressed or demented and cannot request to be terminated (voluntary active euthanasia), but is suffering and in pain from disease, will need a surrogate, and that immediately changes his category to involuntary active, which also has its special problems. It is noteworthy, though, that some degree of depression accompanies some chronic physical illnesses, thus complicating the process of excluding depression entirely from tainting the patient's request for assisted death.

# Active euthanasia

Active euthanasia is the overtly intentional performance of an act with known lethal consequences, e.g., physician-assisted death—impermissible if not requested by the patient, but only sometimes permissible if so requested. It becomes indirect when the termination of life is a covertly intended side effect of treatment that is not ordinarily lethal but is made so by overdosing; e.g., morphine administered in doses greater than ordinarily required for control of pain. Whatever the case, some people think that intentional killing is always wrong on deontological grounds, forgetting that one can justly kill another in self-defence when his own life is in jeopardy.

# Passive euthanasia

Passive euthanasia is the omission of steps and efforts that are necessary to sustain life, with lethal consequences, e.g., withholding antibiotics for a lethal infection from a patient who needs them, and withholding vital assisted respiration, or discontinuing it in a patient who cannot sustain autonomous respirations. It is permissible if the patient requests it or his hopeless condition dictates it, but may be impermissible to the doctor whose oath of duty demands that she not allow harm to come to her patient. Furthermore, it may turn out to be an undesirable, lingering, and miserable process of death that does not do the patient any good. Hence the harm condition is arguable, because it assumes that death is a greater harm than the harm caused by the patient's suffering or his hopeless condition. The more humane approach would be to relieve the patient of his suffering by complying with his voluntary request. Only when the parents of a child with a disability request the withholding of corrective treatment that can otherwise offer the child a reasonably tolerable life free of pain and suffering, in spite of the disability, is the question debatable.

On the surface, it seems that there is an essential difference between active and passive euthanasia, since one category entails action while the other entails inaction. Their moral implications may appear to be different from the deontological perspective, because the motive for active killing in one instance is tainted by an action that is lacking in the other instance of refraining from saving and letting die; but it is, in fact, only the means of satisfying the same motive that varies with each circumstance. The motive may or may bot be be the same in each case but the consequences are certainly the same; and this fact alone presents a challenge, since the right to life is a negative right not to be killed, unless one requests it, but it also does not entail the positive right to be kept alive by others. Furthermore, nothing is said about being left to die, or not being kept alive by the use of respirators, ordinary methods of oral feeding, or exceptional means like parenteral feeding. Letting die is refraining from using one's ability, power, and advantageous position with respect to saving another person's life, and some people believe that it should bear the same moral burden as killing a patient (see page 305). Nevertheless, unlike choice where choosing and not choosing is still making a choice, non-action cannot be classified with taking action as acting. Someone who does not do anything cannot be accused of having done something, in spite of what Mill rightly says, "A person may cause evil to others not only by his actions but by his inaction, and in either case he is justly accountable to them for the injury."[6] (cf. Tooley, p 304).

## Voluntary euthanasia

Voluntary euthanasia occurs when a patient who is suffering from a painful, incurable, and burdensome terminal illness exercises his autonomy to live or die and consistently refuses treatment, directly or through a surrogate, or requests painless termination with another person's assistance of what he considers his worthless life, because he is unable to do so himself (assisted death). It is the undisguised version of the content of the previous discussion. As in all other cases, we must assume that the person's wishes are backed by clarity of reason and absence of duress or trickery about any disvalue or worthlessness of his life and how to resolve the predicament of continuing with it, and that the execution of his request will therefore not constitute a violation of his right to life. It would be wrong to go by the decision of someone who is constantly under heavy sedation, which might impair his judgment, or one who considers himself a burden on the family, friends, and public health systems that are tending him. In that state few people want to wield their prerogative of "entitlement" to health services, regardless of whether they are in the form of mechanical support or palliative care.

These impediments not withstanding, it would not be right for those who are not living that life to dismiss the patient's autonomous decision about the worth of his life, as only he knows what matters to him in his life-as-lived. Many people are still dying in severe pain from their illnesses because of lack of palliative care and facilities for care (personnel with adequate training), and we have no right to compel them to persevere in that state of pointless suffering, a situation that we would not relish for ourselves. A variety of this category is that of physician assisted death in which the doctor intentionally helps the patient to commit suicide, on his own request, by only providing him with the means to carry out the deed. The efforts of several states to provide solutions to the controversies tied to this category of euthanasia are discussed on page 338.

# Involuntary euthanasia

Involuntary euthanasia occurs when a patient's life is terminated without his consent and in spite of his desire not to be terminated, because other persons think his termination will serve his interests. It is always a questionable procedure, because the patient is not the one who is making this value decision on his life, but his lack of appreciation of his spatiotemporal existence and circumstances is proffered as the right of those making this decision to decide his fate for him. Most people have reservations about this category of euthanasia and consider it wrong without qualification.

# Non-voluntary euthanasia

Non-voluntary euthanasia is the termination of the life of a helpless person without his input and consent by reason of inability to do so; e.g., an infant with dim prospects of a quality-filled life and an incurably ill patient without a living will or previously expressed wish, both of whom are presumed not to have the capability to form, and cannot express, a preference for life or death. Such a person may be judged by others to be in a persistent vegetative state, where he is unaware of his surroundings and lacks cortical functions and self-consciousness to the extent that he cannot harbour any interests or entertain any present or future preferences. The question therefore is whether he is still fully a person with a right to life that should not be infringed and a will that should not be violated, or he has lost his personhood to the discretion of other people: close family, friends, physicians, and the medical and legal communities. Furthermore, it seems that euthanasia in such cases opens the door to the elimination, on bogus grounds of mercy, of the socio-economically disadvantaged, the unsophisticated, and the infirm, all of whom may not have someone to champion their rights. The implication is, therefore, that we should have legislation to regulate the use and scope of all categories of euthanasia, bearing in mind the fundamental principle that in all situations the primary aim should be to kill the pain and not the patient, unless it becomes absolutely necessary to terminate him for incontestable reasons.

Even with that proviso, the fear of being eliminated may still persist in those defenceless persons, like the elderly, who will feel guilty for still being around and using up limited health resources that could be allotted to younger and more productive citizens, thus inducing in them health facility phobia while also prompting the bureaucracy to degrade the facilities where terminal care is provided for them to less than optimal, since the option to exercise their choice not to be around is squarely in their court. Other fears relate, as we have seen, to the elimination of social misfits, deformed and severely retarded infants and children, doubts about the genuineness of voluntary consent in cases where communication is sub-optimal and patients' wishes are not adequately conveyed. Fears also relate to taking irrevocable action on the basis of mistaken diagnoses and prognoses that can be minimized by pooling opinions, and hasty decisions about stopping arduous palliative care for this easy and inexpensive but controversial way out of dealing with incurable diseases for which cures may be unavailable at the time, although they may be imminent. Besides, some patients have sometimes been given a few months to live on the basis of mistaken diagnoses that became evident when they outlived themselves by many years.

The overall rapid advance of scientific research has brought with it breakthroughs made in time to save some persons who were written off while others were allowed to die on the eve of the same breakthroughs, because their conditions were believed to be incurable, and recovery impossible. Such breakthroughs are seen in the substitution of damaged genes with functional beta-globin genes in Beta-Thalassemia, enzyme substitution therapy for Gaucher's disease, and the use of tumour vaccines and introduction of foreign genes (cytokine, suicide, tumour suppressor and other genes) into cancers, which kill cancer cells selectively or inhibit the growth of the cancer. Meanwhile, some afflicted persons have lingered for years without any chance of recovery at great financial and emotional cost and other hardships to their families, and to health care systems. In the end, it seems as if we are engaged in gambling for or against the odds of beating the suffering that accompanies terminal illnesses.

## Rationalizing euthanasia

Doctors would like to think that withholding treatment is less culpable than withdrawing it, because one cannot be blamed for what one did not do—not starting treatment, but one can be blamed for what one did—starting and stopping treatment in a patient who subsequently dies. But such thinking is only cold comfort. Failing to treat a patient when circumstances demand it is probably more culpable than treating and withdrawing treatment when the burdens of treatment exceed its benefits, i.e., it is futile. Instead of stopping treatment and thus terminating the life of someone who is suffering excruciating pain from his disease, thereby risking being accused of murder, the doctor thinks that she can simply omit life sustaining actions to accelerate the patient's rate of progressive and painful decline to his demise and thereby escape the charge of murder or culpable homicide that would be levelled against her for active intervention meant to terminate the patient's life prematurely, but painlessly. Ultimately, the same purpose is served by removing the intravenous drip, gastric tube, or the respirator as by administering a lethal injection: the demise of the patient. The intent of active intervention and passive non-intervention is the same, although the methods of achieving it are conveniently different for the sake of the moral comfort of the doctor. She believes that no one can prove decisively that the (undesirable) consequences of non-intervention are worse than those of active intervention, barring the fact that suffering is prolonged in the former case and active intervention is a causal process with instant results, whereas non-intervention does not cause anything. If cessation of treatment is to be equated with intentional termination of life, the intent should be clearly expressed and not implied, because implication may be misinterpreted.

Whatever the case, there is no reason to subject the patient to a lingering death for selfish reasons through a process that displays a sore lack of the element of mercy, but which should instead be a case where *primum non nocere* should come into play. The Hippocratic Oath does state that "I will give no deadly medicine to any one if asked, nor suggest any such counsel; [but it also states that] I will follow that system of regimen which, according to my ability and judgment, I consider for the benefit of my patients"[7]; which means "If it is given me to save a life, all thanks. But it may also be within my power to take a life"[8] for the benefit of the patient who is suffering with no hope of relief. The more merciful action would be to end the patient's life promptly and spare him prolonged suffering, but deontologists defend their position of non-intervention by claiming that intentions are the only factors to be considered in assessing the moral worth of actions undertaken, even if the consequences turn out differently than anticipated. So any action intended to end the patient's life would be immoral and wrong regardless of desirable consequences for any one. But according to Mill, only those always unpredictable immediate and remote consequences which guided the action will bestow moral worth on it; intentions have no role in the attribution of moral praise or blame—a difficult position to reconcile when intent and consequences happen to coincide as in this instance.

Perhaps, then, we should appeal to the doctrine of double effect to resolve the impasse: the death of a patient from overdose of morphine is not wrong in itself as a side effect of the palliation of pain that requires doses of that magnitude, because it was not intended. It is morally neutral in that context, and if the patient dies, he dies as a result of the intended good of making the natural course of the disease tolerable—a good effect that outweighs the bad effect of pain and suffering endured by him. The high doses of morphine probably accelerated his demise, but that is not the issue. The patient's life was already in dire jeopardy before the administration of morphine, which only added an element of unintended premature termination of a life that was already ebbing out. In normal circumstances the administration of morphine has both good (relief of pain) and bad (predisposition to death of the patient) effects; but in a case where the patient requests the doctor to end his life and the doctor is willing to go along, death is the good effect that emancipates the patient from the bad life of pain. The anticipated death of the suffering patient may be rationalized as the foreseen, but unintended, consequence of an action intended to have only the good effect of pain relief, as if the analgesic effect of the overdose can be separated from its foreseeable lethal results in this instance. This last consideration makes the bad effect (death) the intended means of producing the good effect (relief of pain), thereby violating the terms of the doctrine, even though the agent may have good reasons for doing so. Thus she cannot invoke double effect to justify her actions in this paradoxical situation.

Therefore this doctrine is still not a way out for her; perhaps it is a cop out, because a claim cannot be made that the good effect of relief of pain was the intended outcome, rather than the foreseen demise of the patient. Here the good effect intended is the demise of the patient, because his oblivion (death) means the absence of pain (good), while his persistence means the persistence of pain (bad). The presumption is that all other methods for the control of pain and mental misery have been given adequate trial, but have failed to produce relief. If this line of rationalization proves inadequate in this way, then the final refuge may be found in another piece of rationalization that we have already considered: the doctrine of acts and omissions which attempts to provide exoneration from moral culpability by claiming that there is a moral difference between performing an act with certain consequences, such as deliberately giving a large single overdose of morphine with the intent of killing the patient immediately, and performing different acts like omitting active treatment with identical but delayed consequences, except that the overt intent has been submerged in the inactivity of the agent.

In contrast to the dilemmas presented by the foregoing cases, the doctor may think that she faces a simpler choice in connection with her intentional termination of the treatment of a patient who requests this positive inaction and refuses any further active treatment. The patient may be refusing treatment, often unadvisedly and without wanting to die, because he believes that he has the right to control what happens to his body and to prohibit other people from invading it in any manner. But to give in to the patient's refusal of treatment when one is not certain that the patient is not being unrealistic and irrational is to be irresponsible. On the other hand, if a selected course of treatment does not seem to be turning the patient's illness around, a doctor who is concerned more with the person and his illness than with the disease might decide to stop specific treatment and resort instead to palliative care for the sake of keeping her patient comfortable, and also to avoid raising false hopes in her patient. The patient might realize that he is fighting a losing battle, but still entertain hopes for improvement in his condition so that he can return to some form of active life, simply because the doctor is fanning those hopes by rashly pushing ahead against odds with what is ordinarily regarded as curative treatment that is futile in this case, or she is resorting to some extraordinary measures in the hope of improving her patient's condition and saving face. Such selfish attitudes and actions deserve only the utmost condemnation, because they do nothing for the patient. Patients appreciate honest talk.

# More slippery slopes

The previously discussed slippery slope arguments claim without adequate justification that the legalization of euthanasia will necessarily result in progressive abuses until the whole process is out of control, resulting in the corruption of the practice of voluntary euthanasia into involuntary euthanasia and the random killing of people for immoral reasons. The basis of this type of argument is that sometimes unintended and undesirable results follow the best laid plans to achieve a desirable end; but it is also dependent on the method of probabilistic prediction, which is highly liable to error, because as probabilities are predicted on the strength of preceding probabilities, they decrease progressively in strength, thus making succeeding predictions less likely. Therefore it becomes less likely that voluntary euthanasia will of necessity result in involuntary euthanasia in time. Besides, if laws governing the practice of euthanasia are well framed and strictly enforced, no person, regardless of his social, educational, or economic status will be coerced into opting for death, because he feels guilty of depriving a worthier person of privilege—there are no worthier persons; and no one will want to take the chance of abusing the practice, lest they pay the heavy penalty that will be imposed for doing so. The slippery slope argument underrates the force of secular law as against that of divine law, which people instinctively revere. The counterargument is that acquiescence in the request of some terminal patients for the ending of their lives should not be made to appear as discriminatory treatment against those who are unable to make such requests by reason of their severe incapacitation; both categories of patients should enjoy the same privilege, in spite of the difficulty of making decisions where patients are unable to communicate their wishes.

When everything has been decided, the doctor now has to perform the act of euthanasia. How she responds will depend on her moral persuasion and how comfortable she feels with the job that she is expected to carry out, especially if she believes that her sole function is to save lives and never to participate in terminating them. She may be beneficently disposed toward the patient and yet feel obligated to respect his autonomy in wanting to be permanently relieved of his suffering, and she may also feel uneasy about being the instrument of her patient's demise. No one can compel her to act against her conscience in resolving this conflicting duty of now having to terminate the life of the patient who has credibly placed his sacred trust in her as the virtual custodian of his life and not its terminator. These difficulties are further enhanced by the present legal prohibition of any form of assisted death, defined as: causing the death of a human being, directly or indirectly by any means, including acts of omission; and this applies regardless of the person's request for or consent to the termination of his life.

Those who disapprove of assisted death claim that we still can't be sure that such consent is fully informed and voluntarily given and that the patient's decision is irrevocably made—people can be unstable in their decisions and change their minds with changing circumstances. They cite the inadequacy of an advance directive in the absence of the actual experience of the medical problems that the patient is anticipating at the time of issuing that directive and his rational inadequacy during the course of an illness that is weighing heavily on him. Directives may refer to permanent unconsciousness or confusion, complete dependency on others for daily activities of living, and irreversible end-stage illness that does not respond to any treatment. They also cite the vulnerability of chronically ill patients and those who feel isolated and see only a bleak future of despair ahead of them due to the physical or psychosocial problems that have degraded the quality of their lives and destroyed their self-esteem and relationships. They tend to succumb to pressure to terminate their lives, because they feel that they are a burden to those who have to care for them, although it is wrong that any one should feel obligated to choose to die for the sake of increasing universal human happiness —a misapplication of the utilitarian principle and an infringement on patient autonomy and rights. In addition, the relative ease of carrying out assisted death will tend to make it an easy substitute for research into new and more effective ways of treating chronic, terminal, and painful diseases, with resulting demoralization of those who are suffering from those diseases. It will also ease the transformation of assisted death into an additional category of medical treatment for underprivileged sections of the community in the face of limited palliative care facilities. Finally there is always the lurking possibility of mistaken diagnoses, which cloud the patient's outlook on life and result in his mental suffering, even though his physical suffering is

337

not excessive. Such are the fears about assisted dying.

In response to the fears stated above, Abilash Gopal states that "no evidence of heightened risk of death by physician assistance in the elderly, women, uninsured people, the poor, racial and ethnic minorities, people with low educational status, minors, patients with psychiatric illness, and patients with chronic nonterminal illness"[9] was found in the Netherlands and Oregon where assisted dying is legal. Kathryn Tucker also noted that "more than eight years of experience in the state of Oregon has demonstrated that risks to patients are not realized when a carefully drafted law is in place. . . . staunch opponents have recognized that continued opposition to such a law can only be based on personal moral or religious grounds."[10] In any case, people are not so foolish as not to recognize problems arising from the practice and to deal with them rationally. Other jurisdictions that have legalized assisted death are Washington, Vermont, and Montana. Belgium has adopted more lenient conditions than those of the Netherlands, while Switzerland has made provision for outsiders to avail themselves of its liberal services. In Canada patients in circumstances of "grievous and irremediable" medical conditions that cause them intolerable, enduring suffering qualify for physician-assisted death. Safeguards emplaced by all these jurisdictions that should serve as a hedge against sliding down the slippery slopes referred to above include adulthood; patient's voluntarily expressed wish to die that is unequivocal and unwavering, written, signed, and witnessed; suffering due to terminal illness and unendurable suffering with no reasonable source of relief in sight, even if death is not imminent; his judgment is not impaired; he has been advised of all available modalities of treatment; he has the option of withdrawing his request; consultation by the doctor with other professionals and his family. Some people suggest video recording the entire process to ensure prevention of abuse. But in spite of all these measures, physicians, philosophers, politicians, and the legal profession are still wrestling with the task of deciding how to satisfy of all concerned parties about the morality and legality of euthanasia amidst the clamour of some patients for it.

When the patient is not able to make that decision and request, who decides what's best for him, as is the case with neonates who are born with malformations that are clearly inconsistent with self-awareness and rationality at any time in their lives. Previously nature intervened by curtailing their survival, but modern advances in treatment and technology have made possible the rescue and survival of extremely premature infants with their inevitable and often severe neurological problems: more advanced spina bifida, self-mutilating defects like Lesch-Nyhan Syndrome, brain malformations like lisencephaly, and other sources of insurmountable stress and disruption of the lives of their families every day of every year. Are we doing them a favour? So what guidelines should be followed by all concerned in the solution of these perennial problems? How should moral principles be balanced against real life challenges and legal stipulations in setting these guidelines without sacrificing the infant's total interests for the benefit of his caretakers, his family, and costs to society at large?

Plato thought that "the children of the inferior Guardians, and any defective offspring of the others, will be quietly got rid of. . . . if they fail to prevent its birth, to dispose of it as a creature that must not be reared."[11] (= infanticide). Utilitarians advocate a policy of the maximum amount of happiness accruing to all who are affected by the malformed infant's condition, including the termination of his life, actively or passively, on the assumption of the permanence of his burdensome condition, regardless of how long the infant is likely to survive with this burden. As long as the burden remains, the resulting quality of the lives of the infant and those affected by it will also remain always severely challenged, and replacing this unhappy state with a happy one by terminating its life should not result in moral culpability; but it does. Deontologists invoke the sanctity of the life of the infant as a person, which in some cases may be controversial, in light of our previous discussion of extending moral personhood to human beings at this level of lack of self-consciousness. Their stand thus prohibits the active termination of the infant's life, because such action amounts to using the infant as a means to achieving the happiness of other persons. They permit provision of the barest humane needs for survival, like nutrition and pain relief, but hardly any heroic interventions to prolong the infant's life of suffering.

Whatever the case may be, Loretta Kopelman maintains that contrary to Baby Doe rules that mandate provision of nourishment and life-saving treatment to handicapped infants without regard to the quality of life, parents as the infant's proxies have his best interests at heart and can be relied upon to choose the best morally practical actions on his behalf, including his protection from needless suffering. Parents should be wary of ethical theories that harp on maintaining the sanctity of human life in the midst of that suffering, and state legislations that use civil rights arguments to mandate treating all neonates afflicted with life-threatening conditions, even if the treatment will result in a life that lacks any semblance of quality and dignity as defined on page 14. Sometimes, too, the life of the family pet might seem better than that of the severely disabled child who results from these strict legal enforcements. They restrict the physician's expert and reasonable medical judgment to make decisions about relieving or correcting the condition and the liberty to make other choices, except and only, as required by the amended rules, when the infants are irreversibly comatose and treatment would be ineffective, futile for their survival, or would prolong their dying and thus prove to be inhumane. These conditions exclude considerations of the neonate's prospective quality of life. She says,

> Baby Doe Rules should be rejected because they sometimes require actions that violate duties to act compassionately, provide individualized treatment decisions at the end of life, and minimize unnecessary suffering. [They] unfairly single out one group, infants under one, for treatments adults do not want for themselves, violating duties to treat others as we want to be treated.[12]

Concern with such theories and legislation relates to the ambiguity of their operational terms: *futile* and *inhumane*. Proxies who decide what is best for these infants should be guided by a desire to maximize their benefits and minimize their burdens within existing value systems, regard for their short and long term interests, cognizance of the trauma of raising false, unattainable hopes by salvaging lives marked by severe functional shortcomings, and prospects of an abbreviated existence in oblivion for some of these infants (Tooley's non-persons) as in the case of anencephalics or those with Trisomy 18. They should not make unreasonable demands in the guise of exercising their autonomy, and where their demands prove to be selfish, dogmatic, or irrational, the caring physician should act as their trusted guide through the challenges presented by a medical situation on which their technical knowledge may be limited or non-existent, and where their emotional expectations from and demands on the health care system (and themselves) may be unrealistic.

The ultimate relief will come from society's genuine concern for its disabled members and its desire to promote their interests, actual or potential, without submerging them in its many contrary and self-serving agendas. Concern for the theoretical sanctity of, and respect for, human life should not suddenly spring into prominence as an expedient device to advance a religious or pseudo-moral course, but it should be driven by a universal concern for justice, equity, and compassion for the plight of those that we profess to be serving and on whose behalf we are called upon to make momentous decisions. The decisions that we impose on others should also be good enough for us in similar circumstances, and we should be able to live comfortably with them. So, even if no laws are enacted to legalize assisted dying, but consideration is accorded to worthy cases without threat of mandatory punishment of those involved, that would still be a move in the right direction toward justice, equity, and compassion for those who richly deserve it.

340

# The Totality Viewpoint

# 21

# Darwinian Medicine

## Evolutionary aspects of Disease

It is fitting now to make reference to a hitherto neglected aspect of Medicine that should be part of our comprehensive approach to clinical problem solving by asking not only proximate reasons for how disease happens in people or which statistical parameters a condition must satisfy to be called a disease, but also ultimate reasons for why it happens the way it does, and why natural selection has not eliminated the genes that foster habits that are causally related to diseases like lung cancer, viz., smoking. That aspect is the Darwinian or evolutionary approach to understanding of disease and disability, combating and eliminating disease, and promoting good health. It substitutes evolutionary maladaptation for the previously considered mechanisms of disease. It helps us to postulate theories for understanding the competition between us and microbes in the context of evolution and the struggle for survival that has been waged by one group against the other over many centuries; also the favourable and unfavourable adaptations and compromises that each group has had to make in its efforts to prevail in this struggle. Certain characteristics have been selected for from the ancestral gene pool, because they were best suited for enabling each group of organisms to perform those tasks that are necessary for its survival and for the propagation of its good genes into future generations; and that is why we are still here. Still, microbes have gained an advantage in this process by reproducing more rapidly than humans and retaining a simpler genetic structure that is less vulnerable to the detrimental effects of genetic variation seen in our complex physicochemical systems—an evolutionary trade off aptly termed environment discord.

Hence it is that microbes always seem to be ahead of us in this struggle by virtue of their highly adaptive ability to replicate via extremely rapid rates of vertical gene transfer through successive generations and horizontal transfer to existing members of their species. They outstrip our self-protective devices by evolving new methods of attacking us, while defending themselves from our drug-assisted offensive on them, by altering their genetic constitution to develop resistance to the toxic effects of commonly used antibiotics on themselves and their progeny, thus necessitating the use of larger doses of drug and broader spectrum antibiotics. Examples of this phenomenon are the emergence of Clostridium difficile, methicillin resistant Staphylococcus aureus (MRSA), and macrolide resistant bacteria as a result of the indiscriminate use of antibiotics to treat conditions like the simple cold. They also subvert our genes to serve their own ends, as in the case of the human immunodeficiency virus, which replicates by fusing its genome with the cell nuclei of the host. Others like Yersinia pestis release a protein, which in turn activates proteins (capases) of the host cell to trigger breakdown of its nuclear DNA, thereby making it possible for them to escape phagocytosis by the host cell. The results are an inactivation and disintegration of mitochondria, rearrangement and breakdown of the cell's plasma membrane, and release of bacteria into the tissues, which in turn facilitates their propagation and the spread of infection.

To be able to fight off invading microbes, our bodies had to develop adaptive immune mechanisms that rapidly produce enough antibodies at the time of first infection, and on subsequent occasions by virtue of their B and T lymphocyte memory. Still, some viruses envelop themselves in membrane-coated vesicles that have budded off the host cell's plasma membrane and fused with the plasma membranes of new host cells that they infect to protect themselves from these mechanisms. Others (SARS-CoV-2) mutate repeatedly from B.1.1.7 to XAK to avoid destruction by the same immune mechanisms that make us victims of (?infection-caused) immune disorders like auto-immune hemolytic anemia in which certain of our antibodies attack and destroy antigens on our erythrocytes. In other disorders like Systemic Lupus Erythematosus, antibodies are formed against certain nuclear proteins resulting in immune complexes that settle in various organs and destroy them to cause the many phases with which the disease presents. The basis of these reactions is misidentification of native proteins as foreign antigenic proteins. Some invaders like group A Streptococci have evolved an antigenic armamentarium that cross reacts with human tissues, so that the body's immune response to onslaught by these streptococcal antigens produces antibodies that damage the person's kidneys, heart valves, and basal ganglia, resulting in glomerulonephritis, rheumatic fever, and Sydenham's chorea. On the other hand, elevated levels of scavenging uric acid, which saves our tissues from oxidation by free radicals that cause ageing, also predispose us to attacks of gout.

Viewed from this perspective, health and disease lose their normative character by becoming mere cogs in the mechanistic wheel of evolution—a blind process of "descent and modification" that does not concur with the psycho-social approach to the human organism and disease advocated by normativists. Furthermore, it is clear that natural selection and evolution are not geared to ensuring our

344

protection from disease and disability, but their function is solely the non-teleological perpetuation of genes into future generations through reproduction, and that the genes themselves do not possess a teleological streak by which they can select reproductive intent in preference to the prevention of the person's diseased state. Their behaviour is determined by the physicochemical circumstances of the milieu in which they exist, and it differs accordingly. Future genetic representation of the species is only a fortuitous by product of this blind process, which can derail if the colonizing micro-organisms replicate at a rate that outstrips the efforts of the immune system of the host to contain their effects by eliminating them.

One result is that in any conflict between the two ends, adaptations for the facilitation of reproduction will always carry the day at the expense of life span or good health in the later life of the person, as long as function in the reproductive years of life can be maximized. The net result is that we are left disadvantaged and vulnerable to disease and disability, although a case can still be made for the utility of some forms of acute disability in protecting the person from ongoing physical activity that may be detrimental to his welfare and survival, such as inflammation in damaged tissues, which prevents their continued use and further damage. The use of analgesics to relieve the pain of damaged tissues so that the subject can continue with his activity is another example of defying nature's protective mechanisms and thereby causing further damage to those tissues in the absence of pain, and delaying the healing process.

Darwinian Medicine is especially interested in the more natural view of disease by asking why the human body is constituted the way it is, why it works the way it does, and why natural selection has left us vulnerable to infections, fevers, aches and pains, diarrhea, deadly and disabling genetic disorders, and all the other disorders detailed in medical textbooks, when it could have perfected our bodies to work better for the better health and survival of our species. It also reflects on its effects on the human species in its totality, while concerning itself with why individuals or small groups display that vulnerability. In its consideration of why certain diseases appear to be limited to some racial groups while other diseases occur with more frequency in some groups than in others, it directs attention to each group's adaption to environmental conditions in its chosen habitat and the evolution of genes that combat the effects of other (deleterious) genes in the survival of the group within that particular habitat, which is not shared by other groups that necessarily lack those genes, because they have developed genes that are consistent with conditions in their own habitats, e.g., Sickle Cell gene in the inhabitants of the malaria-infested regions of West Africa where it affords protection against the disease, or Tay-Sachs gene in Eastern European Jews who lived in tuberculosis-infested ghettos and selected for the gene as a means of protection from tuberculosis. Furthermore, natural selection is not limited to discovering mechanisms only, but it also ventures into the area of the evolutionary import of clinical phenomena; e.g., E. coli causes diarrhea by enterotoxigenic, entero-adhesive, and enteroinvasive mechanisms that may cause the demise of the host from dehydration or toxicity from overwhelming infection, but its ultimate purpose is to enhance its chances of propagation and survival as a species indirectly through the survival of the individual bacterium as it is passed around in the community on the hands of infected individuals.

As speculative as it is, a Darwinian approach to Medicine also helps us to understand the constraints, compromises, perfections, and imperfections that have evolved through time to regulate the body's functions, including the persistence of deleterious genes that cause old age and deprive us of eternal youth, or those that cause illnesses like arteriosclerosis in old age when they help to heal our bones in our earlier years while they also calcify our arteries; hence the futility of the quest for conditions that will favour longevity. It also provides a logical explanation for the existence of cancers from the fact that the stimulus for the single fertilized cell to multiply, with subsequent organization of the resulting cells into functional organ systems can, in the majority of cases, proceed relatively faultlessly to form a complete human being. So we expect the same stimulus that guides normal development to cause the same cells to proliferate unchecked to form cancers, except that it is now misguided to the extent that the cancer cells invade body tissues other than those from which they arose, eventually causing the demise of the person if he does not receive treatment, or if the treatment fails. This may happen de novo or as a result of adaptation of the cancer cells to the drugs used to treat the patient in the same way as microbes develop resistance to antimicrobial agents.

# Darwinism and the scientific method

The traditional reductionist and scientific approach to disease presents pathophysiological facts from which it draws conclusions about strategies for the appropriate management of the patient's existing clinical problem without regard for its evolution. As noted, the evolutionary approach, tries to unfurl the ultimate reasons for the coming into being of the current clinical situation, its meaning, and how it should be regarded in the context of the phylogeny of the group to which the patient belongs and his ontogeny within that group. The two methods have been styled inquiries into the ultimate and proximate causations of disease, each of which, however, gives less than a complete explanation of the biological processes involved in disease and illness generally and specifically. Accordingly, their

346

roles have to be considered in combination, and in doing so attention has to be given to conditions that may appear to be the detrimental effects of disease, such as fevers and pains, but are, in fact, adaptive mechanisms induced by natural selection. All are non-specific efforts to preserve the integrity and survival of the individual, and hence the species, in that variety of adverse circumstances where the patient's responses are appropriate to that end. Where the body fails to respond or exaggerates its responses, disease ensues.

At all times, it is useful to remember that these speculations cannot be subjected to proof by the methods of science, as previously discussed, because there is no way of telescoping the evolutionary process to be able to observe it over a finite period, or to return it to past eons so as to induce a different path of progression that will favour selected experimental conditions. We have to reason backward from the present situation and try to justify it by postulating probable factors that brought it about as developmental adaptations or maladaptations entailed by this trade-off process. In all cases, genes were initially and subsequently preferentially selected for, because they conferred certain advantages on the organism, in spite of their potential to increase the incidence of disease, disability, or hypofunction in the later life of that organism. In the case of some so-called pleiotropic genes (multifunctional genes that affect more than one physical characteristic) such as the gene of Down syndrome that affects facial features, intelligence, and cardiac morphology, it is difficult to imagine that the advantages of the calm demeanour of these patients is a fair trade-off for these and many other serious problems that they have to face in life.

Such cases notwithstanding, understandable trade-offs occur in other cases such as the strategic protective deployment of the tonsils and adenoids (Waldeyer's ring supplemented by the cervical ring and bilateral vertical chains of cervical lymph nodes, all of which serve to contain microbes that would otherwise enter the body orally an nasally) and their hypertrophy which causes upper airways obstruction and sleep apnoea, corpulmonale and congestive heart failure, both lethal conditions. On the other hand the vermiform appendix, although a vestigial structure with no residual function, becomes potentially lethal when it becomes inflamed and it ruptures, causing acute peritonitis; and our upright posture, while conferring the advantage of dexterity, comes with the cost of undue pressure on our lumbar spines, resulting in lumbo-sacral problems of herniated discs, postural hypotension, and hemorrhoids (if we did not have an upright posture, we would not suffer these back problems—the claim is an instance of counterfactual causation). The amenorrhea of anorexia nervosa also serves the useful function of preventing pregnancy in a state of depressed nutrition when the would be fetus would suffer severe intrauterine malnutrition with its untoward sequelae. In short, every condition has it attendant costs and benefits.

Darwinian Medicine should be regarded, not as a competing discipline, but as a fitting complement to the scientific understanding of human disease processes and their management. It also serves to explain why we have to be careful about trying to indiscriminately eliminate or suppress symptoms such as sneezing, coughing, and runny noses, which act as defence mechanisms in helping us to clear our respiratory tracts of microbes, mucus, and foreign material, even as they facilitate the spread of infection by droplets. At the same time, it rekindles our awareness that the pharmacologically sedated patient who cannot cough up infected sputum postoperatively, or one who is debilitated, will most likely succumb to hypostatic pneumonia. On the other hand, one whose excessive vomiting is uncontrolled, or one who receives drug treatment but does not receive enough fluids and electrolytes for his non-virulent shigella-induced diarrhea, runs the risk of protracted illness with possible complications of dehydration and hypovolemic shock with its lethal sequelae. The down side of these defence mechanisms is that sometimes they seem to come into operation without the sufficient cause for which they exist; but that is the price we have to pay for their efficiency, as long as we do not confuse them with disease *per se*. In this regard, we only need to observe that it is better to suffer the inconvenience of a coughing fit triggered by a bit of phlegm than to succumb to the deleterious effects of an obstructing foreign body, simply because we lack the cough reflex to clear it. The principle is the same as that whereby we tolerate false alarms from smoke detectors, as long as we know that they will alert us in the event of a genuine fire and smoke emergency.

# Logic of Darwinism

By employing evolutionary explanations to attempt to place in its proper perspective the vulnerability to disease of the human species, Darwinian thinking offers logical explications for anomalies such as polycythemia, which results from chronic hypoventilation or pulmonary oxygen diffusion defects that ordinarily severely diminish the total amount of oxygen carried by 5 million desaturated erythrocytes. So 2 million more erythrocytes enter the circulation to carry and deliver more oxygen to the tissues under the influence of a hypoxia-induced secretion of erythropoietin and its resulting erythrogenic effect on the bone marrow. It also involves turning much of our thinking about health and disease on its head. So, instead of saying that the predominant consumption of fruit and vegetables reduces our risks of developing colon cancer, we should rather say that insufficient of these foods versus other foods to which evolution has not yet adapted our systems increases our risk of developing colon cancer. The same argument applies to the consumption of foods that promote obesity and its related problems of diabetes, atherosclerosis, stroke, and coronary heart disease, and to the push for physical exercise in this modern age of sedentary pursuits when a naturally active life would provide the exercise needed to keep us non-obese and healthy—walking instead of riding, playing outside instead of watching television with a packet of potato chips, or playing video games from a couch. It is conceivable that in our primitive state we were very active as we went around looking for food that was different from modern fast foods with their obesity inducing high caloric content, and, of course we did not drive to the corner store. The absence of all these conveniences of modern civilization literally kept us healthy in the same way as it did in our own childhood when we played and ran around outside and ate healthy food. Natural selection has, however, not kept pace with industrial, economic, and social development; as a result our bodies are still running on the primitive program that was admirably suited for a natural outdoors life of hunting and gathering, thus rendering us vulnerable to these modern conditions which entail the consumption of excessive amounts of calories, coupled with voluntary restriction of the means of metabolizing and expending those calories.

With further reference to life-style, Darwinian Medicine explains the increased incidence and prevalence of breast cancer in terms of biochemical changes in breast tissue related to prolonged exposure to estrogen from frequent menstrual cycles, because the modern life-style favours early onset of menarche and later childbirth, fewer children born and breast-fed, if at all, for shorter periods, and the earlier onset of menopause. As a result, one of the solutions to the prevailing high incidence of breast cancer may reside in the introduction of measures for reducing this undue exposure to estrogen. The same frequency of menstrual cycles and blood loss explains the increased prevalence of iron deficiency in some women. Life-style changes are also important in preventing the spread of microbes by interrupting the life cycles of those that cause non-devastating disorders. These microbes terminate their own life cycles by killing their hosts, a process that also allows natural selection to weed out the more virulent strains, which end with the patient, in favour of the ones that can colonize their host without killing him while they produce milder clinical disruptions or even eke out a kind of symbiotic existence with him. Cases in point are those of the elimination of virulent diphtheria strains through immunization, leaving the milder strains whose effect on their host is like that of a vaccine, which lacks the intensity of an infection; and reduction in the virulence of cholera with improved sanitation, which is the certain means of halting the propagation of the bacterium that is favoured by unsanitary conditions, since droplet spread from person to person is not possible with these bacteria.

Equally illuminating is the role of iron in the control of infections. For instance, the anemia that accompanies acute and chronic infections is known to resolve spontaneously when the infection passes. The responsible mechanism for the onset of this reaction is the automatic reduction in the output of erythropoietin, which in turn reduces erythropoiesis, thereby making less iron available to microbes that thrive on free iron from their host. Also postulated, but not yet fully explicated, in the pathogenesis of anemia that commonly accompanies chronic disease is what has been termed "the blockade of reticuloendothelial iron release,"[1] which increases the quantity of hepcidin in the circulation, thereby rendering the hosts iron unavailable to invading microbes. On the other hand, bacterial virulence has been noted to increase with the free availability of iron, a factor that is counterbalanced by the higher content of lactoferrin in breast milk than in cow milk and the postulated presence of conalbumin in the human body, both of which proteins bind iron to make it unavailable to invading microbes for their metabolism. (That is a good reason for promoting infant breast feeding). The pivotal role of iron in the regulation of this aspect of microbial infection is illustrated by the following facts: the patient and his infecting microbes both need iron for their metabolism; the patient's oxygen carriage to his metabolically active tissues depends on the dissociation of iron from hemoglobin; hemoglobin formation depends on adequate supplies of iron; bacteria scavenge their iron supplies from the patient's cells and iron transporting proteins transferrin, ferritin, and lactoferrin. Therefore supplying more iron in the form of iron supplements to the anemic patient who has an infection increases the hazard to the patient and should not be undertaken, because it defeats the body's own mechanisms for thwarting bacterial scavenging. To reduce available iron supplies to bacteria, the body can suppress the following processes:

(1) assimilation of dietary iron up to 80%, by sequestering it in the liver out of the reach of bacteria, thereby complementing the effect of fever in disabling the bacteria, and

(2) neutrophil release of apolactoferrin used to bind iron at septic sites.

It can also enhance the synthesis of nitric oxide by macrophages to cause disruption of the iron metabolism of bacteria and promote hepatic release of hemopexin and haptoglobin to bind hemin and hemoglobin, respectively, making them unavailable for bacterial metabolism. All in all, remarkable processes of natural selection for protecting the person from harm without using drugs,

350

and a tribute to the role of Darwinian Medicine as a complement to, and not a competitor against, traditional medicine.

## Benefits from disease and adverse symptoms

In the midst of these logical explanations for the utility of what appear to be disadvantageous biophysical processes, the case of genetic disorders seems to be an aberration that baffles reason; but once we come to appreciate the reasons for the survival of deleterious genes and why we should not always try to eliminate them completely (heterozygous state), we begin to see the benefits that come with them. Appropriate examples are provided by Sickle Cell and Cystic Fibrosis genes. The Sickle Cell gene results from a substitution of glycine for leucine at the #6 position of the $\beta$ globin chain of the hemoglobin molecule. The resulting erythrocyte in the patient with Sickle Cell trait is thereby rendered able to resist colonization by the Plasmodium falciparum malarial parasite by stunting its growth and development while accelerating its clearance from the circulation and thereby reducing its deposition in deep postcapillary vascular beds that results in ischemic tissue damage. This situation is all very well for the heterozygous sickle cell individual who does not suffer the crises and other ill effects of the disease, but for the homozygous patient it is a marker of an increased rate of early mortality and propensity to hyperhemolytic crises, besides presenting as a case of Hobson's choice between two undesirable diseases, one of which is not better than the other in its accompanying effects. On the other hand, there is a body of opinion which claims that alpha+ thalassemia is associated with a higher incidence of milder forms of malaria like Plasmodium vivax, while protecting the individual from more severe forms like Plasmodium falciparum. Other hemoglobinopathies, like Glucose-6-phosphate dehydrogenase deficiency, have also proved to be protective against malaria.

Cystic Fibrosis (CF) is caused by the loss of a phenylalanine residue at amino protein number 508 of the CFTR (cystic fibrosis transmembrane conductance regulator) protein coded for by the CF gene situated on the long arm of chromosome 7. The CFTR protein regulates the transport of chloride, sodium, and water across epithelium. Abnormal CF genes result from $\Delta$F580 mutation in the CFTR protein. Patients who are hemizygous for $\Delta$F580 gene mutation (carriers) are protected against the deadly diarrhea of Cholera that normally occurs in those who are homozygous (possibly also against penetration of their guts by Salmonella and E.coli). Vibrio cholerae's adaptive mechanisms enable it to adhere to enterocytes without penetrating the mucosal barrier thereby avoiding peristaltic expulsion. It then elaborates an enterotoxin, choleratoxin

351

or choleragen, which places the G protein in the epithelial cells of the intestine in a state of continual activation, causing excessive levels of cellular adenylate cyclase in the mucosa resulting in synthesis of equally increased levels of cyclic adenosine monophosphate (cAMP) that turn on electrolyte secretory pathways by activating protein kinases. The hemizygous state protects against this eventuality by inhibiting total opening of chloride channels in the bowel epithelium that causes profuse loss of water and electrolytes surpassing the absorptive capacity of the colon.

Cholera toxin induces these sometimes fatal profuse water and electrolyte losses as a compensatory evolutionary mechanism for flushing out invading bacteria to protect the host from the effects of further tissue damage and continuing water loss. But the other trade off of this mechanism is to help propagate the bacterium by making it available to more potential hosts out there in the community. In the case of the cystic fibrosis carrier, these losses are minimized by the partial opening of the chloride channels. Preservation of the haploid individual and the species is doubly enhanced by saving the patient from fatal dehydration and from the fatal complications of cystic fibrosis that occur in the individual with a diploid genetic constitution in both cases. The result is the catch22 situation where we are presented with a two-horned dilemma. On the one horn is the survival of the individual, and on the other horn is the survival of the group to which he belongs. Public health measures of isolation and safe disposal of waste products resolve one horn of this dilemma in favour of the group by preventing the spread of infection, leaving resolution of the other horn to vigorous rehydration and partial correction of the patient's adverse clinical state by the advantageous aspects of the pathophysiological mechanisms of cystic fibrosis as detailed above

In these kinds of situations, there is no scope for treatment with anti-diarrheal medications that might otherwise prolong the infectious process; management resides in understanding the under-lying mechanisms and their evolutionary foundations, and providing the kind of treatment that is geared to correcting those abnormalities. In the parallel situation of Shigellosis, where antispasmodic-containing anti-diarrheals are also contraindicated for the same reason, Pepto Bismal relieves the diarrhea without interfering with the cramping mechanism that helps to expel the infecting bacteria, whereas Lomotil inhibits both mechanisms, prolonging the infection. These considerations do not, however, militate against the use of specific antimicrobial agents where septicemia presents a threat to the survival of the victim of infection, but in the regular run of cases where such treatment would prolong the infection, natural mechanisms are left to eliminate the infecting bacteria, and fluids, glucose and electrolytes are administered to restore homeostasis, in the same way that emesis does it by eliminating toxins from the stomach before more of them can be absorbed while they traverse the remainder of the intestinal tract. Here too, the patient receives only the necessary support modalities to maintain his homeostasis.

To return to the intriguing example of the role of fever in infections, fever is the natural mechanism for combating infection by simultaneously disabling the pathogen's physiological mechanisms and enhancing the host's immune mechanisms, although it can also sometimes help some microbes to survive and propagate. It has been proved to enhance neutrophil phagocytic activity and lymphocyte cytotoxicity, thereby helping to eliminate the microbes that cause infection. So our obsession with seeking to eliminate fever in clinical conditions like viral colds is unfounded, ill advised, and largely misplaced, as it tends to disrupt the body's immunological responses and obfuscate those deleterious features of the disease process whose expression is fever dependent. Suppression of fever also prolongs the duration of clinical symptoms in some viral infections. It must be seen as the discomfort that the host has to endure for his survival.

In the same way that analgesics and narcotics may mask important symptoms of potentially lethal peritonitis resulting from a perforated bowel, thus disabling the body's alerting mechanism to some derangement in its function, fever ablation inhibits the manifestation of important pointers to the progress of the infectious process. A low grade fever that suddenly becomes persistently elevated may be an indication of the spread of infection or the development of an abscess in a previously resolving tissue infection; but if it is masked, this important pointer may be missed. Fever lowering efforts are indicated only for preventing simple febrile convulsions and increased intracranial pressure in susceptible infants and other subjects. Fever caused by common viral infections will not cause brain damage as many people falsely believe, although prolonged high fever may cause delirium, stupor, and increased requirements and consumption of oxygen to satisfy the demands of an accelerated catabolic state. On the other hand, the sick patient who does not develop a fever in response to the challenge of his infection may be the critical victim of a failing immunological system.

A final example of the utility of Darwinian thinking comes from the psychosomatic arena. During the state of fear induced by panic or over-reaction to danger, increased adrenaline secretion causes tachycardia, tachypnea, and diaphoresis, making for increased oxygen intake and muscular metabolism, and cooling down of the body in preparation for a fight or the flight to safety in the insecure ancestral environment. Such a mechanism exacts considerable caloric expenditure from the body and is an excessive reaction to common, regular challenges; but by understanding the evolutionary "why" of it more than the mechanistic "how", the patient who is the subject of major panic and phobic attacks will be better able to handle his pathological fear against the background of its concomitant evolutionary life-saving advantages. The patient only needs to understand the "why" of this defence mechanism to be able to deal with his panic and phobic attacks with the confidence of knowing that he is not diseased, but reacting naturally and positively to a challenging situation that can sometimes be life threatening.

Similar considerations apply to people who do not appear to be affected by states of anxiety, or have no fear of fear. Ordinarily and ancestrally, life in trees and high rocks induced fear of falling from these precarious situations, tending to discourage risk taking in favour of cautious behaviour. These persons, however, do not hesitate to take extreme risks, and it is likely that their foolhardy behaviour dates back to infancy and beyond when, unlike most children, they had no fear of falls from heights and of other serious injuries. Psychiatrists believe that these "hypophobics", who are insufficiently anxious, rather than unduly optimistic, also get into trouble with their relationships, because of their lack of the inhibitions that would normally act as their brake mechanisms.

## Epidemiology and Darwinism

The field of epidemiology has highlighted important aspects of evolutionary biology and medicine in the case of varying degrees of microbial virulence and the living human body's responses to the effects of these microbes. It used to be the accepted dogma that microbes adapt to their hosts over time to the point of symbiosis, thereby ensuring their own survival, but also preventing the host from succumbing to their adverse effects. But it is clear now that the abbreviated survival of some of the hosts of the most virulent microbes, e.g. Ebola, contrasts with the prolonged survival of other hosts to microbes that retain low grade virulence over a long time before causing terminal damage to their hosts; e.g., Mycobacterium tuberculosis and Plasmodium species. The highly virulent, rapidly reproducing microbes that kill their hosts in short shrift severely limit their own chances of propagation, as we will see in the case of Ebola, but they also increase these chances by virtue of their overwhelming numbers that facilitate their spread. The slower reproducing, less virulent microbes that can be easily confined through public health measures will depend on the breakdown in these measures, e.g., passive access to public sources of propagation like overcrowding and contaminated food and water supplies, which are still causing many minor, but annoying community infections like rhinoviruses, and

354

more serious infections like E. coli complicated by hemolytic uremic syndrome and its sequelae.

This tug of war between the person and the microbe spirals upward all the time with the microbe always forging ahead of the person, because it evolves faster and acquires the ability to resist different forms of antimicrobials; e.g., emerging tubercle bacilli that are resistant to standard anti-tuberculosis drugs, a property that they acquire from their prolonged survival in the tissues of their victims. An interesting scenario is also provided by the natural control of Ebola virus infection, which has a high fatality rate and for which there is no antimicrobial treatment, but there is now a vaccine. As Lori Oliwenstein has observed, "With a nod to evolution's god, physicians are looking at illness through the lens of natural selection to find out why we get sick and what we can do about it."[2] She quotes an example from biologist Paul Ewald, who has stated that the spread of the virus and the devastation that it causes are limited by its inability to move freely from one person to the next, because it kills its victim in a short time before he has had the time to pass on the infection by contact with other persons. If he could live long enough, he would most likely be able to transmit the viral progeny to other persons and spread the infection. Besides, after the death of its victim, its spread is limited by its extreme vulnerability to sunlight and by its inability to survive for more than a day outside the human body. In this way, natural selection limits an infectious disease that would otherwise wreak widespread havoc on defenceless populations.

From these experiences we learn that we can control infections better if we take advantage of evolutionary interventions in addition to the antimicrobial cudgels that we have come to rely on to eliminate these elusive killers. The use of vaccines, quarantine, and isolation within the context of the public health initiative are good example of this strategy. Also, limiting the use of antibiotics allows commensal, protective microbes to flourish and directly reduce the virulence of more hostile microbes by inducing evolutionary changes in their toxic armamentaria—a kind of Lamarckism in espousing the inheritance of acquired characteristics, but really a form of accelerated evolution by natural selection. A similar process is seen in the attenuation of microbial virulence with the use of public health control measures.

# Expanding horizons—the whole picture

Ontogenetically many human organs recapitulate their phylogenetic history, which also reveals a progression in fractal structure from the single saccular lung of the reptile, through the simple accessory air sacs of the birds, to the to the complex bronchioles and multiple air sacs of the human species; and from the primitive heart and aorta of the insect, which terminates in the head and body cavity, through the single stream, two-chambered heart and circulation of the fish, to the four-chambered double stream heart and circulation of the human being with its complex, arborizing (fractal) Purkinje and vascular systems. (see pages 355 and 228-232).

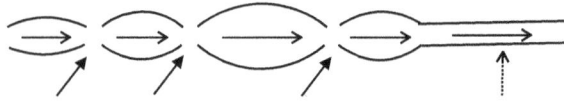

Insect heart with inflow ostia and aorta

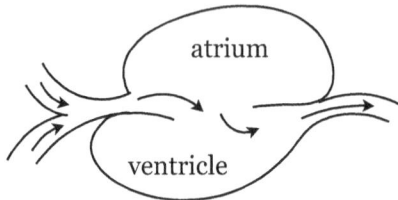

Fish heart

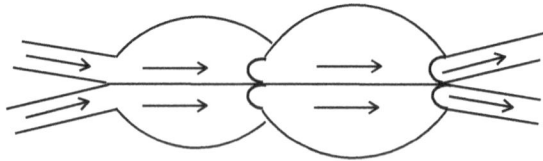

main veins /atria /ventricles with valves/ main arteries

Human heart

These and other evolutionary changes in organs, which enable the individual to survive in the struggle for existence by natural selection, raise philosophical questions about the nature of the designer who is directing the process teleologically, or the blind mechanistic force that is propelling it. They also raise questions about what it is that the surviving genes carry forward, or the contrary. What these particulate genes carry forward is their potential or disposition to endow the individual who results from their union in the zygote with the ability to withstand the buffetings of his intrauterine and extra-uterine environment by dint of a "functional

356

efficiency" that allows him to develop characteristics that befit him for adaptation and survival. With this concept in mind, it seems that the criticism levelled against natural selection that it is tautologous (circular), i.e., natural selection is the survival of the fittest, and the fittest are those who survive, is without foundation, because survivors are not individuals but genes and propensities. Natural selection also dispels the erroneous notion relating to the fixed nature of the species, in accordance with their creation, by demonstrating progressive changes in their nature, which are consistent with their continuing existential struggle—the same kind of mortal struggle that obtains between disease-causing microbes and their hosts.

In light of these universal struggles among all kinds of living organisms, the discourse that pertains to the human organism has been epitomized in the existentialist dictum: "existentialism is about the *struggle* to live"[3]. As Sartre has indicated (page 75), we are not born with fixed and deterministically established natures, but we acquire our essences as we continue to struggle against the odds of living. Further contact between Darwinism and Philosophy emerges in its relationships and parallels with epistemology. The holistic evolution of knowledge brought about by the evolution of successive facets of the hypothetical-deductive method through a process of selection of data that contributes to the formulation of a comprehensive, rational, operative theory finds its parallel in the Darwinian process of the evolution of species through a process of selection that contributes to the emergence of a competent human organism that is able to withstand and adapt to the ever changing circumstances of living.

The facts of Darwinism may have started off as modest linchpins of a grand theory that could not hold its own against creationism, but they have proved over time to be facts on par with the accepted facts of empirical science, against the myth of creationism, and they have also demonstrated its connection with Philosophy. Nevertheless, Wittgenstein, in his role as a linguistic philosopher, disagreed with the idea that Darwinism has connections with Philosophy, based on his contention that "Philosophy is not one of the natural sciences"[4], because it is "an activity [that] . . . aims at the logical clarification of thoughts, . . . [rather than] a body of doctrine [or propositions]."[5]. Contrary to the evidence adduced above and to what the evolution of the Philosophy of Science has come to represent, he states categorically that "Darwin's theory has no more to do with philosophy than any other hypothesis in natural science"[6]. Needless to say, this opinion is subject to serious challenge.

Running practically parallel with the Darwinian approach to health care is the expanding, multidisciplinary, holistic approach to disease, illness, and healing whose surface I have only scraped. This perspective also treats the whole patient in his particular milieu as the victim of the psycho-physiological disruption caused by environment, disease, and illness. In all these holistic disciplines the person is the subject of the healing process, as opposed to that part of him with which the reductionist approach of scientific medicine concerns itself in the hope that all will be well when the particular disrupted, objective function has been restored, as determined largely by quantitative tests and to a minor degree by qualitative assessment. The person as the concern of the physician who promotes his own self-healing should also be seen against the background of the evolutionary process that has played a major role in producing the problems with which he and his physician are confronted.

Both Darwinism and holistic or so-called integrative medicine should be seen as complementary to orthodox medicine, not as its competitors, because Medicine as a genus is holistic and one discipline can never exhaust all its facets; the subject of disease, who is its host and focus and to whom treatment should primarily be directed, is a complex organism that should be approached from all possible angles. That is why allergists can adopt and adapt the methods of homeopathic practice by administering progressively increasing amounts of the substance to which they wish to desensitize someone, starting with what we would call a homeopathic dose. The practice of homeopathy, which claims avoidance of the toxicity of drugs, entails the controversial administration of the smallest doses of the substance that causes symptoms when administered in large amounts, with the sole aim of stimulating the body's defensive and combative mechanisms to fight off whatever is attacking it. As Michael Cohen so aptly observed,

> To shift from an exclusively medical paradigm to a framework that includes touch and other forms of holistic healing does not mean that the insights, discoveries, and therapeutic devices of modern medicine will be discarded or diminished. Nor does the movement from medicine to healing mean returning to the Dark Ages or succumbing to quackery.[7]

# References and Index

# References

Chapter 1. Prolegomena

1.  Paul Thagard, "The Concept of Disease: Structure and Change", http://cogsci.uwaterloo.ca/Articles/Pages/Concept.html
2.  Havi Carel, "Can I Be Ill and Happy?", *Philosophia* (2007) 35:96 DOI 10.1007/s11406-007-9085-5, https://www.researchgate.net/publication/225148136_Can_I_be_ill_and_happy

Chapter 2. Philosophhical tools and concepts

1.  Samuel Cartwright, "Report on the Diseases and Physical Peculiarities of the Negro Race", in *Health Disease and Illness*, eds. Arthur L. Caplan, James J. McCartney, Dominic A. Sisti, (Washington D.C: Georgetown University Press, 2004), 34, 35.
2.  Lewis Carroll, *Through the Looking Glass*, (London: Harper Collins, 2017), 83.
3.  Peter Geach, *Mental Acts*, (London: Routledge and Kegan Paul, 1957), 22-23.
4.  John Wilson, *Thinking with concepts*, (Cambridge: Cambridge University Press, 1963), 10, 11.
5.  Ludwig Wittgenstein, *Tractatus Logico-philosophicus*, tans. D. F. Pears & B. F. McGuinness, (London: Routledge and Kegan Paul, 1963), 5.6.
6.  Wittgenstein, *Tractatus* 5.61.
7.  Wittgenstein, *Philosophical Investigations,* trans. G.E.M. Anscombe, (Oxford: Blackwell, 1963), 43.

Chapter 3. Philosophical and Scientific concepts

1.  Frege, Gottlob, "On Sense and Reference", http://home.sandiego.edu/~baber/metaphysics/readings/Frege.SenseAndReference.pdf
2.  Bertrand Russell, "The Philosophy of Logical Analysis" in *A History of Western Philosophy,* Chapter XXXI, http://www.naturalthinker.net/trl/texts/Russell,Bertrand/Philosophy/Russell,%20Bertrand%20-%20The%20Philosophy %20Of%20Logical%20Analysis.pdf
3.  George Berkeley, "A Treatise Concerning the Principles of Human Knowledge" in *A New Theory of Vision and Other Writings*, (London: J. M. Dent & Sons Ltd, 1938), 94.
4.  Dominic Murphy, "Concepts of Disease and Health", http://plato.stanford.edu/entries/health-disease/#pagetopright
5.  Richard Kamber, "Philosophy's Future as a Problem-Solving Discipline: The Promise of Experimental Philosophy," *Essays in Philosophy:* Vol 12: Iss. 2, Article 7, 2011, 291-311.

6. Ron Amundson, "Homology and Homoplasy: A Philosophical Perspective", http://www.uhh.hawaii.edu/~ronald/pubs/2001-ELS -homology.pdf

Chapter 4. Personhood

1. Keith Campbell, *Body and Mind*, 2d ed., (Notre Dame, Indiana: University of Notre Dame Press, 1984), 14.
2. Jaegwon Kim, *Mind in a Physical World*, (Cambridge, Massachusetts: Massachusetts Institute of Technology Press, 1998), 9.
3. Thomas Nagel, "Conceiving the Impossible and the Mind-Body Problem", *Philosophy* 73, no. 285 (1998): 337-352, http://philosophy .fas.nyu.edu/docs/IO/1172/conceiving.pdf
4. Nagel, *Mind and Cosmos,* (Oxford University Press: New York, 2012), 45.
5. Stephen Priest, *Theories of the Mind*, (London: Penguin Books, 1991), 210-211.
6. John Locke, *An Essay Concerning Human Understanding*, ed. Peter H. Nidditch, (Oxford: Oxford University Press, 1979), 335.
7. Jason Stanley, "Persons and their Properties", *The Philosophical Quarterly* 48, (1998): 149-175, http://w.w.w.rci.rutgers.edu/~jasoncs /persons.pdf
8. John Searle, *Mind*, (New York: Oxford University Press, 2004), 297.
9. Peter Strawson, *Individuals*, (London: Methuen, 1964), 89.
10. René Descartes, "Discourse on the Method of Rightly Conducting the Reason and Seeking for Truth in the Sciences", in vol. 1 of *Philosophical Works of Descartes*, trans. Elizabeth S. Haldane and G.R.T. Ross, (New York: Dover Publications Inc., 1955), 101.
11. Steven D. Edwards, *Disability: Definitions, value and identity*, (Oxford/Seattle: Radcliffe Publishing, 2005).
12. Michael Tye, *Consciousness and Persons*, (Cambridge, Mass: MIT Press, 2003), 143.
13. Warren Bourgeois, *Persons*, (Waterloo, Ontario: Wilfrid Laurier University Press, 2003), 389.
14. Jean-Paul Sartre, *Existentialism and Humanism*, trans. Philip Mairet, (London: Methuen, 1963), 28.
15. John P. Lizza, "Persons and Death: What's Metaphysically Wrong with Our Current Statutory Definition of Death?", https://pmr.uchicago .edu/sites/pmr.uchicago.edu/files/uploads/Lizza-personsanddeath -metaphysicalchallenges-forhigherbrain.pdf Lizza-personsanddeath -metaphysicalchallenges-forhigherbrain-2.pdf
16. Searle, *The Mystery of Consciousness*, (New York: Review of Books, 1997), xiii.
17. Searle, *Mystery of Consciousness,* 4.
18. Searle, *Mystery of Consciousness,* 8.
19. Searle, *Mystery of Consciousness,* 213.
20. Gilbert Ryle, *The Concept of Mind*, (London: Hutchinson & Co., 1962), 15.

21.   Joseph J. Fins, "Rethinking Disorders of Consciousness", *Hastings Centre Report* 35, 2 August 2005: 22-24, http://w.w.w.medscape.com/viewarticle/503854?src=mp

22.   Fins, "Consciousness".

Chapter 5. Psychology and Psychiatry

1.   Jerome C Wakefield, "The concept of mental disorder: diagnostic implications of the harmful dysfunction analysis", *World Psychiatry* 2007 Oct; 6(3): 149–156, http://www.ncbi.nlm.nih.gov/pmc/articles/PMC2174594/

2.   Thomas S. Szasz, "Mental Disorders Are Not Diseases", http//:www.szasz.com/iol10.html

3.   Szasz, "The Therapeutic State", http//:www.iaapa.de/szasz_arcive.htm#13

4.   Clemens E. Benda, "What is Existential Psychiatry", *American Journal of Psychiatry* 123 (1966): 288-296.

5.   Benda, "Existential Psychiatry".

Chapter 6. The patient analysed

1.   *Merriam Webster's Dictionary on line*, s.v. "Homeostasis" http://www.m-w.com/dictionary/homeostasis

2.   Andy Denis, "Methodology and policy prescription in economic thought: a response to Mario Bunge", *The Journal of Socio-Economics* 32 no. 2, (2003): 219-226.

Chapter 7. Causality

1.   Rudolp Carnap, *An Introduction to the Philosophy of Science*, ed. Martin Gardner, (New York: Dover Publications, Inc., 1995), 11.

2.   *Webster's New World Medical Dictionary*, s.v. "Koch's postulates" http://www.medterms.com/script/main/art.asp?articlekey=7105

3.   "Hill's Criteria of Causation", http://www.drabruzzi.com/hills_criteria_of_causation.htm.

4.   David Hume, *A Treatise of Human Nature*, III, xv, (London/New York: Dent /Dutton, 1951), 170-1.

5.   Avi Sion, "The Logic of Causation", http://www.thelogician.net/LOGIC-OF-CAUSATION/JS-Mill-Methods-Appendix1.htm

6.   Joe Lau and Jonathan Chan, "Causal Inferences", http://philosophy.hku.hk/think/sci/inference.php

Chapter 8. Challenges to the person's survival

1.   James Birch, "A misconception concerning the meaning of 'disease'", *British Journal of Medical Psychology* 52 (1979): 367-375.

2.   Charles R. P. George, "Disease Explicated and Disease Defined", http://hdl.handle.net/2123/654

3. Christopher Boorse, "Health as a theoretical concept", *Philosophy of Science* 44 (1977), 567, 562, http://bioetyka.uw.edu.pl/wp-content /uploads/2014/10/06_BoorseHealthConcept.pdf

4. Lennart Nordenfelt, "Understanding the Concept of Health", 12. http://www.fil.lu.se/hommagewlodek/site/paper/NordenfeltLennart.pdf

5. Nordenfelt, "Concept of Health", 10.

6. John G. Scadding "Essentialism and nominalism in medicine: logic of diagnosis in disease terminology", *The Lancet* 348 (1996), 594-596.

7. Michel Foucault, *The Birth of the Clinic,* (New York: Vintage Books, 1994), 3.

8. DEFINING DISEASE TYPES I, II and III, http://www.who.int/phi/3 -background_cewg_agenda_item5_disease_types_final.pdf

9. Karem Sadegh-Zadeh, "Fuzzy Health, Illness, and Disease", *The Journal of Medicine and Philosophy* 25, no. 5 (2000), 605-638.

10. Lester S. King, "What is disease?", *Philosophy of Science* 21 (1954): 193-203.

11. Boorse, "Health as a theoretical concept", 555.

12. Kennth Richman, *Ethics and the Metaphysics of Medicine*, (Cambridge, Mass: The MIT Press, 2004), 56.

13. Ereshefsky, Marc, "Defining 'Health' and 'Disease'", *Studies in History and Philosophy of Biological and Biomedical Sciences* 40 (2009) 221–227, http://www.academia.edu/22337270/Defining_health _and_disease

14. Szasz, "The myth of mental illness", in *Health Disease and Illness*, 46, 48.

15. Eric Juengst, "Concepts of Disease after the Human Genome Project", in *Health Disease and Illness*, 257.

16. Juengst, "Concepts of Disease", 257.

17. Juengst, "Concepts of Disease", 258.

18. Ray Moynihan, Iona Heath, David Henry, "Selling sickness: the pharmaceutical industry and disease mongering", *British Medical Journal* 2002;324:886–891, http://www.bmj.com/content/324/7342/886.1

19. Moynihan, Heath, and Henry, "Selling sickness".

Chapter 9. Identifying and managing disease

1. Edmond A. Murphy, *The Logic of Medicine*, 2d ed. (Baltimore: Johns Hopkins University Press, 1997), 122-127.

2. K. Danner Clouser, "Approaching the Logic of Diagnosis" in *Logic of Discovery and Diagnosis in Medicine*, ed. Kenneth F. Schaffner, (Berkely: University of California Press, 1985), 39.

3. Benjamin Freedman, "Equipoise and the Ethics of Clinical Research", *New England Journal of Medicine* 317, no. 3 (1987): 141-5.

4. Charles Weijer, Stanley Shapiro, "Clinical equipoise and not the uncertainty principle is the moral underpinning of the randomised controlled trial", *British Medical Journal* 2000 Sep 23; 321(7263): 756–758, http://www.ncbi.nlm.nih.gov/pmc/articles/PMC1127868/

Chapter 10. Disability

1.  Jonathan Wolff, "Disability Among Equals", July 2004, rev. January 2005, Homepages.ud.ac.uk/~uctyjow/DAE.doc.
2.  "Constitution of Disabled Peoples' International", http://www .un.org/disabilities/documents/convention/convoptprot-e.pdf
3.  "Convention on The Rights of Persons with Disabilities", http://www .un.org/disabilities/documents/convention/convoptprot-e.pdf
4.  "International classification of functioning, disability and health : children & youth version" 3.2, apps.who.int/iris/bitstream/10665/43737/1 /9789241547321_eng.pdf
5.  "White Paper on The Rights Of Persons With Disabilities 2016", https://www.gov.za/sites/www.gov.za/files/39792_gon230.pdf
6.  Ron Amundson, "Disability Rights: Do We Really Mean It?" *in Philosophical Reflections on Disability*, ed. D. Christopher Ralston and Justin Ho (Dordrecht: Springer, 2010), 170 https://www.researchgate.net /publication/226989443_Disability_Rights_Do_We_Really_Mean_It
7.  "White Paper"
8.  "Constitution of Disabled Peoples' International"
9.  Christopher Newell, "Disability, Bioethics, and Rejected Knowledge", *Journal of Medicine and Philosophy*, 31:269–283, 2006, https://www .tandfonline.com/doi/pdf/10.1080/03605310600712901
10.  Michel Foucault, *Birth of the Clinic*, 33.

Chapter 11. Chronology of disease

1.  David H. Mellor, *Real Time II*, (London: Routledge, 1998), 2.
2.  Saint Augustine, *Confessions*, trans. R. S. Pine-Coffin (London: Penguin Books, 1961), 244.
3.  John McTaggart, "The Unreality of Time", *Mind: A  Quarterly Review of Psychology and Philosophy*, no. 17 (1908): 456-473, http://www.ditext.com/mctaggart/time.html
4.  Isaac Edery, "Circadian rhythms in a nutshell", *Physiological Genomics* 3 (2000): 59-74, http://physiolgenomics.physiology.org/cgi /content/full/3/2/59
5.  Ian B. Hickie, Sharon L. Naismith, Rébecca Robillard, *et al.* Manipulating the sleep-wake cycle and circadian rhythms to improve clinical management of major depression. *BMC Med* **11**, 79 (2013). [https:// bmcmedicine.biomedcentral.com/articles/10.1186/1741-7015-11-79]
6.  Hickie et al., "Sleep-wake cycle and circadian rhythms" .
7.  Hickie et al., "Sleep-wake cycle and circadian rhythms".

Chapter 12. The final phase

1.  Robert M. Veatch, "The Impending Collapse of the Whole-Brain Definition of Death," *Hastings Center Report* 23, no. 4 (1993): 18–24, https://doi.org/10.2307/3562586
2.  Veatch, "Whole-Brain Definition of Death".

3.  James L. Bernat, "A Defence of the Whole-Brain Concept of Death", *The Hastings Centre Report* 28, no. 2 (1998): 14-23, http://philosophy rutgers.edu/COURSE/COURSE-SYNOPSIS/SPRING-2008/SYLLABI /249/Bernat.pdf

4.  David B. Hershenov, "The Death of a Person", *The Journal of Medicine and Philosophy* 31, no.2, (2006):107-120.

5.  Hershenov, "Death of a Person".

6.  Hershenov, "Death of a Person" .

7.  Daniel C. Stevenson, "Letter to Menoeceus by Epicurus", *The Internet Classics Archive 2009,* trans. Robert Drew Hicks, http://classics.mit.edu /index.html

8.  Nicholas D. Schiff and Joseph J. Fins, "Brain death and disorders of consciousness", https://doi.org/10.1016/j.cub.2016.02.027

9.  *Webster's New World Medical Dictionary,* s.v. "Locked-in syndrome", http://www.medterms.com/script/main/art.asp?articlekey =11024

10.  Bernat, "Whole-Brain Concept of Death".

11.  *President's Commission for Study of Ethical Problems in Medicine and Biomedical and behavioural Research*, "Defining Death", July 1981, https://repository.library.georgetown.edu/bitstream/handle/10822 /559345/defining_death.pdf

12.  *Harvard Ad Hoc Committee on Brain Death*, http://euthanasia .procon.org/sourcefiles/Harvard_ad_hoc_brain_death.pdf

13.  *The Canadian Law Reform Commission's definition of death*, 1981, https://bccla.org/our_work/the-canadian-law-reform-commissions -definition-of-death/

14.  Amir Halevy and Baruch Brody, "Brain Death: Reconciling Definitions, Criteria, and Tests", *Annals of Internal Medicine* 119, no. 6 (1993): 519-525.

15.  Paul Tillich, *The Courage To Be*, (New Haven: Yale University Press, 1952), 90-91.

16.  Rihito Kimura, "Death, Dying, and Advance Directives in Japan: Socio-Cultural and Legal Point of View", *Advance Directive and Surrogate Decision Making in Transcultural Perspective* (Baltimore: Johns Hopkins University Press, 1998), http://www.bioethics.jp/licht_adv8.html

17.  John P. Lizza, *Persons, Humanity, and the Definition of Death,* (Baltimore: Johns Hopkins University Press, 2006), 33.

18.  John MacQuarrie, *Existentialism*, (Harmondsworth, Middlesex: Penguin Books, 1977), 196.

19.  MacQuarrie, *Existentialism*, 197.

20.  Stevenson, "Menoeceus".

21.  James Park, *An Existential Understanding of Death: A Phenomenology of Ontological Anxiety*, 5th ed. (Minneapolis: Existential Books, 2006) https://s3.amazonaws.com/aws-website-jamesleonardpark ---freelibrary-3puxk/UD.html

22.  Diane Zorn, "Heidegger's Philosophy of Death." *Akademia*, 2, No.2: 10-11, http//:www.yorku.ca/zorn/files/Phil_of_death.pdf

23.  Frederick Grinnell, "Defining Embryo Death Would Permit Important Research", *The Chronicle Review,* 16 May 2003, https://pdfs .semanticscholar.org/7d28/a323bd71bcf6ca0849574e739c9d51442172.pdf

24.  Donald Landry, "Definition of Embryo Death Criteria May Open Doors For Stem Cell Research", *Stem Cell Research News,* 4 June 2006, https://www.technologynetworks.com/cell-science/news/definition-of -embryo-death-criteria-may-open-doors-for-stem-cell-research-192119

Chapter 13. Logic and scientific method

1.  Gottfried Leibniz, *The Monadology*, trans. Robert Latta, (London: Oxford University Press, 1951).

2.  Colin McGinn, *Logical Properties*, (Oxford: Clarendon Press, 2000), 3.

3.  Herbert Keuth, *The Philosophy of Karl Popper*, (Cambridge: Cambridge University Press, 2005), 12.

4.  W. V. Quine, J. S. Ullian, *The Web of Belief*, 2d ed. (New York: McGraw-Hill, 1978), 72.

5.  Thomas Kuhn, *The Structure of Scientific Revolutions*, (Chicago: University of Chicago Press, 1996), 10.

6.  Kuhn, *Scientific Revolutions*, 6.

7.  Kuhn, *Scientific Revolutions*, 4.

8.  Ary L. Goldberger, "Giles F. Filley Lecture. Complex Systems" *The Proceedings of the American Thoracic Society* 3 (2006), 467-471, http://pats.atsjournals.org/cgi/content/full/3/6/467

9.  Goldberger, "Complex Systems".

10.  Irving Dardik, "The Origin of Disease and Health Heart Waves", *Cycles* 46, no 3 (1996), http://caltek.net/dan/connectivity/phibiz/dardik /index.html

Chapter 14. Epistemology

1.  Descartes, "Discourse on the Method", 101.

2.  Louis Lasagna, "Hippocratic Oath, Modern version", (1964) http:// guides.library.jhu.edu/c.php?g=202502&p=1335759

3.  Hippocrates, "Of the Epidemics" in *The Internet Classics Archive,* Section 2, 5. ed. Daniel C. Stevenson, trans. Francis Adams, http://classics .mit.edu.Hippocrates/epidemics.1.i.html

4.  William Rosenberg and Anna Donald, "Evidence-Based Medicine: an approach to clinical problem-solving", *British Medical Journal* 310, (29 April 1995): 1122.

5.  Ronald De Vera Barredo, "Reflection and Evidence-Based Practice in Action: A Case Based Application", *The Internet Journal of Allied Health Sciences and Practice* 3, no. 3 (July 2005), http://ijahsp.nova.edu/articles /vol3num3/De_Vera_Barredo-1.htm

6.  Leroy Hood, "Systems Biology and P4 Medicine: Past, Present, and Future", *Rambam Maimonides Medical Journal.* 2013 April; 4(2): e0012, http://europepmc.org/articles/PMC3678833

Chapter 15. Ontology and Existentialism

1. Rom Harre, *Theories and Things,* (London/New York: Sheed and Ward, 1961), 8.

2. John F. Sowa, "Ontology", http://www.jfsowa.com/ontology/

3. Alexius Meinong, "Zur Gegenstandstheorie" in *Die Philosophie der Gegenwart in Selbstdarstellungen,* trans. Reinhardt Grossmann, http://www.formalontology.it/meinonga-texts.htm

4. McGinn, *Logical Properties.*

5. Edward Makhene, *Things, Objects, and Persons,* (self-pub, 2022), 6-21.

6. Barry Smith, Igor Papakin, and Katherine Munn, "Bodily Systems and the Modular Structure of the Human Body", *Proceedings of AIME 2002: 9th Conference on Artificial Intelligence in Medicine Europe,* 86-90, http://citeseer.ist.psu.edu/smith03bodily.html

7. Maureen Donnelly, "On Parts and Wholes: The Spatial Structure of the Human Body", *Medinfo* 11, Pt. 1 (2004): 351-5, http://cmbi.bjmu.edu.cn/news/report/2004/medinfo2004/pdffiles/papers/5096Donnelly.pdf

8. Jean-Paul Sartre, *Being and Nothingness,* trans. Hazel E. Barnes, (Pocket Books: New York, 1956), 87.

9. Sartre, *Existentialism and Human Emotions,* (New York: Carol Publishing Group, 1995), 15.

10. Immanuel Kant, *Foundations of the Metaphysics of Morals,* trans. Lewis W. Beck, (Chicago: University of Chicago Press, 1950), 80.

11. Lasagna, "Hippocratic Oath".

12. Robert G. Olson, *An Introduction to Existentialism,* (New York: Dover Publications Inc., 1962), 14.

13. Martin Buber, *I and Thou,* trans. Walter Kaufmann, (New York: Charles Scribner's and Sons, 1970), 54.

14. Lasagna, "Hippocratic Oath".

Chapter 16. Phenomenology

1. Pierre Thévenas, *What is Phenomenology,* (Chicago: Quadrangle Books, Inc., 1962), 19.

2. Lasagna, "Hippocratic Oath".

3. S. Kay Toombs, *The Meaning of Illness,* (Dordrecht: Kluver Academic Publishers, 1993), 117.

Chapter 17. Ethics

1. John S. Mill, *Utilitarianism, Liberty, and Representative Government,* (London: J. M. Dent & Sons Ltd, 1947), 9.

2. Mill, *Utilitarianism,* 9.

3. Kant, *Metaphysics of Morals,* 87.

4. Kant, *Metaphysics of Morals,* 73.

5. William. D. Ross, *The Right and The Good,* (Oxford: Clarendon Press, 1946), 19.

Chapter 18. Right to life

1. Mill, *"Utilitarianism"*, 73.
2. J. David Velleman, "Against The Right To Die", *The Journal of Medicine and Philosophy* 17:665-681,1992, http://www2.warwick.ac.uk/fac/soc/philosophy/undergraduate /modules /ph137/2014-15/velleman-_against_the_right_to_die.pdf
3. Michael Tooley, "Abortion Lecture 2", *Philosophy 11, Introduction to Ethics,* http://spot.colorado.edu/~tooley/Abortion2.html
4. Frederick Schauer, "The Right to Die as a Case Study in Third- Order Decisionmaking", *The Journal of Medicine and Philosophy* 17, no. 6 (1992).
5. Tooley, "An Irrelevant Consideration: Killing versus Letting Die", http://www2.sunysuffolk.edu/pecorip/SCCCWEB/ETEXTS/Deathand Dying_TEXT/An%20Irrelevant%20Consideration.pdf
6. Tooley, 'Killing versus Letting Die".
7. John Hardwig, "Is there a Duty to Die?", *Hastings Centre Report* 27, 2 (Mar. - Apr. 1997):34-42, http://web.utk.edu/~jhardwig/dutydie.htm
8. Peter Singer, "Unsanctifying Human Life", https://philosophy introcourse.files.wordpress.com/2012/10/singer-unsanctifying-human -life.pdf

Chapter 19. Abortion

1. Mary Anne Warren, "On the Moral and Legal Status of Abortion", in *Intervention and Reflection,* by Ronald Munson, (Belmont: Wadsworth, 2000), 101.
2. Barry Smith and Berit Brogaard, "Sixteen Days", *The Journal of Medicine and Philosophy* 28, no. 1 (2003).
3. Gregor Damschen, Alfonso Gómez-Lobo, and Dieter Schönecker, "Sixteen Days? A Reply to B. Smith and B. Brogaard on the Beginning of Human Individuals", *The Journal of Medicine and Philosophy* 31, no. 2 (2006).
4. Judith J. Thomson, "A Defense of Abortion", in *Moral Problems in Medicine,* ed. Michael Palmer, (Toronto: University of Toronto Press, 1999), 42.
5. Carlos A. Bedate and Robert C. Cefalo, "The zygote: to be or not be a person", *The Journal of Medicine and Philosophy* 14, no. 6 (1989): 643.
6. Bedate & Cefalo, "The zygote", 643.
7. Bedate & Cefalo, "The zygote", 644.
8. Bedate & Cefalo, "The zygote", 644..
9. Antoine Suarez, "Hydatidiform Moles and Teratomas Confirm the Human Identity of the Preimplantation Embryo", *The Journal of Medicine and Philosophy* 15, no. 6 (1990): 630.
10. Suarez, "Preimplantation Embryo", 631.
11. Suarez, "Preimplantation Embryo", 633.
12. Thomas J. Bole, "Zygotes, Souls, Substances, and Persons", *The Journal of Medicine and Philosophy* 15, no 6 (1990): 637-652.
13. Carol A. Tauer, "Personhood and Human Embryos and Fetuses", *The Journal of Medicine and Philosophy* 10, no. 3 (1985): 262.

14. Tauer, "Personhood ", 255.
15. Laura Purdy and Michael Tooley, "Is Abortion Murder?", in *Moral Problems in Medicine*, 46.
16. Purdy & Tooley, "Is Abortion Murder?", *Moral Problems*, 47.
17. Purdy & Tooley, "Is Abortion Murder?", *Moral Problems*, 48.
18. Aaron Ridley, *Beginning Bioethics*, (Boston/New York: Bedford /St. Martin's, 1998), 111.
19. *Merriam-Webster's OnLine Dictionary*, s.v. "Life." http://www .merriam-webster.com/dictionary/life
20. *Random House Unabridged Dictionary*, s.v. "Life." http://dictionary.reference.com/browse/life

Chapter 20. Euthanasia

1. Tom L. Beauchamp and Arnold I. Davidson, "The Definition of Euthanasia", *The Journal of Medicine and Philosophy* 4, no. 3 (1979): 294.
2. Carson Strong, "Euthanasia: Is the Concept Really Nonevaluative?", *The Journal of Medicine and Philosophy* 5, no. 4 (1980): 323-324.
3. Bob Lane, "Euthanasia", http://records.viu.ca/www/ipp/euthanas .htm
4. Mill, "On Liberty", 75.
5. Shanda Hastings, "Euthanasia", http://www.sbcc.edu/philosophy /website/STOA_files/STOA%201.2.pdf
6. Mill, "On Liberty", 74.
7. "Hippocratic oath", *Encyclopædia Britannica*, https://www .britannica.com/topic/Hippocratic-oath
8. Lasagna, "Hippocratic Oath".
9. Abilash A. Gopal, "Physician-Assisted Suicide: Considering the Evidence, Existential Distress, and an Emerging Role for Psychiatry", *J Am Acad Psychiatry Law* 43:2:183-190 (June 2015), http://www.jaapl .org/content/43/2/183.ful
10. Kathryn Tucker, "Physician Assisted Death in Oregon: A Success", https://www.utm.edu/staff/jfieser/class/160/6-euthanasia.htm
11. Plato, *The Republic*, Book V, Part VI, trans. H. D. P. Lee (Harmondsworth: Penguin Books, 1964), 216, 217.
12. Loretta M. Kopelman, "Rejecting the Baby Doe Rules and Defending a 'Negative' Analysis of the Best Interests Standard", *The Journal of Medicine and Philosophy* 30, no. 4 (2005),331-352.

Chapter 21. Darwinian Medicine

1. Stanley Zucker, Samuel Friedman, and Rita M. Lysik, "Bone Marrow Erythropoiesis in the Anemia of Infection, Inflammation, and Malignancy", *Journal of Clinical Investigation* 53, no. 4 (1974): 1132–1138, http://www .pubmedcentral.nih.gov/articlerender.fcgi?artid=333099
2. Lori Oliwenstein, "Dr Darwin", http://www.chester.ac.uk /~sjlewis/DM/TEXTS/TEXT5.HTM)
3. Austin Cline, "Existentialism and Darwinism - Evolutionary Theory and Existentialist Philosophy", http://atheism.about.com/od /existentialism/aphilosophies_4.htm

4. Wittgenstein, *Tractatus Logico-Philosophicus*, 4.111.
5. Wittgenstein, 4.112.
6. Wittgenstein, 4.1122.
7. Michael H. Cohen, "A Fixed Star in Health Care Reform: The Emerging Paradigm of Holistic Healing", http://www.rosenthal.hs.columbia.edu/legal/cohen.fixedstar. contents.html

# Index

378

Ockham's razor, 11, 258-259
Ogden, Charles, 5
Oliwenstein, Lori, 355, 370
Olson, Robert, 264, 368
Oregon law, 338
organism, biological, 197
other minds, 76, 89

pain, 270: fibres A and C, 69-71, 79,
    197; protective effect of, 345
Papakin, Igor, 368
paradigm: 225-226; shift, 226
parasite, fetus as, 319-320
parentalism, 283, 303
Park, James, 207, 366
past future, 123
paternalism, 175, 293
pathologist, 103
Pears D.F., 361
Penfield, Wilder, 49
Penicillin, 122
person: definition, 67; fetus as, 82,
    313-316
personal identity, 80-82
personhood: 67, 72-77, 82, 238;
    clinical perspective, 83-84;
    corporeal foundations , 76-77;
    dawn of, 209, 311-313; of fetus,
    313-316; of zygote, 313-314
perspective, bat's, 268
petri dish, 314, 316
pharmacogenomics, 249
phenomenology: essence in, 267
    essence of, 266-267;
    of consciousness, 78; of death,
    195; of disability, 176; of illness,
    152; of pain, 70; phenomena in,
    266
Philadelphia chromosome, 148
physician, earliest, 108
Piaget, Jean, 57-58
pill pedlars, 275
placebo, 166-167, 242
Plato: 339; 370; "The Forms" of, 27
Poincare, Jules H., 223, 227
polarization, 280, 308, 323
Popper, Karl, 38, 46, 56, 124,
    220-223, 225, 243, 367
possibilities, logical, 216
positivism, 38-40, 60

post hoc, 122, 245
potentiality, 316-319, 320
pragmatism, 94-95.
predicates, M and P, 86, 87, 198,
    202, 302
predictive value, 158-159, 162
Priest, Stephen, 71, 362
prima facie, 283, 284, 327
principle, moral symmetry, 305
principlism, 284
probability, 240, 161-163
problem solving: 60-62; and
    Darwinism, 343
properties: 84, 92, 98, 257, 290,
    302; abstracting similar, 29; and
    perspectives, 78; and relations,
    252; bundle of, 83; cannot
    coexist in the same object, 216;
    collections of sensible, 42;
    contradictory, 142; disease A,
    with, 255; emergent, 109;
    emergent functional, 143;
    essential, 24, 27, 82; exactly the
    same in all their, 218-219;
    lacking in specified, 30; logical,
    367, 368; logically incompatible,
    180; mental, 72-74; necessary
    and sufficient, 303; non-physical
    higher order, 69; of their spatial
    cohesion, 253;  Persons and
    their, 362; physical, 72;
    transcendent, 110
propter hoc, 19, 122, 245
proteomics, 250
Psychiatry, existential, 90-91
psychological integration, 314
Psychology, 85-87
psychophysical parallelism, 69
Purdy, Laura, 369
Putnam, Hilary, 43, 49-50
PVS, 22, 67, 189, 192, 196, 197, 198,
    318
Pyrrho of Elis, 32

Quine, Willard van Orman, 223,
    252, 367

rationalism: 47-49; critical, 55-57
Rawls, John, 242-243, 284, 294
RCT, 242, 251

www.ingramcontent.com/pod-product-compliance
Lightning Source LLC
Chambersburg PA
CBHW021547210326
41599CB00010B/347